MOOD DYSREGULATION

Beyond the Bipolar Spectrum

MOOD DYSREGULATION

Beyond the Bipolar Spectrum

Deborah A. Deliyannides, MD

First edition published 2024

Apple Academic Press Inc.
1265 Goldenrod Circle, NE,
Palm Bay, FL 32905 USA

760 Laurentian Drive, Unit 19,
Burlington, ON L7N 0A4, CANADA

CRC Press
6000 Broken Sound Parkway NW,
Suite 300, Boca Raton, FL 33487-2742 USA

4 Park Square, Milton Park,
Abingdon, Oxon, OX14 4RN UK

© 2024 by Apple Academic Press, Inc.

Apple Academic Press exclusively co-publishes with CRC Press, an imprint of Taylor & Francis Group, LLC

Library and Archives Canada Cataloguing in Publication

Title: Mood dysregulation : beyond the bipolar spectrum / Deborah A. Deliyannides, MD.
Names: Deliyannides, Deborah A., author.
Description: First edition. | Includes bibliographical references and index.
Identifiers: Canadiana (print) 20230225535 | Canadiana (ebook) 20230225594 | ISBN 9781774912430 (hardcover) | ISBN 9781774912447 (paperback) | ISBN 9781003340249 (ebook)
Subjects: LCSH: Affective disorders. | LCSH: Affective disorders—Diagnosis. | LCSH: Affective disorders—Treatment. | LCSH: Affective disorders—Etiology. | LCSH: Bipolar disorder.
Classification: LCC RC537 .D436 2023 | DDC 616.85/27—dc23

Library of Congress Cataloging-in-Publication Data

CIP data on file with US Library of Congress

ISBN: 978-1-77491-243-0 (hbk)
ISBN: 978-1-77491-244-7 (pbk)
ISBN: 978-1-00334-024-9 (ebk)

For
my children

About the Author

Deborah A. Deliyannides, MD, graduated from Jefferson Medical College, Philadelphia, PA, and completed her psychiatric residency at the New York Hospital-Cornell Medical Center in White Plains, NY. She is Assistant Clinical Professor of Psychiatry at Columbia University College of Physicians and Surgeons, New York, NY, and has been a research psychiatrist at the New York State Psychiatric Institute since 1990 where she treated a vast array of people with mood disorders at the Depression Evaluation Service. She was medical director of the Zucker Hillside Hospital Methadone Maintenance Treatment Program at the Long Island Jewish Medical Center in Glen Oaks, NY for over 10 years and earned added qualifications in Addiction Psychiatry. She is currently a psychoanalytic candidate at the NYU Postdoctoral Program for Psychotherapy and Psychoanalysis, and maintains private practices in Manhattan and in Darien, CT.

Acknowledgments

First and foremost, I owe a debt of gratitude to all those who have entrusted themselves to my psychiatric care over the years. Their needs and often unrecognized symptoms created the impetus for this book, and their collaboration with me some of its most important insights. I thank each and every one of you for your essential contribution to this work.

My interest in mood disorders began at the Depression Evaluation Service of the New York State Psychiatric Institute where I learned creative approaches to pharmacologic treatment under the mentorship of Patrick McGrath, Jonathan Stewart, and the late Frederic Quitkin, all of whom made critical contributions to my thinking. Yet while treating people with medication, I was always in touch with their psychotherapists, who introduced me to the broader psychoanalytic community. Within this community I found colleagues and mentors who have been invaluable to me in generating this project: you are too many to name, but I trust you know who you are. I am indebted to all of you for your ideas, guidance, critique, and encouragement.

And finally, I am grateful to my family for their enduring love and support, with a special callout to my daughter for recognizing the centrality to this book of Nietzsche's voice: "And those who were seen dancing were thought to be insane by those who could not hear the music."

Contents

Preface

I wish I had a penny for every time a person with mood instability in my psychiatric practice asked me, "Do you think I'm bipolar?" Within the population I see, the answer is typically a resounding "no!" Just because "bipolar disorders" is the only section in our DSM to describe symptoms of unstable mood doesn't mean a person struggling to regulate their mood has a diagnosis of bipolar disorder.

Just as often, I see people who say they are "depressed," but they are often more irritable than sad. They would never regard themselves to have bipolar symptoms, but medications used to treat people with bipolar disorder are often the most effective treatment for them.

Usually, neither person has bipolar disorder: they have what I have come to call "mood dysregulation." Difficulties with mood or emotion regulation characterize nearly 75% of our current psychiatric diagnoses, not just the bipolar disorders. From substance abuse to eating disorders, anxiety disorders to ADHD, mood dysregulation is a common thread that runs throughout many psychiatric disturbances but is often confused with, or conflated with, bipolar disorder.

At present, people with mood dysregulation are not adequately described on the pages of any diagnostic manual. In some way they are "diagnostic orphans." Their symptoms often bear some attenuated resemblance to those with diagnoses on the spectrum of bipolar disorders, and at times they respond to the same interventions as people with bipolar disorders. Thus, they are tethered in some way to disorders on the bipolar spectrum, albeit remotely. But they cannot truly be regarded to have any bipolar disorder: they are better described as beyond the bipolar spectrum.

To understand this relationship, it is important to know something about the way bipolar disorders have been discussed throughout the years. This will involve a historical review of the description and classification of mood disorders. But looking at the current classification system brings to light a number of limitations that obviate the need for more flexible approaches to viewing "mental ill health." This requires seeing behaviors as existing on a continuum rather than describing disorders in categorical terms and understanding symptoms as crossing diagnostic lines. It also requires looking at the etiologic and biological underpinnings of the phenomena in question. This book attempts to understand mood dysregulation within this new context.

Introduction

"When you have got an elephant by the hind legs and he is trying to run away, it's best to let him run."
—Abraham Lincoln

I was unprepared for the reaction I received when I first began assessing depressed people decades ago in a clinical research setting with a structured diagnostic interview. Using a standardized questionnaire, the clinician presents the prospective subject with diagnostic questions that parallel each DSM (*Diagnostic and Statistical Manual of Mental Disorders*) criterion. I naively assumed my interviewees would respond to me in an open, straightforward manner as one might expect from a person in distress seeking help from a professional. But the most curious pattern emerged: whenever I reached the "bipolar disorders" section of the questionnaire, many people, for some reason, became hesitant. My "talker" turned skittish. Where we'd had an easygoing dialogue, they often became cagey. They would shut down. Previously elaborate, thoughtful answers became a terse "no:" they seemed to want to move on to the next section. Often people would tell me outright, "I know where you're going…and no, I'm not bipolar." I heard these disclaimers even if I had a sense there might be some mood instability but doubted there would be a diagnosis of bipolar disorder. This type of response was especially true of people who had a relative with a diagnosis of bipolar disorder.

Bipolar disorder is a biologic mood problem that runs in families and is characterized by cyclical mood swings associated with shifts from low energy to high energy states. In its most extreme form, the manic state of the cycle is characterized by psychosis. While mood dysregulation is different from bipolar disorder, the vast majority of the hundreds of people I have seen with some form of mood dysregulation have a family member with a diagnosis (or suspected diagnosis) of a bipolar disorder. Too often, however, that family member is NOT remembered kindly. He or she is often remembered as being wildly eccentric.

For the sake of illustration, let's make a composite such relative and call her Aunt Sadie. Here is a typical story I might hear about Aunt Sadie:

My mother didn't like to talk about Aunt Sadie...she was an embarrass-
ment to the family. My dad told me most of what I know. She had highs
and lows in the extreme. At times she would be curled up in bed for days
on end like a total recluse; she didn't eat or bathe. My uncle played Mr.
Mom for our cousins. I really felt sorry for them. Nobody saw her, not
even the family. Then, out of nowhere, she'd get a burst of energy and go
crazy: she would wear outrageous clothes, start calling everyone up, and
spend money like there was no tomorrow. My uncle had to take all the
credit cards away. At times she left the house for days on end: we know
she cheated on my uncle. She would spread crazy rumors about people:
mom could never talk sense into her head when she was like that. Then
she would get so testy she got into arguments with everyone. At other
times she was downright paranoid. Once she thought the neighbors were
taking her things and went over to their house to get them: she got into a
fight with them that landed her in the hospital. She was supposed to take
medication, but I don't think she ever did. She was a totally different
person when she was like that: unrecognizable. Dad said if you tried
to point out that she wasn't herself when she acted like that, she didn't
know what you were talking about. There were periods of time when she
seemed totally normal, but she had no memory of her other self. Then out
of nowhere she'd get depressed again. I know she got shock treatment
a few times when she was hospitalized. Whenever my brother and I get
moody we tease each other, "You're going to become Aunt Sadie" but
somewhere deep down we're scared it's not a joke.

Bipolar Aunt Sadie is the family member nobody wants to be. Not
surprisingly, the person I am interviewing fears he or she will become
Aunt Sadie. If they haven't already told me, they are surely thinking, "I
am not bipolar," so they never quite disclose their full history to me, the
history I need in order to determine the appropriate treatment. For many
people struggling with depression who have a relative like Aunt Sadie,
my questions seem threatening. If I address things head-on by asking,
"Are you worried you might become like your Aunt Sadie?" the answer is
inevitably "No."

Living with someone who has severe bipolar disorder can be quite
scary. The seemingly inexplicable mood changes can be very disturbing.
The thought of being like that person can simply be too difficult to process.
"I am not bipolar" is an understandable response. It is common, almost a
cliché, for people who have a family member with bipolar disorder to be
vigilant about their own behavior and symptoms, wary that they may be
exhibiting or come to develop symptoms of mania or severe depression.

The fear of a bipolar diagnosis and the stigma associated with it leads to underreporting or denial of mood instability, which prevents clinicians from recognizing the attenuated symptoms of bipolar disorder often seen in mood dysregulation. Underreported signs of mood dysregulation such as temper flares, overreactivity, rejection sensitivity, road rage, or problems with impulse control including eating disorders, gambling, and sex additions can lead clinicians to make inaccurate diagnoses, resulting in treatment mistakes that can end up in costly and discouraging treatment failures.

Though mood dysregulation as we are discussing it is not a bipolar disorder, it often responds to the same medication Aunt Sadie uses. How to explain this to a person with mood dysregulation who has Aunt Sadie in mind is often a challenge, and feeds back into the problem of underreporting. If mood dysregulation were better understood, clinicians would be more effective with their treatment recommendations, and those with symptoms would be more open to interventions that might otherwise be stigmatized as belonging to those with bipolar disorders.

The very word "bipolar" can evoke such awful images that many people avoid its discussion. The stigma associated with bipolar disorder is so great that relatives or friends of those with the diagnosis often refuse to consider taking a medication they associate with that person, even if they do not have bipolar disorder, though it might be the most appropriate treatment for them. Many times, mood dysregulation happens to respond to the same medications as bipolar disorder, despite the fact that it is not bipolar disorder or even on the bipolar spectrum of disorders: it is beyond the bipolar spectrum, though related to it in certain ways. Unfortunately, the bipolar association stigmatizes treatments often most effective for mood dysregulation, depriving many from getting the help they need.

To be able to say, "I am not bipolar," one needs to know something about what bipolar disorder is. Chapter 1 is a short historical review of the identification of cyclical mood disorders throughout the centuries. Chapter 2 elaborates on what we mean today when we use the term "bipolar disorder" by reviewing the DSM-5 diagnostic descriptions of bipolar and depressive disorders. The bipolar spectrum is described as it is currently conceived including efforts to extend that spectrum beyond the bounds of DSM nomenclature to more fully capture clinical phenomenology.

Chapter 3 picks up on the topic of the limitations of the DSM classification system and introduces emerging thinking about new diagnostic systems that address the problems associated with the old. Chapter 4, the most central chapter in the book, introduces the concept of mood dysregulation, locating it beyond the bipolar spectrum and defining it in terms of the literature on "affective temperaments." A hallmark symptom of mood dysregulation, dysphoria, is defined and distinguished from depression, with which it is often confused. Mood dysregulation is identified as a transdiagnostic process and compared with emotion (dys) regulation and affect (dys)regulation. In Chapter 5, the transdiagnostic nature of mood dysregulation is elaborated as it is discussed in the context of other conditions. Chapter 6 reviews normal human physiology and the process of emotion regulation before discussion of the pathophysiologic origins of mood dysregulation based on attachment theory in the setting of relational trauma, elaborated in Chapter 7. Chapter 8 reviews treatments for mood dysregulation before a presentation of clinical examples in Chapter 9.

CHAPTER 1

Historical Background

Symptoms, then, are nothing in reality but a cry from suffering organs.
—Jean Martin Charcot

ABSTRACT

Mood disorders have been documented for over 4000 years. What we think of today as "depression" was described by ancient Greek physicians (who regarded these disorders as related to imbalances in bodily fluids) to be "melancholia." Aristotle was the first to document a relationship between alternating states of sadness and cheerfulness. It was then recognized that there existed a single phenomenon characterized by the cycling of these mood states: this condition came to be known as "bipolar disorder." The conceptualization of bipolar disorder as a single disorder came to be known as the "unitary hypothesis," articulated by the German psychiatrist Emil Kraepelin, and was incorporated into the first major diagnostic compendium of mental health in the mid-twentieth century. But Kraepelin's idea was challenged by Karl Leonhard, who argued that depression was a distinct disease entity from bipolar disorder. This resulted in revisions to our current diagnostic nosology, which now distinguishes unipolar depression from bipolar depression, a distinction increasingly viewed as problematic by leaders in the field. One might argue it obscures symptom recognition and poses problematic treatment implications for those with mood dysregulation.

Disturbances of emotions have been described for millennia. We read in the Bible that David eased a melancholic King Saul with his harp over 3000 years ago. Melancholia, from the Greek "melaina khole" means "black bile" one of the four "humors" considered to be in excess in the body in Hippocrates' (460–370 BC) conception of disease. For Hippocrates, disease was the result of the imbalance in the humors of black bile, yellow bile, blood, and phlegm. We may think of this as a primitive conceptualization

Mood Dysregulation: Beyond the Bipolar Spectrum. Deborah A. Deliyannides, MD (Author)
© 2024 Apple Academic Press, Inc. Co-published with CRC Press (Taylor & Francis)

of disease, but it was quite advanced for his time: Hippocrates, himself the son of a physician, was the first to understand that diseases had natural causes and were not the result of superstitions, the gods, or demonic possession (Garrison, 1996; Simon, 2008). Hippocrates' followers are thought to have been the first to document mental illnesses.

Aristotle (384–322 BC), like Hippocrates, considered melancholia as a condition of excess black bile. But he described a spectrum of severity associated with black bile from a normal melancholic temperament to the disease of melancholia (Hett, 1965). Aristotle's distinction foreshadowed our modern notion of the continuum between depressive temperaments (dysthymia) and major depressive disorder (Akiskal, 1983a).

Descriptions of melancholia, first called "depression" in the 18th century, dominated the discussion of mood disturbances throughout the ages. But it was clear that disorders in mood were not always in the direction of depression.

Documentation of a cyclical mood disorder appears for the first time in the writings of Aristotle. Aristotle understood that melancholia was associated with mania in a cyclical way and was the first to record this association over two millennia ago (Pies, 2007) in *Problemata:*

> If [black bile] is in excessive quantity in the body [it] produces apoplexy or torpor, or despondency or fear; but if it becomes overheated, it produces cheerfulness with song, and madness...those...in whom the bile is considerable and cold become sluggish and stupid, while those with whom it is excessive and hot become mad, clever or amorous and easily moved to passion and desire, and some become more talkative... many, because this heat is near to the seat of the mind, are affected by the diseases of madness or frenzy...now when the mixture of black bile becomes colder it gives rise...to all kinds of despair, but when it is hotter, to cheerfulness. That is why the young are more cheerful and the old are less so. For the former are hot and the latter cold (Hett, 1965).

Note that from the outset, what came to be known as "bipolar disorder" was understood to be cyclical in nature. Recognition of the cyclical nature of mood states brought an understanding of their biologic rhythmicity, their internal, endogenous underpinnings. Aristotle recognized cycling moods but did not recognize the association of one state with the other.

Aretaeus did, however. Aretaeus was another celebrated Greek physician, six centuries after Hippocrates, who lived in Cappadocia (now Turkey) and practiced state-of-the art medicine. He is known as the father of bipolar disorder as he was the first to document mania and melancholia as two different faces of the same disease (Marneros & Angst, 2000). He described a group of euphoric people who would "laugh, play, dance night and day, and sometimes go openly to the market crowned, as if victors in some contest of skill," only later to appear "torpid, dull, and sorrowful." Aretaeus' observations enabled him to identify and describe the successions of manic and depressive states separated by periods of lucidity in what became known as the "circular disease" (McGrew, 1985):

> …it appears to me that melancholia is the commencement and a part of mania. but those affected with melancholy are not every one of them affected according to one particular form; but they are either suspicious of poisoning, or flee to the desert from misanthropy, or turn superstitious, or contract a hatred of life. Or if at any time a relaxation takes place, in most cases a hilarity supervenes, but those persons go mad (Aretaeus, 1856).

For centuries, the notion of a singular, cyclical mood disorder prevailed. But different perspectives began to emerge (and have been vying for prominence to this day). At some point, depression and mania came to be regarded as two distinct illnesses. Marsilio Ficino (1433–1499), a Neoplatonic philosopher of the Medici court writes:

> …the melancholic humour lights and burns, producing that excitement which the Greeks call mania and we, furor. But when it dies out, only a black soot is left…which makes people foolish and sluggish. This state of mind is properly called melancholia, dementia and madness (Ficino, 1995 trans).

Similarly, Thomas Willis (1676) says, "These two, melancholy and mania, mutually exclude and replace each other like smoke and flame."

With a return to the notion of a cyclical mood disorder, a bit of drama emerged in the mid-19th century between two French psychiatrists, both of whom claimed to have discovered bipolar mood disorder. On January 31, 1854, the psychiatrist Jules Baillarger gave a lecture to the French Academy of Medicine entitled "On a form of insanity characterized by two regular periods, one of depression and the other of excitement," proposing the name "Insanity with a double form" ("Folie a double forme"). Two weeks later, Jean-Pierre Falret gave a lecture to the same audience "On a form of mental disease characterized by the successive and regular reproduction of the manic state, the melancholic state, and a more or less lucid interval, 'Circular insanity' or 'Folie circulaire.'" His lecture was based on a prior lecture he had given at the Salpetriere hospital in 1850, subsequently published in 1851 (Pichot, 2006).

Who had priority in the description of this disorder? The psychiatrist Morel, who had been a pupil of both Baillarger and Falret, attempted a conciliation, suggesting that the differences between the two conceptions were more apparent than real, even as Baillarger and Falret continued arguing with one another. Two years before Falret's death in 1870, Baillarger was elected president of the French Academy of Medicine and became the most influential French psychiatrist, claiming priority to the description of the disease. Nevertheless, "an objective study leaves no doubt about Falret's priority in the description of 'folie a double forme,'" which he had originally called "folie circulaire" and which came to be called "bipolar disorder" (Pichot, 2006).

Falret believed heredity played an important role in the etiology of bipolar disorder. His clinical descriptions of the syndrome are very close to the present conceptions of the diagnostic entity. Falret's observations of individuals with "folie circulaire" reflect important details of the condition:

> …one sometimes sees manic paroxysms and there is even a condition of anxious melancholia characterized by constant pacing and inner turmoil, which incapacitates these patients so that they cannot concentrate, and this state sometimes ends up in manic agitation (Falret, 1864).

Jean-Pierre Falret was actually describing mixed states ("etat mixte"), which were formally recognized and elaborated by his son Jules. Among his clinical descriptions, Jules Falret observed:

> … "mixed states" characterized by "predominant ideas, often of sad nature, in the middle of an excitation state, simulating true mania," for which one is forced "to use the hybrid and contradictory expressions of manic melancholies or melancholic manias" (Falret, 1861).

Jean-Pierre Falret's concept of "folie circulaire" influenced the German psychiatrist Karl Kahlbaum (Pichot, 2006), who was the first to use the term "cyclothymia" for cyclical mood disorders with milder symptoms in his classification of mental illnesses. Kahlbaum's thinking, in turn, impacted the ideas of the German psychiatrist Emil Kraepelin.

Emil Wilhelm Georg Magnus Kraepelin (1856–1926) was a German psychiatrist. He is regarded by some to be the founder of modern scientific psychiatry, in part because he was the main architect of modern psychiatric nosology, the medical classification of diseases.

Kraepelin famously separated psychiatric illnesses into two categories: dementia praecox ("premature dementia" or "precocious madness") and affective psychoses. In Kraepelin's diagnostic system, all psychiatric disorders were psychoses. Psychotic disorders were either schizophrenia (dementia praecox) or bipolar disorder (affective psychoses).

Of interest to us is the designation of affective psychosis. With this diagnosis, Kraepelin describes a manic–depressive illness, including depression and mania within a singular disorder. This is an important if not controversial formulation of the clinical entity, reflecting a "unitary" view of an affective disorder in which a single underlying process is responsible for mania, depression and all intermediate states (Kraepelin, 1921). Kraepelin envisioned a broad spectrum of manic-depressive illness which encompassed attenuated forms of severe symptoms as well as major depressive states (Kraepelin, 1899a). In Kraepelin's "unitary" view, a person with major depression and no evidence of mania would still be regarded to have a bipolar disorder as depression is included within the same diagnostic entity as manic-depressive illness. Kraepelin's ideology and nomenclature were so compelling and influential that they were incorporated into the first DSM (1952).

Meanwhile, Karl Leonhard, another German psychiatrist interested in psychiatric nosology in the tradition of Kraepelin, but 50 years his junior, disagreed with Kraepelin's unitary conceptualization of manic-depressive psychosis. He distinguished two depressive disorders in contrast to Kraepelin's unitary view of depressive disorders (Leonhard, 1999), arguing in his description of the presentations of "Mania and Melancholia" that depression was symptomatic of two different illnesses, one in which melancholia cycled with mania and one in which it did not. He introduced the term "bipolar" for the form of depression which cycled with mania and "unipolar" for the form of depression which did not, thus dichotomizing depression into two types. His argument for dichotomizing the diagnosis was based, in part, on his observation that familial frequency of depression was greater in the bipolar form of the illness than in the unipolar form (Leonhard, 1957).

The shift from "manic depression" to "bipolar disorder" was not simply a change in nomenclature. It was an ideological shift from Kraepelin's vision of a unitary spectrum of disorders to Leonard's vision of a dichotomy between unipolar and bipolar depression.

Leonhard's nomenclature had a lasting impact on the broader psychiatric community. Several others adopted his position that there were three distinct syndromes: unipolar depressed, unipolar manic, and bipolar (Angst, 1973; Angst & Perris, 1972; Perris, 1966; Winokur et al., 1969). In 1980, the bipolar/unipolar nomenclature was incorporated into DSM-III (1980): it is still used today. If Aunt Sadie had been treated in 1980, her diagnosis would have been bipolar affective disorder.

Despite the fact that DSM has adhered to Leonhard's nomenclature since 1980, the research community has become increasingly uncomfortable with the dichotomized conception of cyclical mood disorders that began with his use of the term "bipolar" (Akiskal, 1983b; Gershon, 2000; Goodwin & Jamison, 1990; Maj et al., 2002). In a move toward Kraepelin's conceptualization, "bipolar spectrum" terminology (Klerman, 1981) was introduced by Dunner and colleagues (1970) to describe varying degrees of mania and hypomania (subthreshold mania) as researchers began elaborating a spectrum of cyclical mood disorders subsumed under the bipolar umbrella. Note again that cyclicity is the hallmark of bipolar disorders.

Remarkably, Jules Angst, who had once held Leonhard's bipolar v. unipolar position on depression, came to embrace the broadest concept of the bipolar spectrum in keeping with Kraepelin's unitary view of depression. The broadening of the bipolar spectrum of diagnoses is reflected in DSM-IV and DSM-5, which allow for diagnostic inclusion of the greatest range of cyclical mood disorders we see to date. Despite the expanded bipolar diagnostic criteria, there remain people with symptoms that cannot technically be included on the bipolar spectrum, for example, people with treatment resistant depression, yet they have symptoms related to bipolar disorders and features in common with Aunt Sadie, impossible as it may seem. Some of these people we will describe as having mood dysregulation, which falls beyond the bipolar spectrum. We will trace the evolution of some of the more recent thinking about mood disorders to understand this connection.

While depression may often be thought of as the major mood disturbance, cyclical mood swings have been described for over 2500 years, during which time the pendulum has swung back and forth between the notion that depression and mania represent two faces of the same illness and the notion that depression is a separate illness. Until half a century ago, the former view (Kraepelin's "unitary" view), that depression and mania are regarded to be different aspects of the same illness, dominated. This conceptualization of the illness was incorporated into the first DSM in 1952.

But Kraepelin's view was supplanted in the second half of the 20th century by the thinking of Karl Leonhard, who believed that depression was a distinct illness: that there was unipolar depression and bipolar disorder. Leonhard's dichotomized conceptualization was incorporated into DSM-III in 1980 and remains a part of the DSM classification system to this day.

The impact of this shift cannot be overstated. The DSM is the didactic foundation for the clinical training of virtually all mental health professionals and becomes the lens through which they assess those they treat. As we will see, DSM's failure to account for the range of symptoms encountered in clinical reality has led clinicians to misinterpret the presentation of "depressed" individuals that may have mood dysregulation, a symptom cluster with important treatment implications distinct from those of depression. With clinical problems such as these, cracks in our current diagnostic system have begun to make themselves known.

KEYWORDS

- **melancholia**
- **"folie circulaire"**
- **"etate mixte"**
- **unitary hypothesis**
- **unipolar depression**
- **bipolar depression**

CHAPTER 2

DSM Mood Disorders

*Madness is to think of too many things in succession too fast,
or of one thing too exclusively.*
—Voltaire

ABSTRACT

This chapter looks more closely at our current psychiatric diagnostic systems, the ICD and, in particular, the DSM, which is the most widely used diagnostic compendium in the US. Mood disorders in the DSM are divided into bipolar disorders and depressive disorders a la conceptualization of Leonhard. People with bipolar disorders cycle between LOW ENERGY depressive states and HIGH ENERGY hypomanic/manic states. While there is only one mood phase associated with the low energy depressive state (the negative mood of depression), there are two mood phases associated with the high energy hypomanic/manic state (the positive mood of euphoria and the negative mood of dysphoria). It is crucial to distinguish between the negative mood of depression and the negative mood of dysphoria, both for diagnostic and treatment purposes. Bipolar disorders in the DSM are categorized along a spectrum of symptom severity. Recognizing the numerous iterations of mood disorders not described in the DSM, Hagop Akiskal and his colleagues elaborated an array of related mood syndromes known as the "soft bipolar spectrum." A newcomer to the group of depressive disorders in the latest edition of the DSM is "disruptive mood dysregulation disorder" (DMDD), which does not require depression as a diagnostic criterion. We will discuss this curious addition further.

Mood disorders have been classified since the time of Hippocrates when mania, melancholia, phobias, and paranoia were described. With the emergence of the medical model in the 20th century, two major compendia of psychiatric nosology were published: the International Statistical

Mood Dysregulation: Beyond the Bipolar Spectrum. Deborah A. Deliyannides, MD (Author)
© 2024 Apple Academic Press, Inc. Co-published with CRC Press (Taylor & Francis)

Classification of Diseases and Related Health Problems (ICD) and the Diagnostic and Statistical Manual of Mental Disorders (DSM).

The ICD is a compilation of medical disorders first published by the World Health Organization (WHO) in 1900 and is used worldwide. In 1949, 3 years before the publication of the first DSM (1952), the WHO published its sixth revision (ICD-6) which contained, for the first time, a section on mental disorders (Chapter V). ICD chapter V parallels the DSM system largely but not altogether. Since October 1, 2015, ICD-10 has been the required code set for use in all health care insurance transactions in the US. For this reason, corresponding ICD-10 codes are included for diagnoses in the DSM-5. The most recent version of ICD is ICD-11, published in May 2019.

The DSM was first published by the American Psychiatric Association in 1952 and has been revised six times since: DSM-II (1968), DSM-III (1980), DSM-III-R (1987), DSM-IV (1994), DSM-IV-TR (2000) and the most recent version, DSM-5 (2013). The DSM is used primarily in the US and has been the main diagnostic resource for mental health practitioners, researchers, pharmaceutical companies, the legal system, and (until 2015) health insurance companies. Because it is the primary clinical diagnostic resource for psychiatric disorders, we will refer to it here.

DSM mood disorders have been classified over the years according to the prevailing zeitgeist shared by the manual's authors and the psychiatric community at large. Thus, in the first DSM (1952), major mood disorders were all both psychotic and bipolar. As per Kraepelin's unitary conceptualization of mania and depression, the diagnosis of major depression as a separate category did not exist in this first DSM: there was no unipolar depression, and all major affective disorders were psychotic disorders. In 1952, Aunt Sadie's diagnosis would have been "psychotic disorder with manic depressive reaction." Milder depressions (dysthymia/cyclothymia) were either neuroses or personality disorders. When it appeared as a separate entity in DSM-II (1968) as "involutional melancholia," what we might now call "major depression" could still only be a psychotic disorder.

In 1952, when DSM was the diagnostic "bible," Aunt Sadie's diagnosis of "psychotic disorder with manic-depressive affective reaction" was a "disorder of psychogenic origin without clearly defined physical cause or structural change in the brain." Her problem would not be thought of as biologic or brain based, rather, it would be considered mental or mind based: in other words, Aunt Sadie didn't have a biological problem, she was "mentally ill." Unfortunately, nearly two decades passed before the APA recognized her problem as a brain based medical issue, reclassifying manic-depressive disorder as manic-depressive illness in DSM-II (1968). In 1968, her diagnosis would have been "manic-depressive illness, circular type." Now Aunt Sadie had a medical problem.

In DSM-III, we see the influence of Leonhard's dichotomization of bipolar and unipolar illness. Major affective disorders were now divided into bipolar disorder or major depression AND were no longer necessarily psychotic in nature. In addition, dysthymia and cyclothymia were no longer neuroses or personality disorders but classified for the first time as affective disorders.

In DSM-III-R and DSM-IV, Leonhard's conceptualization was further emphasized. The "major affective disorders" category of DSM-III which housed both unipolar and bipolar depression disappeared in DSM-IV and was replaced by two separate categories: "bipolar disorders" and "depressive disorders."

DSM-5 maintains this bifurcated (bipolar disorders/depressive disorders) diagnostic categorization and amplifies the depressive disorders section, replacing dysthymia with persistent depressive disorder and adding disruptive mood dysregulation disorder and premenstrual dysphoric disorder. Refer to Appendix C to follow this evolution and compare the complete diagnostic criteria for mood disorders from DSM through DSM-5. We will now look at DSM-5 mood disorders in more detail.

2.1   DSM-5 BIPOLAR DISORDERS

DSM-5 has seven categories of bipolar disorder of which three are given the most attention. These are (1) bipolar I disorder, (2) bipolar II disorder, and (3) cyclothymia. Other bipolar disorders are (4) substance/medication induced bipolar disorder, (5) bipolar disorder due to a medical condition, (6) other specific bipolar disorder, and (7) unspecified bipolar disorder.

2.1.1   *BIPOLAR DISORDER*

Bipolar I disorder is hard to miss. A person in a manic state is usually talkative and gregarious, but what they say often sounds outlandish and grandiose if not unhinged. Confrontation with their reality may be met with a flare of anger: they typically have no insight into their condition. They can become agitated and at times violent. The activated energy states, the reactivity, and the distorted, often paranoid cognition that typically lead to impaired functioning are all characteristic of mania.

The sine qua non of bipolar I disorder is a manic episode. A manic episode is an energized/activated state in which mood can be euphoric (positive phase) or dysphoric (negative phase). Cognitive activity is increased, frequently with "jamming" or "racing" thoughts. Psychosis is usually present with delusions (often paranoia), and at times, auditory hallucinations ("hearing voices"). Symptoms cause functional impairment, typically creating the need for hospitalization.

The Psychodynamic Diagnostic Manual (PDM) (Lingiardi & Mc Williams, 2017), a diagnostic manual compiled by psychoanalysts, follows DSM criteria and amplifies the description of manic states as follows:

> **Affective states:** These may include feelings of intense pleasure or euphoria. They may also involve intense irritability, often accompanied by transient anxiety, agitation, hypersensitivity to, and expectation of insult and rejection. Mania is characterized by excessive energy that may be experienced negatively as a disruptive and distracting internal pressure, or positively as a sense of infinite power, ability, and creativity. In either case, the internal hyperarousal is often accompanied by impulsive behavior. Patients experiencing manic states often report increased desire–even desperate cravings–for others, frequently accompanied by intense, constant sexual desire and social disinhibition.
>
> Individuals with mania may alternate between feeling frayed, fractured, and anxious and feeling perfectly complete and elated. The quick fluctuations in their moods are accompanied by equally rapid fluctuations in the sense of self. One minute the individual feels sullen, useless and agitated: the next minute like a conquering hero. These changes in mood are sudden, unpredictable and uncontrollable. There can be feelings of loss when the mania dissipates – a yearning for the emotional intensity, ecstasy and productivity. Many people with manic tendencies seek out "upper" drugs such as cocaine or methamphetamine to bring back or intensify the euphoria of mania.
>
> **Cognitive patterns:** These may include fantasies of invincibility and exceptional talent; a sense of capacity to succeed at any task regardless

of preparation or training; wishes for fame or adoration; and difficulty thinking clearly, linearly or logically. Individuals may fear that they cannot hold onto their thoughts, which seem flighty and ungraspable. At times, individuals in manic states can feel highly disoriented as they are often unable to identify which of their racing thoughts are important or relevant. At other times they may experience the welter of thoughts as freeing and joyful. They frequently express the contents of their mind without self-censorship and inhibition.

Somatic states: These may include restlessness and sleeplessness. Some people in manic states insist that sleep is a needless waste of time. They may feel physically invigorated and aroused and describe a need to "keep going"! Don't Stop!". Sexual desire is frequent and intense. Physical exhaustion, which may evoke suicidal depression, is a serious danger.

Relationship patterns: These are often unpredictable, chaotic, impulsive, and sexualized. Some people with manic tendencies inspire followers and proteges whose mood may be elevated by sharing in the grand schemes of the manic person.

Clinical symptoms of bipolar I disorder are the most dramatic of the bipolar diagnoses because the bipolar I diagnosis requires the occurrence of a manic episode. A manic episode requires functional impairment. One lifetime manic episode is sufficient to make a diagnosis of bipolar I disorder (unless the episode has been precipitated by medication or a substance). It is hard to be in a state of mania, which requires functional impairment, without also being psychotic, so mania is usually associated with psychosis. Milder episodes of activation not associated with functional impairment, hypomanic episodes, are common in bipolar I disorder. Depression is not required for a diagnosis of bipolar I disorder but may be present.

2.1.2 BIPOLAR II DISORDER

Both girls walked into the fraternity, exuberant, enthusiastic, gregarious, the life of the party. They were fun-loving if not a bit brash. They partied all night and drank a few guys under the table, then got up early the next day and went to class without missing a beat. For one girl, this was a normal, stable personality. For the other, it was the face of hypomania.

What was the difference? The first girl was exuberant and gregarious all the time: this wasn't just her party face or the time of the week. She was naturally a high-energy and sunny person. Her personality was stable: the person she was at a party was the person she was while sitting in class at school. For the girl in a state of hypomania, these behaviors represented a change from her normal function, which she could not sustain, and her hypomanic (party) state cycled, unprovoked, into a depressed state. When depressed, her friends noticed she came to class in sweatpants and no makeup for a few days (instead of her usual stylish and impeccable look), then disappeared for a few days.

A diagnosis of bipolar II disorder requires a history of only one episode of hypomania and one episode of major depression. Hypomania, using the Greek prefix "hypo" meaning "under," describes a state which is less severe in intensity than mania. Psychosis and functional impairment cannot be symptoms of hypomania.

During hypomanic episodes, people typically take risks they would not take in a normal or depressed state. Reckless driving, unsafe sex, unrestrained spending, and overuse of substances that would otherwise not occur in a "baseline state," all are features of a hypomanic episode. Need for sleep can be diminished, pace of speech can pick up, thoughts can race, self-esteem can soar: hypomania is an activated state. The changes must be noticeable and a distinct departure from a baseline level.

During episodes of hypomania, mood does not have to be euphoric. though euphoria can alternate with dysphoria. The words "euphoria" and "dysphoria" come from the Greek "eu-" meaning "good" and "dys" meaning "bad," and the word "phero," "to bear or carry. Euphoria is a part of our everyday language but dysphoria is not. One literal translation of the word dysphoria is "difficult to bear," which is an apt description of the feelings of people whose moods are dysphoric. But it does not clearly differentiate the negative phase of dysphoria from the negative phase of depression, which is a crucial distinction. Dysphoria is the negative mood that accompanies the activated hypomanic/manic state. It is a bad mood but not a depressed mood. It is usually associated with irritability, restlessness, and agitation. It is a negative mood that accompanies the activated state but is distinct from a negative mood of sadness and hopelessness that accompanies the slowed, depressive energy state.

Bipolar II disorder is typically characterized more by bad moods than by good moods. Hypomanic states alternate with depressive states but negative moods tend to predominate, be they dysphoria or depression.

Note that a major depressive episode is different from major depressive disorder. Major depressive disorder, as we will see, requires a major depressive episode AND no evidence of symptoms of hypomania/mania. A major depressive EPISODE that occurs during the course of bipolar disorder and a major depressive EPISODE that occurs during the course of major depressive disorder share the exact same criteria. Leonhard or Kraepelin?

2.1.3   CYCLOTHYMIA

When moods cycle from energized states (less severe than in mania) to slowed/depressed states (less severe than in major depression), a diagnosis of cyclothymia may be present. Cyclothymia is a bipolar diagnosis with attenuated cycles of energy states. People with cyclothymia may not always have visible mood disorders to those who do not know them well.

2.1.4   BIPOLAR SPECTRUM

If one accepts as a working model the DSM classification of bipolar affective disorders, then one might imagine a range of mood disturbances arrayed linearly along a spectrum of severity from bipolar I disorder to cyclothymia. One can see the evolution of this spectrum since 1952 from DSM in which there was only one category, "affective reaction" (manic or depressive) to DSM-III with the separation of bipolar disorder and major depression and the introduction of cyclothymic disorder, to DSM-IV with the distinction between bipolar I and bipolar II disorders.

A visual model for thinking about the bipolar spectrum along such an axis might be that of a prism casting off a rainbow of light. On the RED end of the spectrum, we find people with symptoms like Aunt Sadie. Like all individuals with cyclical mood disorders, shifts in her mood are generally triggered by biologic, internal, endogenous rhythms, not by people or circumstances. Aunt Sadie's highs are energized and frequently (but not always) happy, elated, and euphoric. Lows are often so extreme suicide is a high risk. Because her behavior creates so much impairment in her life and is associated with psychosis, she has symptoms of what we call bipolar I disorder. In the ORANGE zone of the spectrum, we might find people without psychosis but whose symptoms are still so functionally debilitating, they require hospitalization: they also have a diagnosis of bipolar I disorder. People without psychosis but with protracted episodes of hypomania cycling with severe major depression might be found in the YELLOW area of the spectrum with a diagnosis of bipolar II disorder. Those with brief and less severe episodes of hypomania cycling with less severe or perhaps very infrequent major depression might be found in the GREEN area of the spectrum, also with a diagnosis of bipolar II disorder. And those whose moods cycle without symptoms of hypomania or major depression, diagnosed with cyclothymia, would be found in the BLUE region of the spectrum.

2.1.5 "SOFT" BIPOLAR SPECTRUM

While this model might be useful in visualizing the diagnostic system as it is represented in DSM-5, the DSM-5 descriptive system does not adequately account for our clinical reality. When one encounters the vast range of people with mood disorders, there are many whose symptoms do not fit neatly into DSM criteria for any particular diagnosis. There are a number of descriptive limitations in the DSM bipolar classification spectrum of diagnoses to capturing the range of symptoms people with mood disturbances describe:

1) variations in severity of symptoms
2) variations in duration of episodes
3) mixtures of hypomanic/manic and depressive states
4) non-cyclical mood shifts

Spearheaded by the work of the late Hagop Akiskal, who advanced our thinking about bipolar disorders, an enormous body of literature has been contributed to the field over the past four decades. Akiskal and colleagues greatly expanded the range of DSM bipolar diagnoses, proposing a "soft" bipolar spectrum to much more accurately reflect the plethora of mood disorders seen in clinical reality. Some of the proposed syndromes on the soft bipolar spectrum of disorders not listed in the DSM are more severe and lie toward the RED end of the spectrum in our model, but much of what has been written about the soft bipolar spectrum pertains to milder syndromes which we might say fall in the VIOLET area of the spectrum, adjacent to the BLUE area where DSM cyclothymia is situated.

The soft bipolar spectrum proposed by Akiskal is an expansion of the DSM bipolar disorder criteria and includes some of the following not found in the DSM:

Bipolar ½: Schizomanic or Schizobipolar Disorder
                An extreme form of manic psychosis in which the psychotic features are mood-incongruent and persist between mood episodes (RED end of the spectrum)

Bipolar I: (DSM) Classic Mania
                Classic bipolar I disorder. In comparison with a state of hypomania, in mania, meaningful conversation is difficult to sustain for any length of time; euphoric or ecstatic mood can deteriorate into belligerence if thwarted; frank delusions of grandiose ability, of identity, of persecution, or of love may exist; loss of insight and judgment may lead to frenzied activity and serious social impairment (Akiskal et al., 1977).

Bipolar I ½: Depression with protracted hypomania
                Hypomanic episodes are generally considered to last only days; in this situation, hypomania is more intense and of longer duration than in bipolar II disorder. Hypomanic symptoms are cheerfulness and jocularity, gregariousness, and people-seeking, heightened sex-drive, talkativeness, and eloquence, overconfidence, over-optimism, disinhibition, and carefree attitudes, little need for sleep, and overinvolvement in new projects (Akiskal et al., 1979).

Bipolar II: (DSM) Depression with hypomania
                Major depression is moderate to severe in impairment and hypomanic episodes are at least 4 days in duration. Judgment is relatively preserved compared with mania.

Bipolar II ½: Cyclothymic depression
                Short periods of hypomania cycling with mini periods of depression. Mood often changes from happy to sad without a sense of why. There may be frequent ups and downs without apparent cause, frequent feelings of guilt without reason, pessimism about the future, feeling easily hurt by others, trouble falling asleep because of thoughts running

through the head about the day, and frequent feelings of disgruntlement (Akiskal et al., 1995).

Bipolar III: Antidepressant-associated hypomania
Hypomania (and mania) can emerge spontaneously and unexpectedly during antidepressant treatment. Careful questioning usually reveals a history of a family member with a bipolar disorder (Rosenthal et al., 1981).
(Though Akiskal did not account for this in his soft bipolar spectrum, we typically associate bipolar III with any medication induced hypomania, not just antidepressants, thus, I would add ADHD-associated psychostimulant medication to the bipolar III category, e.g., Adderall, Ritalin, as the use of these medications can precipitate hypomania/mania as well.

Bipolar III ½: Bipolarity masked and unmasked by stimulant abuse. Abuse of stimulants (cocaine, crack, crystal meth) to heighten the effects of moods on the upswing. Withdrawal leads to dysphoria and depression, driving the need for increased stimulant use. Stimulant abuse destabilizes an underlying mood dysregulation.

Bipolar IV: Hyperthymic depression
An episode of clinical depression, usually later in life, superimposed on a lifelong hyperthymic temperament. The hyperthymic temperament is upbeat and exuberant; articulate and jocular; overoptimistic and carefree; high energy/ full of plans; versatile with broad interests; overinvolved and meddlesome; uninhibited and risk-taking; habitual short sleeper (less than 6 h/night) (Akiskal, 1992a).

The incidence of bipolar disorder increases significantly if the spectrum is widened to include these soft spectrum syndromes that are not recognized in DSM-5. Proponents of the soft bipolar spectrum model estimate that 50% of all mood disorders are on the bipolar spectrum (Benazzi, 1997; Cassano et al., 1992; Manning et al., 1997; Perugi et al., 1998). The lifetime incidence of bipolar I and bipolar II disorders has been estimated to be 1.3% (Angst et al., 2002; Strakowski et al., 2011) while the incidence of a bipolar disorder broadly defined on the soft spectrum is as high as 5% (Akiskal et al., 2000).

Clearly, as the soft bipolar spectrum expands, there are many more people on the opposite end of this mood spectrum from Aunt Sadie whose resemblance to her is increasingly remote. We do not see people with syndromes in this VIOLET area of the spectrum reflected in our current nomenclature, but Akiskal's notion of a soft bipolar spectrum is helping us recognize the clinical presentations unaccounted for in our current diagnostic system by addressing:

1. Severity of symptoms: symptoms of "hyperthymic depression" are less severe than those in hypomania but are reflected on the soft bipolar spectrum as BIPOLAR IV.
2. Duration of episodes: hypomanic episodes are often shorter than the 4-day DSM criterion. There is discussion regarding whether this should be revised to 1–3 days (Wicki & Angst, 1991). There is no specification for duration of hypomanic episodes in the soft bipolar spectrum diagnoses of BIPOLAR III and III ½.
3. Mixed states of depression and hypomania/mania are unaccounted for on the soft bipolar spectrum. DSM-5 makes a somewhat unsatisfying attempt to account for them.
4. Non-cyclical mood shifts are characteristic of mood dysregulation and will be discussed in Chapter 4.

2.1.6 MIXED STATES

Depression was classically described as a melancholic state characterized by depressed mood, psychomotor retardation, slowed thinking, and diminished energy: the focus was on **reduction** in the dimension of **energy**. Mania was classically thought of as an activated state characterized by agitation, excess energy, flight of ideas, and euphoric mood: the focus was on **activation** and increased energy. *Energy states* were the fault line around which the two poles of the disorder were described.

But many people describe feeling depressed yet appear to be agitated. And there are people who are agitated but not euphoric. Symptoms do not always fall along such clear-cut fault lines, raising the question of "mixed states."

Mixed states have been described since the time of Hippocrates, but interest in them was renewed in the mid-19th century when Frank Richarz (1858) wrote about "melancholia agitans" and Jules Falret (1861) about "etat mixte." Kraepelin was the first to elaborate on mixed states in a systematic way in his presentation of "manic-depressive insanity." He described the manic mixed state as "mania with inhibition of thought, manic stupor, querulous mania, states of transition, and depression with flight of ideas" (Kraepelin, 1899b) and the depressive mixed state as characterized by "depressive agitation and mania with poverty of ideas" (Kraepelin, 1904).

Interest in mixed states continued through the early 20th century with Weygandt (1899) before subsiding for half a century (Marneros, 2001). The question of how to classify states of "agitated depression" in the DSM sparked a resurgence of interest in the last quarter of the century (Himmelhoch et al., 1976; McElroy et al., 1992): should agitated depression be considered a subtype of major depression or should it be considered a mixed state?

The DSM has dealt inconsistently with the question of mixed states in the last four decades. The "mixed" diagnostic concept was first incorporated into DSM-III in 1980 as a subtype of bipolar disorder: bipolar disorder was classified either as manic, depressed, or mixed, depending on the nature of the most recent episode. These restrictive DSM criteria failed to conform to clinical reality, nevertheless, only modest changes were made to the DSM description of a mixed state over the course of nearly 35 years until the publication of DSM-5 (2013).

In DSM-5, mixed states were defined separately for manic episodes and depressive episodes, and criteria were loosened to reflect clinical reality more realistically, though DSM-5 still results in a grossly under inclusive diagnostic system (Perugi et al., 1997), which is inconsistently used by everyone. In short, DSM-5 requires a mixed state in major depression to reflect at least three concurrent symptoms of hypomania/mania, and a mixed state in hypomania/mania to reflect at least three concurrent symptoms of major depression. To this day, there remains no consensus regarding an operational definition of a mixed state.

By some estimates, 40% of people with bipolar disorder will develop mixed states at some point in their lives (Goodwin & Jamison, 1990). Johns Hopkins professor of psychiatry Kay Jamison writes in her poignant memoir *An Unquiet Mind*, documenting her personal struggle with bipolar disorder:

> The word "bipolar" seems to me to obscure and minimize the illness it is supposed to represent… and it minimizes the importance of mixed manic and depressive states, conditions that are common, extremely important clinically, and lie at the heart of many of the critical theoretical issues underlying this particular disease (Jamison, 1996).

2.2 DSM-5 DEPRESSIVE DISORDERS

There was a major revision in the depressive disorders section in DSM-5 since the publication of DSM-IV. In DSM-IV, there were three diagnoses for depression: (1) major depression, (2) dysthymia, (3) depressive disorder not otherwise specified. This became expanded in DSM-5 to eight diagnoses, of which three are the subject of greatest focus: (1) disruptive mood dysregulation disorder, (2) major depressive disorder, (3) persistent depressive disorder. In addition, diagnoses have been added for (4) premenstrual dysphoric disorder, (5) substance/medication-induced depressive disorder, (6) depressive disorder due to a medical condition, (7) other specified depressive disorder, and (8) unspecified depressive disorder.

2.2.1 *DISRUPTIVE MOOD DYSREGULATION DISORDER*

Disruptive mood dysregulation disorder (DMDD) is quite an interesting addition to the depressive disorders section of the DSM, given the fact that the impetus for this new diagnosis was to find a way to reduce the frequency of diagnosing bipolar disorder in children. Depressed mood is not a criterion for DMDD, but angry/irritable outbursts (as often seen in dysphoric hypomania) are the cardinal feature of a DMDD diagnosis. Dysphoric hypomania is frequently seen in mood dysregulation, suggesting a connection between DMDD, mood dysregulation, and bipolar disorder. DMDD appears to rest somewhat uneasily among the depressive disorders. We will discuss it at length in Chapter 5 on diagnoses related to mood dysregulation.

2.2.2   MAJOR DEPRESSIVE DISORDER

In DSM-5, the hallmark of a major depressive (slowed) episode is depressed (sad, low, blue, hopeless, down, and pessimistic) mood most of the day, nearly every day, or loss of interest in most things for at least 2 weeks. Associated disturbances in appetite, sleep, energy, concentration, and self-esteem must be present as well. A depressive EPISODE with no history of hypomania or mania can be diagnosed as major depressive DISORDER.

Major depressive disorder as it stands under depressive disorders is conceived of as Leonhard's unipolar depression. Naturally, to add a "mixed features" specifier betrays ambivalence about Leonhard's conception, but for some of us, the presence of mixed features locates a major depression on the bipolar spectrum (Akiskal, 1996a; Akiskal & Pinto, 1999; Sato et al., 2003). We would certainly not treat "depression with mixed features" as a "simple" (unipolar) depressive disorder but would approach a mixed feature depression with mood stabilizers in hand.

A weakness in the DSM diagnostic criteria for major depressive episode/ disorder is the failure to contrast depression with hypomania/ mania along the dimension of energy. In other words, while increased energy/activity is a requisite for a diagnosis of a hypomanic or manic episode, decreased energy is NOT a requisite for a diagnosis of a depressive episode. The dimension of energy was historically a demarcation between states of depression and states of hypomania/mania, and lack of DSM recognition of this fault line, I believe, is responsible for much of the current confusion regarding what constitutes a mixed state.

2.2.3   PERSISTENT DEPRESSIVE DISORDER

Persistent depressive disorder (PDD) is new in DSM-5 and is a conflation of dysthymia with major depressive disorder. Until DSM-5, depressive diagnoses were bifurcated in terms of severity, and episodes of major depression were diagnosed separately from chronic, low-grade depression. In reality, people with chronic, low-grade depression who never really feel well experience periods of episodic worsening of symptoms that meet criteria for major depression from time to time. DSM-5 makes provision for the description of this clinical reality with the PDD diagnosis.

Our primary diagnostic classification system at present is DSM-5, which subdivides mood disorders into bipolar disorders and depressive disorders according to the scheme of Leonhard.

The three primary DSM-5 bipolar disorders, bipolar I disorder, bipolar II disorder, and cyclothymia, form a spectrum of disorders in order of severity. Currently, there are three major depressive disorders: disruptive mood dysregulation disorder (which we will see behaves much like mood dysregulation) major depressive disorder, and persistent depressive disorder (which is dysthymia with or without major depression).

The dramatic state of mania is the hallmark of bipolar I disorder, associated with functional impairment and often with psychosis, generally requiring hospitalization. While euphoria is a stereotypic symptom, restless agitation, paranoia and lack of insight into the disturbance are also typical of the state. Switches from a stable baseline mood of depression to states of mania can be sudden and unpredictable. Depression need not be a feature of bipolar I disorder.

Bipolar II disorder is much quieter. Hospitalization is generally not necessary unless indicated for the depressive phase, and if there is psychosis in bipolar II disorder, it is only in the setting of major depression. Where mania is the more disruptive state in bipolar I disorder, depression is more disruptive in bipolar II disorder: most people with bipolar II disorder spend the preponderance of their time in negative mood phases, either dysphoria or depression. Cyclothymia involves cycles of elevated mood which are never hypomanic or manic, and depressed mood which are never as severe as in major depression. The only other bipolar diagnostic categories in DSM-5 are unspecified.

Clinical reality, however, exposes limitations to the DSM-5 classification of mood disorders. Failure of the existing diagnostic categories to capture the range of natural phenomena led Akiskal and colleagues to define a much broadened "soft" bipolar spectrum of diagnoses. As clinical reality diverges from our classification systems, authors of the DSM are showing increasing ambivalence about the dichotomization of affective disorders and struggling with the question of "mixed" hypomanic/manic and depressive states. Further, naturalistic presentation of symptoms demands us to challenge the categorical "present v. absent" requirements of our current diagnostic criteria, necessitating an updated approach to extant classification systems. We will look more closely at these systems in the next chapter.

KEYWORDS

- ICD (International Statistical Classification of Diseases and Related Health Problems)
- DSM (Diagnostic and Statistical Manual of Mental Disorders)
- dysphoria
- bipolar spectrum
- soft-bipolar spectrum
- mixed states

CHAPTER 3

Diagnostic Systems

*I was much crazier than I had imagined. Or maybe it was a bad idea
to read DSM-IV when you're not a trained professional. Or maybe the
American Psychiatric Association had a crazy desire to label all life a
mental disorder.*

—Jon Ronson
The Psychopath Test: A Journey Through the Madness Industry

ABSTRACT

Inadequacies in our current diagnostic classification systems are not limited to the
realm of affective disorders but pervade our approach to behavioral disorders in
general. Converging factors have given rise to an overhaul in our current diagnostic
system, including growing appreciation of the multiplicity of determinants to
behaviors the broad spectrum along which disturbances lie, the co-occurrence of
various disturbances, and the need to destigmatize behavior. A crucial recognition
in the evolution of rethinking our diagnostic systems was an understanding that
symptoms need to be untethered from particular diagnoses as they so frequently
cross diagnostic lines. This transdiagnostic paradigm has become the foundation for
a new research initiative underway at the National Institute of Mental Health to study
and classify behavioral disorders, known as Research Domain Criteria (RDoC). The
RDoC classification system aims to describe behaviors along a continuum in relation
to their biological underpinnings and environmental influences. This approach would
seem to promise a home for today's "diagnostic orphans" such as those with mood
dysregulation.

A number of factors gave momentum to the push for an overhauled
diagnostic approach in the psychiatric field. Diagnoses in the DSM and
in Chapter V of the ICD were designed to rely on clinical observation,

Mood Dysregulation: Beyond the Bipolar Spectrum. Deborah A. Deliyannides, MD (Author)
© 2024 Apple Academic Press, Inc. Co-published with CRC Press (Taylor & Francis)

to be behavior based, and to be agnostic with regard to etiology. With the publication of DSM-III, diagnostic classification was supposed to be focused on validating clinically observed symptoms by expert consensus. But 50 years later, no single categorical diagnosis had fulfilled all these criteria (Sanislow et al., 2015). Furthermore, the pace of research has been unacceptably slow with its reliance on the categorical, symptom based diagnostic systems found in DSM and ICD (Morris & Cuthbert, 2012).

In addition, while the DSM was designed to capture clinical phenomenology and to facilitate research, it has not only failed at this but has begun to serve as a source of stigmatization, at times causing the very individuals it attempts to help, avoid seeking treatment for fear of being labeled (Ben-Zeev et al., 2010). Carrying a diagnostic stigma undermines self-esteem and impairs social functioning (Link & Phelan, 2001; Perlick et al., 2001), and caregivers of people with psychiatric diagnoses often cope with stigmatization by withdrawing (Hawke et al., 2013). Diagnostic stigma is also associated with discontinuation of medication, which only exacerbates the problem of the person seeking help (Kamaradova et al., 2016). In countless articles I reviewed regarding psychiatric nomenclature, I found only one in which there was acknowledgment of concern for the potential stigmatizing implications of a diagnosis (Nusslock & Frank, 2011).

Another limitation of our current diagnostic manuals is that they convey the impression that disorders have discrete causes as do so many physical illnesses. The reality is that most mental health issues have causes that are complex, multiple, and poorly understood (Kendler, 2008), emerging from an interplay between biological, behavioral, psychosocial, and cultural processes that do not respect established boundaries.

A further limitation is the binary notion of "present" versus "absent" required by DSM and ICD to make a diagnosis (Regier et al., 2013) and the imposition of artificial categories on symptoms which, in actuality, lie on a continuum, sacrificing the richness of clinical reality. This leads to massive heterogeneity (and loss of data) within any one diagnostic category. For example, two individuals meeting DSM-5 criteria for major depression might have only one symptom (out of nine possible) in common (Dalgeish et al., 2020). In such a system, many individuals in psychological distress fall short of meeting the criteria for any diagnosis despite a need for care (Kotov et al., 2018).

Some researchers have already incorporated a diagnostic shift from the categorical to the dimensional framework. Recall our discussion of mixed affective states. A group in France, recognizing that affective states exist along a continuum, has been looking at mixed states from a dimensional framework around the scale of increased/decreased energy. They have developed a self-rating instrument (the Multidimensional Assessment of Affective States or MAThyS) to identify a range of mixed states using indices of activation/inhibition and emotional reactivity (Henry et al., 2008).

An increasingly recognized problem is that of comorbidity, the co-occurrence in an individual of two or more diagnoses. Comorbidity is the rule rather than the exception and single, uncomplicated clinical presentations are relatively scarce. This suggests that the normative existence of psychiatric disorders must, to a great extent, be an artifact of the categorical classification system (van Loo & Romeijn, 2015). In a research world driven by our current diagnostic system, clinical intervention is diagnosis driven, based on evidence from clinical trials. In reality, however, evidence-based treatment is inadequate to address the comorbid conditions encountered on the ground, necessitating the increasing use of off-label treatments.

These and other inadequacies in the DSM and ICD diagnostic systems paved the way for a new research framework aimed at classifying human behavior in ways that more accurately describe the world of "mental ill health," accounting for a continuum of phenomenological symptoms as well as appreciating the intricacies of their etiologic underpinnings. Such a system would recognize that symptoms fall along a spectrum of severity, would acknowledge the etiologic complexity of various behavioral disturbances, and would aim treatment at factors deemed to be alterable. It would also reflect the fact that many symptoms co-occur, and those co-occurring symptoms can appear across a range of diagnostic categories. This "trans-diagnostic" perspective became the basis for the evolving classification system underway at the National Institute of Mental Health, the Research Domain Criteria (RDoC).

The RDoC paradigm is a new approach to research in which human behavior is studied within six different neuropsychological domains, each on a continuum from deficit to an excess of function. In the RDoC system, mental illness is regarded to be a dimensional extreme on a spectrum of normal human emotion, cognition, or behavior. Where DSM diagnoses are designed to be agnostic with regard to etiology, the RDoC system seeks deliberately to uncover the neurophysiologic, environmental, and genetic underpinnings of behavioral disturbances.

The six RDoC domains are:

1) negative valence systems (e.g., responses to fear, anxiety, loss)
2) positive valance systems (e.g., reward-seeking behavior)
3) cognitive systems (e.g., attention, perception, memory, language)
4) social processes (e.g., attachment, self/other perception)
5) arousal and regulatory systems (e.g., sleep, energy)
6) sensorimotor systems (e.g., motor function) (Ross & Margolis, 2019)

The RDoC initiative now guides the National Institute of Mental Health (NIMH) in the funding of research grants utilizing genetics, neuroscience, and behavioral science to understand, describe, and define mental illness. Thomas Insel, who oversaw the RDoC project at NIMH, clarified some of its assumptions as follows (Insel, 2013):

1) Diagnoses must be based on biology as well as symptoms.
2) Mental disorders involve brain circuits and are therefore biological illnesses.
3) Levels of analysis must be considered across dimensions of function.
4) Mapping of different aspects of mental disorders will aid the development of targeted treatments.

RDoC is a different initiative altogether from the DSM and ICD. The DSM was conceived as a taxonomy of behavioral disorders. It fails to capture the infinite iterations of symptom configurations we see clinically, and it does not account for the underlying pathophysiology of behavior. RDoC aims to address these concerns and to provide a platform to guide the research to answer these questions. In addition, RDoC is a dimensional system rather than a categorical one, describing behavior along a spectrum of normal to abnormal rather than in an either/or fashion. Third, RDoC starts from the ground-up, linking brain processes to clinical symptoms where DSM and ICD work from the top-down, starting with diagnoses and seeking to fit behavior into those categories. And finally, RDoC incorporates a wider range of data, including genetics, biology, and physiology into its diagnostic formulations whereas DSM only incorporates symptom reports or observations. As described by Insel,

RDoC aims to transform diagnosis by incorporating genetics, imaging, cognitive science, and other levels of information to lay the foundation for a new classification system.

Perhaps one of the greatest weaknesses of the DSM system is its agnostic approach to the etiology of disorders. RDoC's framework provides an important corrective to this in that its six domains are motivational systems. This may be the first systematic attempt by psychiatrists to look at human behavior in terms of biologically based motivational systems. Freud had an instinctual drive theory but did not systematize motivation. Psychologist Silvan Tomkins proposed a motivational theory deriving from nine innate affects based on his observation of infant behavior (Nathanson, 1987; Tomkins, 1980). These are: (1) distress/anguish, (2) interest/excitement, (3) enjoyment/joy, (4) surprise/startle, (5) anger/rage, (6) fear/terror, (7) shame/humiliation, (8) disgust, and (9) dissmell.

Psychoanalysts Lichtenberg and colleagues (2011) posited seven motivational systems which, interestingly, roughly parallel the six RDoC domains (RDoC domains in CAPS):

1) the need for psychic regulation of physiological requirements (Arousal/Regulatory Systems)
2) the need for attachment (Social Processes)
3) the need for exploration/assertion (Sensorimotor Systems)
4) the need to react aversively through antagonism or withdrawal (Negative Valence Systems)
5) the need for sensual enjoyment/sexual excitement (Positive Valence Systems)
6) the need for affiliation (Social Processes)
7) the need for caregiving (Social Processes)

These motivational systems are biologically hardwired and self organizing, exerting regulatory effects on one another (Fosshage, 1995). But Lichtenberg and colleagues view their motivational systems as tools to guide treatment with the (perhaps appropriate) traditional psychoanalytic caution toward diagnostic interest.

The late psychobiologist Jaak Panksepp, who spent his life studying the neuroscience of emotion, identified affective/motivational circuits in the mammalian brain and attempted to associate them with behavioral disturbances. Based on decades of research with animals and humans, Panksepp identified seven core affects or motivations and proposed corresponding diagnoses that might be associated with disruptions in these biologic/functional circuits (Panksepp, 1998).

SEEKING (+ & -)	Interest	Obsessive/Compulsive
	Frustration	Paranoid schizophrenia
	Craving	Addiction
RAGE (+ & -)	Anger	Aggression
	Irritability	Psychopathic tendencies
	Contempt	Personality disorders
	Hatred	
FEAR (-)	Anxiety	Generalized anxiety
	Worry	Phobias
	Psychic trauma	PTSD and variants
PANIC (-)	Separation distress	Panic attacks
	Sadness	Pathological grief
	Guilt/shame	Depression
	Shyness	Agoraphobia
	Embarrassment	Social phobia
PLAY (+)	Joy and Glee	Mania
	Happy playfulness	ADHD
LUST (+ & -)	Erotic feelings	Fetishes
	Jealousy	Sexual addictions
CARE (+)	Nurturance	Dependency disorders
	Love	Autistic aloofness
	Attraction	Attachment disorders
(SELF (core consciousness)	(a mechanism for all emotion)	(Dissociative disorders?)

It is easy to appreciate how the American Psychiatric Association had "a crazy desire to label all life a mental disorder" in its publication of the DSM as Jon Ronson experienced, and how intimidating the DSM can be when one approaches it out of context. It is a two-edged sword, after all, because while the DSM serves to guide research and treatment decisions, it also adds to the stigmatization of the very individuals we seek to help and leaves others without help if their symptoms fail to meet diagnostic criteria for a reimbursable condition. The limitations of our current nomenclature are increasingly evident as binary classification leaves a wealth of clinical information unaccounted for while naturalistic phenomena, which generally occur on a spectrum, are overlooked in the service of textbook categorical diagnoses. These and other problems have obviated the need for a new approach to understanding "mental ill health." Such an initiative is underway at the National Institute of Health, utilizing the new RDoC to describe behavior, employing a dimensional approach that recognizes symptoms across a spectrum while attempting to understand their etiology and neurobiological underpinnings. The RDoC approach accounts for the transdiagnostic nature of many behavioral phenomena which co-occur across definitional boundaries.

KEYWORDS

- **stigmatization**
- **comorbidity**
- **transdiagnostic**
- **dimensional framework**
- **continuum**
- **RDoC (Research Domain Criteria)**

CHAPTER 4

Mood Dysregulation

And those who were seen dancing were thought to be insane by those
who could not hear the music.
—Friedrich Nietzsche

ABSTRACT

The implications of our diagnostic tools are not inconsequential. If a psychiatrist unthinkingly uses the DSM to find a diagnosis for a person consulting for "depression," a mistaken diagnosis of a depressive disorder is often chosen, leading to problematic treatment. Looking at symptom patterns however, and recognizing dysphoria in particular, can reveal previously unseen mood dysregulation, leading to more appropriate treatment with improved outcome. People with mood dysregulation experience negative moods which, for want of a better description, they report as "depression." But these people are typically irritable, agitated, reactive, and anxious: these are the negative symptoms of dysphoria characteristic of the activated/energized state that points toward hypomania. Dysphoria is a clue to the relationship between mood dysregulation and bipolar disorders. Such a similarity in phenotype (the visible expression of a gene) suggests shared biologic underpinnings. Some have proposed that all mood disorders arise from "affective temperaments," genetic vulnerabilities that are variously expressed depending on environmental influence. The commonality between bipolar disorders and mood dysregulation might be on the subterranean, genetic level of affective temperaments. In this invisible way, mood dysregulation might be tethered to bipolar disorders in the ultraviolet space just beyond the visible bipolar spectrum. Yet mood dysregulation should not be regarded as a diagnosis but as a characteristic pattern of symptoms that is present across a spectrum of diagnoses: it is a transdiagnostic phenomenon.

It took me some time to recognize because I was so well indoctrinated in the DSM that I rather automatically sought to find a diagnostic home for

Mood Dysregulation: Beyond the Bipolar Spectrum. Deborah A. Deliyannides, MD (Author)
© 2024 Apple Academic Press, Inc. Co-published with CRC Press (Taylor & Francis)

each individual who came to me for consultation. But there seemed to be no home for one particular group. They did not fit neatly into any familiar diagnostic category. People in this group came in for the treatment of "depression," typically describing themselves as "treatment resistant." But they were mostly irritable and angry. Their negative moods were usually not "depressed" in the classical sad way but in an agitated, activated way, reminiscent of people with hypomania or mania. They were volatile and exquisitely reactive to interpersonal slights or to small disappointments which triggered temper flares that were grossly disproportionate to the provocation. Sometimes they had the more classical sadness of depression but activated states and slowed states did not cycle endogenously as in bipolar disorders. Instead, they shifted when provoked by external factors such as relationship problems or upsetting events in their lives. The overall picture was not typical of a depressed person but of one with a diminished capacity to regulate emotion. They neither met criteria for a bipolar disorder nor a depressive disorder. I came to describe their symptoms as "mood dysregulation."

4.1 NOMENCLATURE

A review of some terms as I use them will help clarify the discussion of mood dysregulation. Symptoms here are assessed along three dimensions: energy, mood, and cognition. The dimension of energy is the fault line around which states of depression and hypomania/mania have classically been defined when diagnosing bipolar disorders. Thus, when we use the term "bipolar," we think of two opposing poles not in terms of mood or affect but in terms of energy. The "poles" of "bipolar" are opposing energy states, not moods (or cognition).

The depressive state, by definition, is a state of decreased energy. People in a depressive state show psychomotor retardation and decreased arousal. In contrast, the hypomanic/manic energy state, by definition, is a state of increased energy, hyperarousal, and activation.

Within these two energy states are the dimensions of mood and cognition.

Typical mood symptoms in the depressive/slowed energy state are depression, sadness, hopelessness, and pessimism. Typical mood symptoms in the activated hypomanic/manic energy state are more complicated because they may be either positive (euphoric) or negative (dysphoric). Euphoria is associated with happiness, optimism exuberance, and excitement, whereas dysphoria is associated with irritability, anxiety, and feelings of upset.

The typical cognitive symptoms in the depressive/slowed energy state are mental slowing and poor concentration. The typical cognitive symptoms in the activated hypomanic/manic state, whether euphoric or dysphoric, are "racing" thoughts, which are often reported as "inability to concentrate" (and which can often be perceived as "anxiety" as well.)

Bipolar Disorders are defined by the cycling of energy states. When we think of bipolar disorder, we envision a horizontal line intersected by a sine wave. As the sine wave travels below the line, we enter the negative (depressive/slowed) energy state of decreased arousal. As the sine wave travels above the line, we enter the positive (activated/hypomanic/manic) energy state of increased arousal. Bipolar disorders may be characterized by the cycling of moods, but they are DEFINED by the cycling of ENGERY states.

Mood Dysregulation is defined by shifting energy states that do not cycle. These changes occur in response to outside triggers, such as perceived rejection, disappointments, stimulant medication, etc. Exogenous (external) factors rather than endogenous (internal, biologic) factors trigger shifts in energy states in mood dysregulation. Again, mood instability may be characteristic of mood dysregulation, but mood dysregulation is defined by shifts in energy states.

Because people with mood dysregulation typically do not have symptoms severe enough to meet criteria for hypomania or mania, I do not refer to the "hypomanic/manic state" but to the more general "activated state" in reference to the positive pole of the energy axis when discussing mood dysregulation.

The typical dimensions of mood and cognition in activated and slowed states are as given below:

Energy of activated state	= Increased arousal
Energy of depressive/slowed state	= psychomotor retardation
Mood of activated state	= euphoria OR dysphoria
Mood of depressive/slowed state	= depression/sadness
Cognition of activated state	= racing thoughts
Cognition of depressive/slowed state	= slowed thinking

Inability to concentrate is reported both in activated and depressive states. In the activated state, racing thoughts and an overall sense of agitation often interfere with one's focus, where in the depressive state, impaired concentration is related to overall physiologic retardation.

Understanding bipolar spectrum disturbances and mood dysregulation within the energy/mood/cognition framework clarifies several important issues.

1. **Depression is not the opposite of euphoria.** The term "depression" is associated with the depressive/slowed energy state (or pole) as well as with the mood of depression typically associated with the depressive/slowed energy state. Dysphoria is more aptly the opposite of euphoria: both dysphoria and euphoria occur in the activated state.

2. Mood is negative both in depressed mood and in dysphoric mood. This causes major diagnostic and treatment confusion. Recognition of the energy state often clarifies the difference. Activation/high energy accompanying a negative mood typically signifies dysphoric activation. Hypoarousal/low energy accompanying a negative mood typically signifies depressed mood.

3. Energy is increased both in euphoria and in dysphoria. In euphoria, increased energy is welcome while in dysphoria it is not: dysphoric activation is experienced as anxiety, restlessness, and agitation.

4. Mixed states can and do occur frequently with mood dysregulation. Though definitions vary widely as to what constitutes a mixed state, the common denominator involves symptoms of the activated energy state co-occurring with symptoms of the depressive/slowed energy state.

5. Dysphoria is a mood (the negative mood of the activated state) whereas mixed states are States with co-occurring symptoms from activated and slowed (depressive) states. This helps clarify the difference between dysphoria and mixed states.

4.2   MOOD DYSREGULATION

Behavioral disorders are increasingly seen to exist as occurring along a continuum as the new RDoC system attempts to reflect. Accordingly, one might think of mood dysregulation as on a continuum with the bipolar spectrum of disorders, albeit beyond its current boundaries and definition.

Mood dysregulation is related to bipolar disorders in some manner of appearance but is different in two principal ways: (1) mood shifts in mood dysregulation are not cyclical as they are in bipolar disorders and (2) mood shifts in mood dysregulation are not endogenously triggered as they typically are in bipolar disorders but require exogenous triggers.

4.2.1 PHENOTYPE

While their bipolar association with Aunt Sadie may not be readily apparent, people with mood dysregulation are conceptually related to her by shared phenotype. "Phenotype" is a genetic concept referring to the observable traits of an organism that are a composite expression of its genes ("genotype") and its environmental influences: phenotype refers to "what it looks like." Hair color is a phenotype. If a phenotype is an appearance, why should we expect to see people with mood dysregulation on the outer fringes of a spectrum defined by people with bipolar disorders?

First, the symptoms of people with mood dysregulation involve states of hyperarousal or hypoarousal as do those of people with bipolar disorder, though not in cyclical fashion. People with mood dysregulation may live in chronic states of irritable hyperarousal with no excursions into well regulated states of normal energy. The person who is constantly flying off the handle too easily has a mood dysregulation. The same person may then shift into a hypoaroused state associated with depressed mood. While not in a cyclical fashion, at some point, they shift from one state to another but not in response to endogenous, biological factors.

Second, the family histories of people with mood dysregulation, like those of people with bipolar disorder, have individuals diagnosed with or suspected to have bipolar disorder or depression. Akiskal (2002) reports that a bipolar family history is the most important validating criterion of a diagnosis on the bipolar spectrum.

And third, medications effective for people with mood dysregulation, characterized in part by mood reactivity and difficulty with impulse control, are the same medications that are effective in the treatment of people with bipolar disorder (Fava et al., 1990; Linden et al., 1986; McCormick et al., 1984).

To suggest that individuals with mood dysregulation are phenotypically related to those on the bipolar spectrum because of similar symptoms, similar family histories, and similar treatment response suggests that mood dysregulation is genetically associated with bipolar disorder. Perhaps, mood dysregulation represents partial gene penetrance or a "forme fruste" of bipolar disorders. In clinical practice, the vast majority of those we see with symptoms of a mood dysregulation have a first-, second-, or third-degree relative with bipolar disorder or suspected bipolar disorder.

4.2.2   AFFECTIVE TEMPERAMENTS

There may be another hidden commonality that explains why mood dysregulation lies remotely on a continuum with bipolar disorders as a sub-syndromal manifestation of the diagnosis. As Akiskal and colleagues developed their work on the soft bipolar spectrum, they came to focus on genetic predispositions which might become the precursors to a mood disorder. Based on the work from the 19th century, they elaborated the concept of "affective temperaments" to capture the nature/nurture character of mood disorders.

Relational patterns in an individual's upbringing often illustrate this concept. One of the people I treated told me he had no reason to be in my office. After all, nothing traumatic had ever happened to him. He could not understand why he was anxious, lonely, and sad. He just knew he was chronically on edge, sort of a worrier. As we pieced things together, it became clear that his somewhat anxious predisposition affected his mother as he was growing up: it created an aversive reaction in her that led her to avoid him, often walking away from him when he most needed her. A vicious cycle ensued between the two of them in which he would reach out to her in his anxiety and she, in reaction to this anxiety, would withdraw, exacerbating his anxiety. He was unaware of this cycle and the tension it created in him and would never have thought to tell me that he grew up in a situation of chronic stress. He had no idea of the effects that chronic stress could have on his body and on his capacity to regulate himself, including his moods, but this stress resulted in biologic changes that diminished his capacity to regulate his emotions.

We are all born with a hard-wired, biologically based temperament that is relatively independent of learning and environmental influences. This man was born with an anxious temperament. But personality develops as one's temperament interacts with the environment: this man's mother responded to his anxious temperament in ways that amplified his anxiety in an escalating cycle that reified it. Nature in tandem with nurture is determinant in the development of a mood dysregulation.

From the time of Hippocrates, with his humoral theory of disease (Akiskal, 1996b), until the middle of the 20th century, focus was on the biologic determinants of behavior. With the rise of psychotherapy in the 20th century, psychologists and psychoanalysts turned their attention to environmental influences on personality development. The nature / nurture dialectic continues to this day.

In the 1970s, psychodynamic formulations of mood disorders were favored as various authors documented the difficulty of distinguishing "temperament" from "personality" in certain characterologically expressed low-grade affective conditions such as "emotionally unstable character disorder" (Rifkin et al., 1971), "intermittent depression" (Spitzer et al., 1977), "hysteroid dysphoria" (Liebowitz & Klein, 1979), and "charactero-logic depression" (Akiskal et al., 1980).

Psychiatrists succeeded in shifting the focus from nurture back to nature in the 21st century. Akiskal and others, advancing their work on bipolar spectrum disorders, developed the concept of "affective temperaments," (Rihmer et al., 2010; Vahip et al., 2005), challenging the DSM conceptualization of personality disorders.

Kraepelin (1921) described four "sub-affective temperaments" derived from preexisting biologic traits. He regarded these temperaments to be life-long, early-onset, attenuated, sub-clinical forms of manic-depressive psychosis. He believed these sub-affective temperaments were the biological bedrock from which full-blown psychopathology emerged. His four temperamental types were:

1. Manic: emotionally intense
2. Cyclothymic: emotionally unstable
3. Depressive: low energy
4. Irritable: excessively restless

Kretschmer (1936) summarized Kraepelin succinctly: "the endogenous psychoses are nothing but exaggerated forms of normal temperament." But these temperaments need not manifest in psychopathology: if unperturbed, sub-affective temperaments are not associated with mood disorders. Environmental stressors are necessary to cause mood disorders, which assume a complex epigenetic "nature with nurture" origin.

Based on Kraepelin's sub-affective temperaments, Akiskal and Mallya (1987) developed a scheme with four affective temperaments:

1) Hyperthymic

    2) Cyclothymic
    3) Depressive
    4) Irritable
       An additional temperament was later added:
    5) Anxious/Fearful (Akiskal, 1998)

Kraepelin's sub-affective temperaments were thought of as *predispositions* to mood disorders (Akiskal et al., 2005a), not as pathological in and of themselves. Mood disorders are regarded to be on a continuum with the normal affective temperaments from which they emerge. From an evolutionary perspective, affective temperaments are regarded to have adaptive value. Thus, mood dysregulation might be found somewhere along the continuum from "normal" to full-symptom bipolar disorder.

For example, the people-seeking, extroverted, uninhibited, high-energy **hyperthymic** affective temperament is important for territoriality and leadership. A person with such an affective temperament will defend the home from challenges within and without the social group. Their warm, gregarious nature helps build social networks. The hyperthymic affective temperament only becomes problematic when exuberance leads to overoptimism and bad judgment. Naivete, overconfidence, bombastic nature, grandiosity, impulsivity, overtalkativeness, overinvolvement, and lack of inhibition are symptoms of **hyperthymia** (hyperthymic affective temperament gone awry). At its extreme, the hyperthymic temperament is associated with bipolar I disorder (Akiskal et al., 2005b).

Individuals with a **cyclothymic** affective temperament promote romance, creativity in all spheres, and procreative transmission of desirable traits (Akiskal & Akiskal, 2005a). Disturbances in the cyclothymic affective temperament result in **cyclothymia,** characterized by dysregulation in a number of dimensions including work, energy, and mood. Individuals with cyclothymia may swing from being lethargic to energetic, from being mentally confused to being creative and sharp, to having low self-esteem to being overconfident, then back to brooding. Cyclothymia is a DSM-5 diagnosis and is associated with bipolar II disorder (Akiskal et al., 2003; Hantouche et al., 1998; Perugi & Akiskal, 2002).

The individual with a **depressive** affective temperament reflects a sensitivity to suffering, tends to be self-denying, and is inclined to devote herself to others, family, institutions, and the helping professions. Maladaptation in this affective temperament leads to **dysthymia,** seen in people who are gloomy, pessimistic, and unable to have fun. Such people are often hypercritical and given to complaining, brooding and worrying, are self-critical, and are preoccupied with inadequacy, at times displaying morbid enjoyment in their failures. Dysthymia is a DSM-5 diagnosis. Although the DSM assigned dysthymia to the category of affective disorders, its authors expressed their ambivalence about the "state" versus "trait" matter with the parenthetical designation of dysthymia as a depressive neurosis until DSM-IV (Akiskal, 1992), describing it as a personality disorder rather than as a biological mood disorder. The dysthymic temperament is associated with major depressive disorder (Akiskal et al., 2005b).

The adaptive nature of the anxious affective temperament leads individuals to worry about their kin, serving to preserve the survival of the family. Maladaptation in this affective temperament leads to the anxiety disorders characterized by excessive apprehension, gastrointestinal over-stimulation, autonomic over-arousal, and tension.

Akiskal does not outline any adaptive value for an irritable affective temperament type. He describes its associated maladaptive behavior as moodiness, dysphoria, impulsivity, brooding, and a hypercritical/complaining attitude. Individuals with irritable affective temperaments can show transient flares of temper that appear to be overreactions to situations. People with irritable affective temperaments may present with chronic dysphoria. They tend to brood, are prone to anger, are hypercritical and complaining, and have a penchant for ill-humor and joking. The ill-humored joking often gives rise to biting remarks that seriously offend loved ones and associates and may further degenerate into obnoxious or overtly abusive behavior. Such people may direct foul epithets at individuals that only hours or days earlier they had adored with eternal passion (Akiskal & Mallya, 1987). Some may be easily provoked into activated states such as reactive arguments with family members or episodes of road rage. This type of irritable dysphoria is characteristic of many people with mood dysregulation.

In the past few decades, several groups have attempted to validate the temperament types of Akiskal and Mallya (1987) and to study their epidemiology and course. A number of instruments have been developed including the Temperament Evaluation in Memphis, Pisa, Paris and San Diego (TEMPS), the TEMPS-A (autoquestionnaire), the Temperament and Character Inventory (TCI), and the Munich Personality Test (MPT). Of these, the TEMPS-A has been the most widely validated and used. A review of six studies using the TEMPS-A from around the world showed significant gender differences in temperament distribution with men scoring higher in irritable and hyperthymic temperament types and women scoring higher in anxious, depressive, and cyclothymic temperament types (Vasquez et al., 2012). In a number of studies using versions of the TEMPS, researchers have demonstrated some association between affective temperaments and the later development of major mood disorders (Rihmer et al., 2010).

4.2.3 BEYOND THE BIPOLAR SPECTRUM

The importance of the work on affective temperaments is the recognition that affective disorders have symptom manifestation and severity that lie on a much broader spectrum than has been accounted for by our current diagnostic systems. Many mood disturbances arising from affective temperaments are mild and non-cyclical and would not be located on the bipolar spectrum in our prism model. Yet, they share a phenotype with bipolar disorders and are thus tethered to the bipolar spectrum. Such mood disturbances I call mood dysregulation and would locate them in the ultraviolet end of a spectrum of light cast by our prism, invisible in its association with the entire spectrum of bipolar disorders (Figure 4.1).

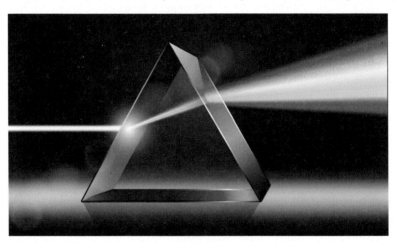

FIGURE 4.1   Bipolar disorders spectrum.

People with mood symptoms in this area of the spectrum have affective disturbances that have enough similarities with bipolar disorders to be seen as on a continuum with the bipolar spectrum, but (1) *they are not biologically cyclical*. They may shift from well-regulated states to states of hyperarousal or hypoarousal but not in any rhythmic, or cyclical manner. Rather, (2) *their associated state shifts are triggered by exogenous factors*, not by endogenous or biologically driven factors. Although exogenous factors may trigger destabilization in people with DSM-5 bipolar disorders, bipolar disorders all have some underlying endogenous cycling.

Syndromes in this far-reaching "invisible" ultraviolet end of the spectrum have no visible presence in the DSM: People with symptoms in the ultraviolet range of the spectrum have symptoms of activation but typically do not meet the criteria for DSM hypomanic episodes. Many such low-grade mood disturbances unaccounted for on the DSM bipolar spectrum fall even beyond the expanded soft bipolar spectrum of mood disturbances (Akiskal, 1986; Giuseppe, 2010; Goto et al., 2011). When activated, they are seldom euphoric: rather, they are dysphoric. Their symptoms are usually milder than those seen with bipolar disorders. On the outside, they do not behave like those with bipolar disorders but "under the hood," they are tethered to disorders on the bipolar spectrum by their excursions from depressive energy states to activated energy states. This type of liminal mood disorder in the ultraviolet region of the bipolar spectrum is characteristic of mood dysregulation.

Akiskal's earliest formulations of the soft bipolar spectrum described syndromes I would regard to be "beyond the bipolar spectrum" in the ultraviolet area in our prism model because they are not cyclical mood disorders. For example, Bipolar III and Bipolar III ½ represent mood disorders in which people switch from the depressive state to the activated state only when triggered by antidepressant or stimulant use: these are not cyclical mood disorders.

The soft bipolar spectrum expanded to include syndromes I would regard to be mood dysregulations in the ultraviolet area beyond the bipolar spectrum. This shift emerged out of the work on temperament types. Such disturbances in the invisible/ultraviolet region of the bipolar spectrum typically present as mild perturbations in a particular temperament type, not in classical symptoms of a mood disorder (Akiskal, 1984). Thus, while Perugi and Akiskal (2002) originally defined the soft bipolar spectrum (Akiskal & Mallya, 1987) around bipolar II diagnostic phenomenology, which is associated with cyclothymia (Akiskal, 1996a), they argue there is

....a spectrum" of related clinical conditions emerging from instability of the cyclothymic temperament characterized by *mood dysregulation*,

anxiety, eating disorders and impulse control (Perugi & Akiskal, 2002).
[italics mine]

While some of their proposed soft bipolar spectrum disorders are
cyclical and might fall on the visible spectrum, I have described above,
many of the syndromes they presented in their later literature do not. In
particular, they regard eating disorders and impulse control disorders,
which are clearly not cyclical, to be on the soft bipolar spectrum in what
we have clearly defined as falling in the invisible ultraviolet area (Perugi
& Akiskal, 2002). In essence, the syndromes Akiskal and colleagues
described the more they elaborated the soft bipolar spectrum essentially
appear to be mood dysregulations.

Even in the earliest iterations of the soft bipolar spectrum, a person
with a diagnosis of bipolar IV on the soft bipolar spectrum might have a
hyperthymic affective temperament, living primarily in activated states,
but, triggered by situational adversity or loss, develop depression, shifting
to a low energy, depressive state. We would regard this to be a mood
dysregulation, not a bipolar disorder.

4.2.4 DYSPHORIA

Although dysphoria is seen in other conditions, it is a red flag for mood
dysregulation. I started seeing it when I stopped making assumptions.
Though people were coming to me for the evaluation of "depression," I
stopped assuming they felt depressed. Instead, I just said, "Describe what
you are feeling." I often heard, "I just feel awful." Or "I'm irritable all the
time." Or "It's not so much depression…I just don't feel good." Or "I get
angry at the drop of a hat." It was always a negative mood, to be sure, but it
was not the sadness I knew to be associated with depression. It was irritable,
restless, angry, agitated. It was a state of activation, not the slowed state typically associated with depressed mood. I recognized it as dysphoria (since
it was the activated/negative counterpart of the activated/positive state of
euphoria) and came to see as a salient feature of mood dysregulation.

The literature about dysphoria is sparse and there is virtually no agree-
ment on what the term means (Starcevic, 2007), although "anxiety," and
"irritability" are the most consistent descriptors (Andreasen & Black,
2002; Berner et al., 1987; Bertschy et al., 2008; El-Mallakh & Karippot,
2005; Starcevic, 2007). Dysphoria becomes clear once one recognizes

irritability, agitation, and activation in the context of a very unpleasant mood that is typically not characterized by sadness. Quite often, dysphoria is triggered by sensitivity to rejection, which causes intense pain followed by reactivity. The observation that this negative mood state was associated with activation became the basis of my understanding of dysphoria and its association with the bipolar spectrum.

Dysphoria is the negative mood phase of the activated state. It is the negative counterpart of the positive "euphoria." Although, like depressed mood, dysphoria is a negative mood, it is not characterized by the sadness, the sense of pessimism, the sense of loss, or the sense of hopelessness associated with depressed mood. Dysphoria is a mood of irascibility, emotional restlessness, and irritability, not tearfulness, sadness, or grief. As a negative mood in an agitated state, it is most often confused with anxiety.

Dysphoria is a mood, not a behavior. But because it is defined by its association with the activated energy state of hypomania/mania, we often speak about "dysphoric hypomania" and "dysphoric mania" even if a full-blown hypomanic or manic episode is not present. "Dysphoric hypomania," used in the context of mood dysregulation, implies an activated state not necessarily meeting the criteria for a hypomanic episode, characterized by dysphoric mood and associated symptoms of activation, such as irritability, agitation, short fuse, anxiety, etc. As a rule, I prefer to use the term "dysphoric activation" to "dysphoric hypomania" to avoid the implication that criteria are met for a full hypomanic episode.

I understand the term dysphoric hypomania in quite a different manner from Akiskal and Benazzi. Akiskal and Benazzi (2005) regard dysphoric hypomania to be a mixed state, that is, one in which symptoms of depression and hypomania/mania are commingled. In my clinical experience, dysphoric activation is not a mixed state but a description of the negative mood phase that occurs during activated states. There are others who agree and do not regard dysphoria to be a mixed state (Berner et al., 1987; Koukopoulos & Koukopoulos, 2000; Stransky, 1911). In addition, Akiskal and Benazzi maintain strict criteria for the presence of dysphoric hypomania which I do not use (Akiskal & Benazzi, 2005; Benazzi, 2006, 2007). Their scheme requires full criteria for a hypomanic episode and full criteria for a major depressive episode to be met to diagnose dysphoric hypomania. I use the term dysphoric hypomania (dysphoric activation) to refer to any activated negative mood phase, regardless of whether or not full criteria for hypomania are met.

The problem with Akiskal and Benazzi's definition of dysphoric hypo-mania, as I see it, is that it does not account for those people with negative moods in activated states who do not describe feeling depressed per se. Their assumption appears to be that there is only one negative mood, depression. In their scheme, when the sad mood of depression occurs in the context of hypomania, it becomes renamed as "dysphoria." They do not seem to distinguish between the two negative moods of dysphoria and depression. I think the confusion is resolved listening closely to the clinical data which reflects two distinct negative moods: the sad, hopeless mood typically seen in those presenting in a slowed energy state (depression) and the irritable mood of those presenting in an activated energy state (dysphoria).

Both euphoria and dysphoria occur in activated states and are associated with physical symptoms of high energy, hyperarousal, and excitation. Hallmark symptoms of dysphoric activation are irritability, restlessness, and agitation (Malhi et al., 2013). Anger, excitability, reactive mood, anxiety, inability to concentrate because of "flight of ideas," and distractibility may also be present. In this activated state, many people become impulsive and do things they might otherwise not do. Cutting and other self-mutilating behavior are not uncommon. Binge-eating and other impulsive eating habits occur frequently during episodes of dysphoric activation. Outbursts at significant others are characteristic of the state as is self-medication with substances, all in a pattern of poor impulse control. Anxiety may be the single most self-reported symptom associated with states of dysphoric activation, most likely because such individuals do not realize that dysphoric activation is a state of heightened arousal. The experience of dysphoric activation, both physically (respiratory, cardiac, gastrointestinal or headaches) and cognitively (racing thoughts or the sense of "not being able to turn off my mind") is one of anxiety.

Dysphoric activation is often mistaken for (agitated) depression. One common example of this occurs when people are evaluated in the emergency room after cutting themselves. An attending clinician will often assume that the individual is depressed and has made a suicide attempt. Such behavior, however, is seldom a suicide attempt or the result of depression. It almost invariably emerges out of irritable, agitated activated dysphoria. Such people rarely describe feeling suicidal when cutting. Most describe feeling overwhelmed with negative feelings in an agitated state that is relieved after they cut. Cutting behavior is generally a red flag for dysphoria. Suicidal thoughts and behavior should not be defined solely in terms of the depressive state or depressed mood per se, but can occur both during depressive states and activated states. An individual in a state of agitated dysphoric activation can feel suicidal, not necessarily out of despair or depression but in an activated state of anger and irritability.

Dysphoria is the single most misdiagnosed mood. Because of the under-recognition of dysphoric activation, it can take 10–25 years for people with mood disorders related to the bipolar spectrum to receive a correct diagnosis (Ghaemi et al., 2000; Lisj et al., 1994; McCombs et al., 2007; McCraw et al., 2014), and proper treatment. Instead, people are diagnosed with depression, anxiety, ADHD, or borderline personality disorder.

4.3 EMOTION V. AFFECT V. MOOD

The words "emotion," "affect," and "mood" are not used in consistent ways and their meanings vary from author to author such that one need pay attention to each author's particular definition of terms to orient oneself to the discussion at hand. Often the terms are simply used interchangeably unless they are the focus of a particular literature. In the literature on emotion regulation, the term "emotion" is typically used to characterize temporally limited feeling states, while in the literature on affect regulation, the term "affect" refers predominantly to self-regulation. The word "affect" is often used when the author's focus is on the biological underpinnings of a mood or emotion. I use the word "mood" rather than emotion or affect when talking about mood dysregulation in keeping with terminology predominant in the psychiatric literature. While it has been said that a mood is a sustained emotional state, moods, in the psychiatric sense, can certainly be short-lived, as in DMDD.

There appear to be distinct literatures on "emotion regulation" (and emotion dysregulation), "affect regulation/dysregulation," and "mood dysregulation." One thing all authors agree upon, however, is that regulation of emotion/affect/mood is all inextricably linked to and dependent upon healthy early attachment relationships (Cassidy, 1994; Roque et al., 2013; Schore, 2003b; Schore, 2016; Waters et al., 2010), which foster normal brain development and are the foundation of self-regulation. This will be discussed in greater detail in Chapter 7 when we discuss the etiology of mood dysregulation.

Between emotion, affect, and mood, the greatest contributions by far have been in the field of emotion regulation, which began to emerge in the developmental literature in the 1980s and have been burgeoning since. In some ways, this work has outpaced itself as there is "conceptual and definitional chaos" (Buck, 1990, p. 330), readily admitted by most authors in the field. Developmental researchers disagree over the core elements of emotion regulation and its definition (Campos et al., 2004; Gross & Thompson, 2007).

The most defining feature of emotion regulation is its focus on the way in which cognitive processes govern the way we regulate our emotions. The most significant of these processes is *reappraisal*, by which we modulate our emotional responses by thinking differently about a stimulus. For example, if one recoils in dread after seeing a snake in the grass, reappraising the situation might lead to the conclusion that a benign garter snake is nothing to fear. Therapy for people with emotion dysregulation in this paradigm is typically cognitively oriented and aimed at helping individuals reappraise stimuli in a self-regulating manner. This may involve learning cognitive strategies that reduce rumination, master thought suppression, and contain or redirect maladaptive thoughts.

Another consistent feature of the literature on emotion regulation is the discussion of its transdiagnostic nature. In a transdiagnostic model, biological, socioenvironmental, and psychological processes are recognized as operating across diagnostic boundaries (Buckholtz & Meyer-Lindenberg, 2012). In such a model, there is a hierarchy of symptoms, each of which is envisioned to lie on a continuum of severity as they cross diagnostic boundaries. Cludius and colleagues (2020) think of a transdiagnostic process as having (1) increased symptom severity compared with the normal population and (2) contributing to the development or maintenance of psychological problems. The RDoC approach to research emerged largely out of transdiagnostic models. Fernandez and colleagues (2016) have actually posited emotion regulation as a new RDoC domain, explaining how it might be integrated with the existing domains.

Over half of individuals with one diagnosis meet criteria for at least one other diagnosis (Kessler et al., 2005). Transdiagnostic models of psychopathology focus on the fundamental processes underlying multiple disorders and help explain their co-occurrence, ultimately leading to more effective assessment and treatment (Barlow et al., 2004; Ehring & Watkins, 2008). Emotion regulation is now typically discussed as a transdiagnostic process as disturbances in regulation of emotion are common among many diagnoses (Aldao & Nolan-Hoeksema, 2010; Kring & Sloan, 2010).

Problems with emotion regulation are referred to as "emotion dysregulation" or as "emotion regulation deficits" (ERDs) in the literature. Treatment approaches for emotion regulation deficits are varied and are often diagnosis specific. For example, a cognitive behavioral approach is often used when anxious people have emotion regulation deficits (Kovacs et al., 2006; Suveg et al., 2006) and a dialectical behavioral approach is often used when people diagnosed with borderline personality have emotion regulation deficits (McMain et al., 2001; Nieacsiu et al., 2014a, b). "Affect Regulation Training" (ART) is described as a transdiagnostic treatment for emotion regulation (Berking & Lukas, 2015; Berking & Whitley, 2014). Again, most of these treatments are largely cognitively oriented.

There is a much smaller body of literature on affect regulation with a number of notable contributors to the field (Fonagy et al., 2018; Mikulincer et al., 2003; Bradley, 2000), but the major contributor to the literature on affect regulation is neuropsychologist Allan Schore (2003a, 2003b, 2014, 2016). His focus is on the effects of attachment relationships on brain development. He details the neurobiological basis of affect regulation and emphasizes what he believes to be the unique role of the right hemisphere in the regulation of affect. Schore addresses not only affect regulation but also affect dysregulation, detailing the way in which relational trauma impacts the development of brain structure and function in ways that interfere with the individual's capacity to regulate affect. While I am not aware of discussion in the affect regulation literature of its transdiagnostic nature per se, affect dysregulation is discussed in the literature as a problem common to multiple diagnoses.

The literature about mood regulation is sparse and disjointed compared with that on emotion regulation and affect regulation. The literature on mood dysregulation, however, is ample but does not refer to mood dysregulation as we have been discussing it here. It refers, rather, to the mood dysregulation defined in DSM-5 disruptive mood dysregulation disorder (DMDD). The two share important features, as we will discuss. Quite possibly some of the literature on DMDD might be applicable to mood dysregulation as described here.

Mood dysregulation, as I have presented it here, describes a syndrome of unstable shifts in energy states from hypoarousal to activation, absent any particular biologic rhythmicity, typically in response to external triggers. The underlying instability has its biological roots in a vulnerable affective temperament, a predisposition shared with many of the bipolar disorders. Because of this phenotypic association with bipolar disorders, mood dysregulation is tethered to but beyond the bipolar spectrum. Thus, those with mood dysregulation are often responsive to the same medications that are effective for those with bipolar disorder. I am not aware of any discussion of medication treatment for emotion dysregulation or affect dysregulation in the literature, but I regard this simply as a reflection of the fact that its authors are typically not physicians. But this should have no bearing on the essential similarities between these three topics of discourse.

We have situated mood dysregulation as being beyond the bipolar spectrum in large part because of the alterations in *energy* states of arousal and hypoarousal it shares with the bipolar disorders, though in mood dysregulation, these shifts are not cyclical or biologically driven. Shifts in biological states of arousal are also recognized in the literature on emotion regulation (Goldin et al., 2008; Ochsner et al., 2002; Thompson, 2011; Thompson et al., 2009) as well as that of affect regulation/dysregulation (Schnell & Herpertz, 2007; Schore, 2003). Schore frames affect regulation in terms of opposing energy states, contrasting the hyperarousal state with the conservation/ withdrawal state (Schore, 1994) and asserts *"Indeed, energy shifts are the most basic and fundamental features of emotion"* (Schore, 2003).

Mood dysregulation shares with emotion regulation and affect regulation/ dysregulation, the feature of being a transdiagnostic process. As we will see in the next chapter, mood dysregulation crosses diagnostic boundaries and co-occurs with numerous other diagnoses. Diagnoses frequently discussed in the literature as co-occurring with emotion regulation and affect regulation/dysregulation are anxiety disorders, eating disorders, substance use disorders, ADHD, and borderline personality disorder. Each of these diagnoses frequently co-occurs with our mood dysregulation as well.

In addition to sharing transdiagnostic features, emotion dysregulation, affect dysregulation, and mood dysregulation all see attachment disturbances as etiologic underpinnings. Compromise in the integrity of brain structures is seen as common to emotion/affect/mood dysregulation.

Would emotion dysregulation and affect dysregulation, like mood dysregulation, be beyond the bipolar spectrum? Quite likely they are all the same phenomena, albeit viewed through different lenses. Whether discussing emotion regulation from the perspective of cognitive psychology or affect regulation/dysregulation from the perspective of relational influences on the brain substrate or mood dysregulation from the perspective of affective temperaments, focusing on shifts in energy states, and the parallels these present with bipolar disorders (including treatment implications with medication), emotion/affect/mood dysregulation may simply be facets of the same clinical phenomenon.

"I'm depressed" they say when they come in. But they don't look as much depressed as irritable, anxious, or agitated, and they never really have a moment of exuberance. They are exquisitely sensitive to rejection and very reactive. They have mood shifts but no real mood cycles. And they haven't responded to the usual antidepressant medications. Dad has a temper. Mom is always down. It's not really a depression and it's not bipolar: it's somewhere in between. I would call it a mood dysregulation.

One of the reasons mood dysregulation is tethered to bipolar disorders is that, like bipolar disorders, mood dysregulation emerges from perturbation in an affective temperament. Affective temperaments are biological predispositions to mood disturbances that may never blossom into mood disorders. But under sufficient stress, affective temperaments can develop into a spectrum of mood disturbances that range in severity from mood dysregulation to bipolar I disorder.

Just like that, "depression" emerges as a misnomer, masquerading for dysphoria, the hallmark of mood dysregulation, in which "depression" can appear somewhat like it does in bipolar spectrum disorders. Mood dysregulation is associated with the bipolar spectrum because of its shared phenotypic features with DSM-5 bipolar disorders, such as heritability and medication response. But mood dysregulation is not a bipolar disorder: it is beyond the bipolar spectrum and distinct from it because (1) it is not a cyclical mood disorder and (2) its mood shifts are triggered by exogenous (situational, not biological) factors.

The negative mood state of dysphoric activation is often confused with depression, resulting in misdiagnosis and treatment with antidepressants, which are often ineffective and result in "treatment resistant depression" when recognition of mood dysregulation and treatment with mood stabilizers could lead to treatment success. Because dysphoria is so often confused with depression, people with misunderstood symptoms of "depression" can bounce around for years before they are recognized to have the dysphoria of mood dysregulation.

While there is essentially no literature on mood dysregulation as we have described it here, it is a feature shared by many diagnoses and is thus a transdiagnostic process like emotion regulation, on which there is a great deal of literature. The literature on emotion regulation and affect regulation appears to be describing the phenomena associated with mood dysregulation, albeit from differing perspectives. In the next chapter, we will look at the intersection of mood dysregulation with various other diagnoses.

KEYWORDS

- **mood dysregulation**
- **activated state**
- **phenotype**
- **affective temperament**
- **soft bipolar spectrum**
- **emotion regulation**

CHAPTER 5

Co-Occurring/Related Conditions

The cure of many diseases remains unknown to the physicians of Hellos because they do not study the whole person.

—Socrates

ABSTRACT

As a transdiagnostic constellation of symptoms, mood dysregulation co-occurs with many other conditions. Borderline personality disorder, with its hallmark symptom of irritability, shares important overlap with mood dysregulation, which is defined largely by irritable dysphoria. The activation and impulsivity associated with ADHD can signal an associated mood dysregulation, but co-occurrence of bipolar and anxiety disorders with ADHD can present diagnostic complications. Eating disorders have long been associated with bipolar disorder and with dysregulation of mood. Post-traumatic stress disorder or PTSD as outlined in DSM-5, does not adequately capture the more common range of symptomatology we see in clinical practice, but a diagnosis proposed by Bessel van der Kolk, developmental trauma disorder (DTD), more fully does. The cardinal feature of DTD, as van der Kolk explains, is mood dysregulation. Anxiety and panic disorders both co-occur with mood dysregulation but the agitation and restlessness of dysphoria are often misinterpreted as anxiety. Substance use/abuse disorders intersect with mood dysregulation in complex ways, both because there is often confusion between cause and effect and because each substance has different physiologic effects on the body. The newest DSM-5 diagnostic addition to the category of depressive disorders, disruptive mood dysregulation disorder (DMDD), in some ironic way, gives mood dysregulation a footprint in the DSM as this diagnosis shares much in common with mood dysregulation, including its hallmark symptoms of irritability and angry mood and its transdiagnostic nature.

Mood dysregulation as a transdiagnostic phenomenon typically co-occurs with one or more extant diagnoses. Below is a discussion of the most

Mood Dysregulation: Beyond the Bipolar Spectrum. Deborah A. Deliyannides, MD (Author)
© 2024 Apple Academic Press, Inc. Co-published with CRC Press (Taylor & Francis)

common DSM-5 diagnoses with which mood dysregulation has interface. DSM-5 diagnostic criteria for these disorders can be found in Appendix C.

5.1   OTHER SPECIFIED BIPOLAR AND RELATED DISORDER

Other specified bipolar and related disorder (OSBD) is a DSM-5 diagnostic category. It applies to

> …presentations in which symptoms characteristic of a bipolar and related disorder that cause clinically significant distress or impairment in social, occupational, or other important areas of functioning predominate but do not meet full criteria for any of the disorders in the bipolar and related disorders diagnostic class (DSM-5, 2014).

Examples of these presentations might include the following:

1.   Short-duration hypomanic episodes (2–3 days) and major depressive episodes
2.   Sub-threshold hypomanic episodes with major depressive episodes
3.   Hypomanic episodes without prior major depressive episodes
4.   Short-duration cyclothymia (<24 months).

Other specified bipolar disorders are visibly on the bipolar spectrum because of their cyclicity but they are typically categorized as "other" because one or more symptoms associated with their presentation are too mild or too short in duration to meet the criteria for a full bipolar diagnosis.

5.2   UNSPECIFIED BIPOLAR AND RELATED DISORDER

Unspecified bipolar and related disorder (UBD) is also a DSM-5 diagnostic category. A PubMed literature search of "unspecified bipolar disorder" yields no results. The category is meant to apply to:

> …presentations in which symptoms characteristic of a bipolar and related disorder that cause clinically significant distress or impairment in social, occupational, or other important areas of functioning predominate but do not meet full criteria for any of the disorders in the bipolar and related disorders diagnostic class (DSM-5, 2014).

Note the "unspecified" and the "other specified" bipolar and related disorder criteria are virtually the same. However, DSM-5 notes that the unspecified bipolar and related disorder category is used

> …in situations in which the clinician chooses not to specify the reason that the criteria are not met for a specific bipolar and related disorder and includes presentations in which there is insufficient information to make a more specific diagnosis (e.g., in emergency room settings) (DSM-5, 2014).

The "other specified and related bipolar disorder" and "unspecified bipolar and related disorder" categories in DSM-5 replace the "bipolar disorder not otherwise specified" diagnostic category in DSM-IV. There is no literature on the epidemiology or treatment of individuals in either of these diagnostic categories or on those with "bipolar disorder not otherwise specified." While all of these diagnoses are ambiguous, they all require symptoms "characteristic" of a bipolar disorder, **cyclicity** being the hallmark.

One often hears the term "rapid cycling bipolar." This term is a colloquialism, not a formal diagnosis, and is generally used in reference to individuals whose moods change far more frequently than months, weeks or days as described in the DSM criteria for bipolar disorders. Rather, they change within a day, sometimes hour to hour or moment to moment.

It may be easy to see how people with the frequent changes associated with rapid cycling bipolar symptoms might be diagnosed with "other" or "unspecified" bipolar disorder, but if one looks at the two distinguishing features of mood dysregulation as we have contrasted it with bipolar disorders, lack of cyclicity and lack of endogenous triggers, the "rapid cycling bipolar" construct raises important questions.

People who describe themselves as having rapid mood shifts typically report frequent and brief but random changes in mood, not necessarily changes in energy state. In such cases, interpersonal disappointment or upset in a biologically predisposed individual may lead to rapid shifts in energy states (from depressive to activated) but more frequently lead to changes in mood WITHIN the activated energy state (from dysphoria to euphoria). Rapid mood shifts within one energy state are different from the rapid cycling of energy states: they are not cyclical and are not driven by endogenous triggers. This type of mood instability appears to be a manifestation of a mood dysregulation, not a bipolar disorder.

Quite likely, the designation "rapid cycling bipolar" leads to erroneous diagnoses such as "other specified bipolar disorder" or "unspecified bipolar disorder." This misconstrual may be at the heart of why those with mood dysregulation struggle to understand themselves when given a bipolar diagnosis, for they know "I am not bipolar." Mood dysregulation is distinct both from "other" and "unspecified" bipolar disorder, both of which require endogenous mood cycling.

5.3   BORDERLINE PERSONALITY DISORDER

One might say there has been a tug of war, of sorts, between psychologists and psychiatrists, as to whether "borderline personality disorder" is a personality disorder or an affective disorder. For many years, opinions fell largely along predictable fault lines, psychologists arguing that borderline personality disorder was a disturbance in personality organization (George et al., 2003; Gunderson et al., 2006; Kernberg, 1967; Koenigsberg et al., 2002; Paris, 2004, 2007a) and psychiatrists regarding it to be a bipolar disorder of one type or another (Akiskal, 2004; Magill, 2004; Mackinnon & Pies, 2006; Smith et al., 2008). There were also those who focused on the overlap between the two diagnostic categories (Henry, 2001; Stone, 2006).

As Michael Stone writes, "Borderline personality disorder enjoys the...distinction of being...the only condition of the nearly 400 in the DSM whose label provides no hint...as to what sort of condition it is." (Stone, 2005). Border? In the late 19th century, Irving Rosse (1890) used the term "borderland insanity" to refer to people on the border between psychosis and normalcy. This was followed by descriptions of "borderline schizophrenia" for people who were "less ill than psychotic but more ill than neurotic." The "borderline patient" was then identified as "too well for the hospital but too ill for the couch" (Stone, 2005). The term "borderline personality disorder" was first used in 1938 by Adolf Stern to describe people who tended to regress into "borderline schizophrenia." In the 1960s, Otto Kernberg described "borderline personality organization" in people whose personality structure lay between psychosis and neurosis and who had primitive defenses such as "splitting" and "projective identification" (Stone, 2005).

Borderline personality organization and borderline personality disorder should not be confused. Borderline personality organization is a psychoanalytic description of psychological defense mechanisms and primitive personality structure. Borderline personality disorder is a DSM diagnostic description of the behaviors associated with the psychoanalytic construct of borderline personality organization.

Kernberg's description of borderline personality organization with disturbances in identity and reality testing capacity did not lend itself well to operationalized diagnostic criteria. Several groups worked to develop diagnostic instruments and to narrow down a diagnosis to the most essential criteria. Gunderson and Singer (1975) published an influential paper that helped pave the way for borderline personality disorder as a DSM diagnosis. Their formulation was made in collaboration with Kernberg. Interestingly, a candidate diagnosis which was not used was "emotionally unstable personality disorder" (Stone, 2005). Emotionally unstable personality disorder is now the ICD-10 diagnosis [F60.3] under which borderline personality disorder falls as a subtype. Interestingly, there is a current trend in this country to use the diagnosis "emotionally unstable personality disorder" in lieu of "borderline personality disorder" to avoid the stigma associated with the latter.

In 1980, borderline personality disorder (BPD) became an official diagnosis in DSM-III. Rapidly fluctuating mood swings—or mood dysregulation—was one of the defining symptoms of the diagnosis from the outset (others were unstable self-image, fear of abandonment, and tendency for self-harm/suicidal thinking). *Michael Stone regards **irritability** to be "the red thread" that runs through the borderline condition* (Stone, 1988).

While many symptoms associated with BPD are highly affective in nature, a DSM-III diagnosis of borderline personality disorder precluded a diagnosis of cyclothymia and vice versa. Borderline personality disorder, despite its dominance by affective symptoms, was not regarded to be an affective disorder in DSM-III.

Yet Akiskal regards some personality disorders to be sub-affective illnesses (Akiskal, 1979). Borderline personality disorder, in particular, he argued, might better be considered to be a sub-affective disorder (Akiskal, 1981). It shares features of affective lability, impulsivity, and emotional lability with the cyclothymic affective temperament and might be thought of as being biologically based rather than as a dynamically driven personality disorder (Akiskal et al., 1985).

Perhaps it was his biting editorial (Akiskal, 2004), challenging psychoanalysts' research conclusions and arguing so persuasively that borderline behavior is **affect driven,** that spearheaded a change reflected in DSM-5, the first DSM to acknowledge the affective nature of the borderline personality disorder diagnosis. In particular (1) borderline personality disorder could now co-occur with cyclothymia, dysthymia, and other affective disorders and (2) a multidimensional trait-based assessment module was added to the personality disorder section of the manual: while cumbersome, this gave more emphasis to the affective/biological aspects of borderline personality disorder (Shedler et al., 2010).

These changes in DSM-5 represent an ideological shift that brings us to a Rashomon moment in which we come closer to "something akin to the blind men and the elephant, with each feeling a particular part of the animal's anatomy and coming to a different disembodied conclusion" (Birnbaum, 2004).

Five of nine DSM-5 symptoms must be positive to meet the criteria for borderline personality disorder. Of those nine, at least three are overtly affective and could be regarded as symptoms of mood dysregulation: (1) impulsivity; (2) affective instability/mood reactivity; (3) intense anger/ frequent temper displays. Suicidality might be added to the list. Akiskal's point is well-taken that borderline personality disorder symptoms are inescapably affective in nature.

Akiskal sees cyclothymia, dysthymia, and bipolar II disorders as being in the borderline realm (Akiskal, 1981), consistent with the notion that borderline personality disorder is a sub-affective state. Evidence for this sub-affective status (i.e., biological underpinnings) is the emergence of angry, irritable dysphoric hypomanic states in individuals with borderline personality disorder treated with antidepressant medication (Levy et al., 1998; Solo et al., 1986), an example of treatment emergent hypomania discussed above. For example, an individual diagnosed with borderline personality disorder who presents with depression and is treated with antidepressant medication might very well become agitated and irritable only after treatment with medication.

Akiskal has also described the affective temperaments as pre-morbid predispositions to various disorders (hyperthymic for bipolar I disorder, depressive for major depression, cyclothymic for bipolar II and anxious for generalized anxiety disorder) (Akiskal, 2007). Why has he elaborated affective disorders based on every affective temperament except the irritable affective temperament?

Stone, noting that "irritability is the red thread that runs through the borderline conditions" may recognize that what is called borderline personality disorder is an environmentally stressed manifestation of the irritable affective temperament. In fact, Walsh and colleagues (2013), in a study of the four affective temperaments, found that those with irritable temperaments had increased levels of anger and irritability and were more likely to engage in risky behaviors compared with normal controls, a population suggestive of people diagnosed with borderline personality disorder as well as those we describe with mood dysregulation. Stone (2005) suggests that this population may have subsyndromal symptoms of DSM borderline personality disorder.

The incidence of BPD has been estimated to be between 1% (Coid et al., 2006) and 4% (Grant et al., 2004) of the general population. Borderline personality disorder runs in families like affective disorders on the bipolar spectrum. Studies have estimated an 11.5–12% risk in first degree relatives (Loranger et al., 1982; Smith et al., 2004) and a 35% incidence of bipolar spectrum disorders in monozygotic twins of people with borderline personality disorder (Smith et al., 2004). In a review of an inpatient population at Westchester Cornell Medical Center, the co-occurrence of narrowly defined bipolar disorders (bipolar I and bipolar II) among people with borderline personality disorder was 44%. The co-occurrence of a broadly defined spectrum of bipolar disorders (bipolars I and II, pharmacologic hypomania, cyclothymia and family history for bipolar disorder) was 81% (Deltito et al., 2007).

Critics have identified two important reasons to clarify diagnostic confusion between bipolar spectrum disorders and borderline personality disorder. One is that the overdiagnosis of bipolar disorder among people with borderline personality disorder will deprive people with borderline personality disorder treatment tailored to their needs. The other is that overdiagnosis of borderline personality disorder among people with bipolar disorder will deprive them of medication if so indicated.

The first problem is of decreasing concern as more treatment modalities have demonstrated efficacy for the treatment of borderline personality disorder. Treatments shown to be effective for BPD are dialectical behavior therapy (Gunderson et al., 2000; Neacsiu et al., 2010; Ruggero et al., 2010; Zanarini et al., 2003), EMDR (particularly when there is a history of trauma) (Allen, 2003; Mosquera et al., 2014), affect regulation therapy (Clarkin et al., 2001; Gabbard, 2001; Kernberg et al., 1989), mentalization treatment (Bateman & Fonagy, 2010), cognitive behavioral therapy (Linehan, 1993), partial hospitalization (Bateman & Fonagy, 1999), and psychoanalytic psychotherapy (Applebaum, 2006; Kernberg, 2003).

A more significant problem is the avoidance of psychopharmacologic intervention in people with borderline personality disorder who could be helped by medication (Paris, 2007b). Some have claimed that medication is less effective in the presence of a personality disorder (Zanarini et al., 1990) but, as Stone (1988) notes, "it is a fallacy that medication does not work in the treatment of borderline personality disorder." In fact, there is a striking overlap in medication response between those with bipolar disorder and those with borderline personality disorder (Smith et al., 2004). Evidence shows with remarkable consistency that impulsivity and anger/irritability in those with borderline personality disorder are responsive to mood stabilizers (Akiskal, 2004; Fuerino & Cowdry, 2011; Gardner & Cowdry, 1986; Links et al., 1990; Mercer et al., 2009). In an interesting study, Akiskal and colleagues (1979), treating people with borderline personality disorder (including symptoms of "weakened sense of self") found that after 1 year of lithium treatment, 7 of 11 reported a "much improved sense of self, which they felt came from achievement of significant control over unpredictable mood swings. When interviewed, these individuals reported, "I have a much better hold of myself." This is a typical response from someone who has had successful treatment with mood stabilizers, and certainly reframes the discussion about whether or not a biologic disturbance in self-regulation impacts one's sense of self and, in turn, personality structure.

A number of people I treat have asked me if I think they are "borderline," a diagnosis given them by other mental health professionals which they appropriated in some pejorative way. Most of the time, once treated with mood stabilizers, self-harming behavior, reactivity, impulsivity, interpersonal sensitivity, and other behaviors often associated with borderline personality disorder seem to melt away. Personality organization as described by Kernberg is subsequently difficult to identify in their interpersonal relationships, arguing for the sub-affective nature of BPD as put forth by Akiskal. Many such people labor under the stigma of BPD when, in fact, they have unrecognized mood dysregulation that has never been properly treated.

Akiskal (2004) suggests the diagnosis borderline personality disorder might better be renamed and reclassified as an affective disorder while psychoanalytic formulations be afforded an entirely separate dimension in the diagnostic system. We may soon be approaching a time when our classification system can account for the complexities of symptom development, one in which the role of dynamic and broader environmental forces upon one's genetic substrate can be better understood and described.

5.4 ATTENTION-DEFICIT/HYPERACTIVITY DISORDER

Core features of attention-deficit/hyperactivity disorder (ADHD) are (1) inattention and/or impulsivity with several symptoms that are (2) present prior to age 12 and (3) present in at least two settings (home/work/school) and (4) interfere with social/academic/occupational functioning. ADHD is highly associated with mood reactivity, a feature of mood dysregulation (Graziano & Garcia, 2016). Though mood instability is not explicitly listed in the DSM-5 diagnostic criteria for ADHD, the manual does describe emotional instability/dysregulation as an associated feature of ADHD.

Mood dysregulation is prevalent in people with ADHD (Biederman et al., 2012; Sobanski et al., 2010; Surman et al., 2011) and is an important transdiagnostic factor relevant to the diagnosis (Barkley, 2010; Reimher et al., 2005; Corbisiero et al., 2013). Affective symptoms are associated with ADHD throughout the lifetime (Shaw et al., 2014) including in pediatric populations (Graziano & Garcia, 2016; Laufer & Denhoff, 1957), adolescent populations (Bunford et al., 2015), and adult populations (Reimher et al., 2005). The dysphoria associated with rejection sensitivity, which we have discussed as a feature of mood dysregulation, has increasingly been recognized as associated with ADHD, though a number of authors regard it to be a feature of ADHD rather than a co-occurring symptom (Dodson, 2021; Wilson & Cooperman, 2021). Dodson (2021) describes rejection sensitive dysphoria (RSD) as a manifestation of emotional dysregulation, and Barkley (2015) posits that emotional dysregulation is a core feature of ADHD.

Irritability is often seen with ADHD. At times such irritability results from the frustration associated with the inability to focus, disruptions in focusing, and one's inefficiency in completing tasks. At other times it may be a symptom of the dysphoric hypomania of co-existing mood dysregulation. The severity and the context in which irritability exist helps to determine its meaning. While irritability may be part of the mood dysregulation associated with ADHD, it is not always indicative of the dysphoric hypomania which can be associated with mood dysregulation.

Problems with mood dysregulation appear to be associated with some of the social impairments seen in people with ADHD. In a sample of young adolescents with ADHD, three specific facets of emotion dysregulation predicted social impairment: (1) low threshold for emotional excitability, (2) behavioral dyscontrol in the face of strong emotions, (3) inflexibility/ slow return to baseline (Bunford et al., 2014). College students with significant ADHD symptoms report more difficulty providing emotional support to friends and overall lower quality of friendships (McKee, 2017).

The close link between mood dysregulation and ADHD has important treatment implications. Two common (related) clinical problems we see are (1) the "self-diagnosed" individual who comes in for a consultation complaining of inability to concentrate (usually having done research on the internet, reaching the "realization" that they have ADHD), and (2) the individual whose therapist has suggested they have ADHD and should speak with a psychiatrist about obtaining psychostimulants. In both cases, our capacity to properly diagnose and treat can be short-circuited by an individual's pre-determination that they need psychostimulants when, in fact, concentration problems are symptomatic of many issues other than ADHD, and psychostimulants may exacerbate symptoms which can often be obscured when one presents with such a pre-determination. A misdiagnosis can lead to mistreatment.

Problems with executive function and concentration are present in many conditions other than ADHD. Impaired executive function is seen in people with emotion dysregulation (Sudikoff et al., 2015) who do not have ADHD. Attention is compromised in bipolar disorder (Normala et al., 2010; Robinson et al., 2006), in part because the anatomical structures found to be abnormal in bipolar disorder are the same structures necessary for normal maintenance of attention (Camelo et al., 2013). Concentration impairment is also a cardinal symptom of depression (DSM-5) and is routinely seen in depressive disorders (Marazziti et al., 2010; Porter et al., 2007; Zakzanis et al., 1998) as well as in anxiety disorders (Bishop, 2009; Braunstein-Berkovitz, 2003; Cushman & Johnson, 2001).

When considering treatment for concentration problems it is particularly important to rule out any mood disorder on or tethered to the bipolar spectrum, such as mood dysregulation, before prescribing psychostimulants. People with such mood disorders are very vulnerable to mood destabilization and can develop hypomania or mania with psychostimulant use. Indeed, these are the individuals Akiskal describes as having bipolar III disorder on his soft bipolar spectrum. I regard such a predisposition to be a mood dysregulation which has an "invisible" association with the bipolar spectrum that would never be manifest if not unmasked by the use of stimulant medication.

There are no reliable predictors of the likelihood of such stimulant destabilization (Ross, 2006), though I frequently see a history of bipolar spectrum disorder in the families of those who become activated with stimulant use. Unfortunately, many times the problem occurs in teenagers diagnosed with ADHD in whom a co-occurring mood dysregulation or bipolar spectrum disorder diagnosis has been missed until they have been hospitalized for a manic episode precipitated by stimulant medication prescribed for ADHD. This trend has accelerated in the era of the fashionable ADHD diagnosis, preferred to the stigmatized bipolar diagnosis with which it is often confused. Comorbid ADHD and mood dysregulation or bipolar disorder generally requires the use of mood stabilizers if psychostimulants are prescribed (Asherson, 2005; Ceraudo et al., 2012).

There is a sub-group of people with ADHD that can safely be treated with psychostimulants without the use of mood stabilizers and will not develop treatment emergent hypomania/mania (Faedda et al., 2004; Ross, 2006; Schaller & Behar, 1998). On this basis, Faraone and colleagues (1997) suggest the possibility of two types of ADHD: one "pure" type and one that is comorbid with bipolar spectrum disorders, identified by familial transmission and a more severe clinical course. If this is so, I would expect those with mood dysregulation to fall into the latter group.

5.5 EATING DISORDERS

Eating disorders (EDO) are a large category of syndromes including avoidant/restrictive food intake disorder, anorexia nervosa, bulimia nervosa, and binge eating disorder among others. Emotion dysregulation is seen across the entire spectrum of eating disorders (Brockmeyer et al., 2014; Danner et al., 2014; Svaldi et al., 2012) as a transdiagnostic phenomenon.

Reviewing comorbidity, symptom overlap, course of syndrome, family history, biology, and treatment response, researchers posit that eating disorders, particularly bulimia nervosa and binge eating disorder, may be a subset of sub-threshold bipolarity (Lunde et al., 2009; Perugi et al., 2006), that is, mood dysregulation. Perhaps one of the most compelling arguments for the association of eating disorders with bipolar spectrum disorders is the response of binge eating to the mood stabilizer lithium. In the first open trial of lithium in women with eating disorders, 12 of 14 women reported decreased binge-purge episodes as well as increased mood stability (Hsu, 1984). In the second open trial, 65% of people reported a 75% or greater reduction in binge-purge episodes with lithium and cognitive behavior therapy (Hsu, 1987). Lithium was reported to be effective in people with anorexia nervosa (Barcai, 1977; Hudson et al., 1985; Stein et al., 1982) and in people with bulimia nervosa (Leyba & Gold, 1988; Pope et al., 1986; Shisslak et al., 1991). There have also been case reports of the successful use of atypical neuroleptics with and without lithium to treat eating disorders (Hansen, 1999; Malina et al., 2003; Newman-Toker, 2000; Pope et al., 1986).

Greenberg and Harvey (1987) found that mood lability was a better predictor of the severity of binge eating than the interaction between dietary restraint and depressed mood. People with anorexia nervosa have been described as having irritability, hyperactivity, restlessness, pacing, insomnia, and hyper-talkativeness, the behavioral activation signs at the core of the hypomanic activation of mood dysregulation (Brambilla et al., 2001; Casper, 1998; Kron et al., 1978; Winokur et al., 1980).

Dysregulation of emotions in eating disorders has been found to involve suppression of negative emotions (Brockmeyer et al., 2013; Corstorphine et al., 2007), inability to identify emotions (Evers et al., 2010), presence of negative affects (Engel et al., 2013), and alexithymia (Tchanturia et al., 2013). Consistent with these patterns of emotion dysregulation, people with eating disorders have been found to be more likely to avoid emotional experiences and to engage in avoidant behaviors (Wildes et al., 2010).

5.6   POSTTRAUMATIC STRESS DISORDER

Posttraumatic stress disorder (PTSD) is associated with disturbances in mood as well as in cognition and behavior. When trauma is circumscribed and acute, the associated affective disturbances do not typically fall into the pattern of mood dysregulation. Chronic trauma, which has the capacity to affect brain development and function, however, typically results in mood dysregulation.

DSM-5 has recategorized PTSD from "anxiety disorders" to "trauma and stress-related disorders." This change was made on the basis of the phenotypic characteristics of individuals with trauma diagnoses (symptoms of anhedonia, dysphoria, aggression, anger, and dissociation) as opposed to those seen in people with anxiety disorders (fear based symptoms). This shift reflects the mood dysregulation associated with PTSD. Among the detailed criteria for this diagnosis are a number of symptoms involving disturbances of affect: re-exposure distress, anxious avoidance of places, experiences, thoughts and generalized triggers, negativity, anhedonia, irritability, and reactivity.

Early trauma impacts brain development and gene expression, potentially fostering mood dysregulation in predisposed individuals (DeBellis et al., 2005; Nelson & Carver, 1998; Teicher et al., 1990; Yehuda & Lehrner, 2018; Yorke, 2010). We also know that interpersonal stressors are more detrimental to the organism than non-social stressors (Rodgers & Cole, 1993). Child abuse and neglect are social traumas with biobehavioral consequences, yet they have no representation in the DSM.

Bessel van der Kolk, in his book *The Body Keeps the Score* (2014) argues that child abuse, a hidden epidemic, is a developmental trauma. Yet 82% of the traumatized children screened in the National Child Traumatic Stress Network failed to meet DSM criteria for PTSD (Spinazolla, 2005). To help bring the issue of this type of trauma to light he proposed a new DSM diagnosis of "developmental trauma disorder" or DTD (van der Kolk, 2005). A cardinal feature of his proposed diagnosis is mood dysregulation. One of his typical cases might be a 13-year-old girl with developmental trauma and severe mood dysregulation who comes in with a history of treatment resistance and numerous diagnoses including bipolar disorder, intermittent explosive disorder, reactive attachment disorder, attention deficit disorder, oppositional defiant disorder, and substance abuse (van Der Kolk, 2014, p.153). His description exemplifies the transdiagnostic nature of mood dysregulation and its emergence in the context of relational trauma.

Exposure of children to physical, emotional, educational neglect and maltreatment or physical/sexual abuse or violence within the context of their caregiving system during childhood is regarded by workers in the field as "complex trauma" (van der Kolk, 2005). While the DSM criteria for PTSD are exhaustive, the PTSD diagnosis fails to capture the developmental effects trauma has on these children: disturbed attachment patterns, rapid shifts in emotional states, aggressive behavior against self and others, and emotional and physical dysregulation.

Van der Kolk's proposed diagnostic criteria capture the experience of many of the people we see with mood dysregulation that have been subjected to developmental trauma in a way that the DSM-5 PTSD diagnosis does not. Whether predisposed by a vulnerable affective temperament or not, exposure to developmental trauma has far-reaching effects on neurophysiologic development, which can hardly be overstated.

Divergent target symptoms are associated with PTSD, not all of which respond to each treatment modality. Overall, there are many effective treatments for PTSD. Eye Movement Desensitization and Reprocessing (EMDR) shows rapid response for depression and anxiety (Marcus et al., 1997). Imaginal flooding reduces acute symptoms of state anxiety but not depression and trait anxiety (Cooper & Clum, 1989). A randomized trial comparing cognitive behavior therapy to exposure therapy found both to be equally effective (Paunovic & Ost, 2001). In a randomized trial comparing affective/interpersonal regulation skills training followed by exposure treatment with cognitive behavioral therapy, 46% of the regulation skills/exposure treatment group reported feeling significantly better compared with 4% of the control group (Cloitre et al., 2002).

Medications have been of help in treating some of the depressive and anxiety symptoms but virtually all controlled studies conducted have used standard antidepressants (SSRIs in particular): a few have used anxiolytics (Friedman & Davidson, 2007; Friedman et al., 2003; Mohamed & Rosenheck, 2008; van der Kolk et al., 1994). There have only been a few controlled studies using mood stabilizers, most of which have been inconclusive (Ravindran & Stein, 2009).

5.7   GENERALIZED ANXIETY DISORDER

Generalized anxiety disorder (GAD) is patterned on the normal behavior of those who are temperamentally well organized, always thinking ahead, planning for contingencies, and taking precautions. But on a spectrum, such behavior becomes a disorder when it is amplified to a disruptive degree. People with GAD are chronic worriers: they fret about routine daily events (finances, being late, health of family members), even when the likely impact of the consequences is greatly overexaggerated. Try as they may, they find it difficult to control their worries, which disrupt the task at hand. Their worries interfere with functioning, are pervasive, and are often associated with physical symptoms: this distinguishes the anxiety of those with GAD from the "worried well." Clearly, GAD and mood dysregulation intersect in a transdiagnostic manner, but is mood dysregulation a feature of GAD or a co-occurring disorder?

Researchers feel that one of the main mechanisms for an individual's failure to regulate emotions associated with anxiety is the inability to articulate feelings (Mennin, 2004; Suveg & Zeman, 2004). Menin and colleagues (2005) have argued that people with GAD have difficulties in four particular areas of emotional functioning:

1. Heightened intensity of emotions compared to people without GAD
2. Marked difficulty in identifying, describing and clarifying their emotional experiences
3. Proneness to greater emotional reactivity to emotions by holding catastrophic beliefs about the consequences of both positive and negative emotions
4. Struggles to manage or self-soothe in the face of negative emotion

Researchers have posited that emotion dysregulation predicts a diagnosis of GAD (Menin et al., 2005; Salters-Pedneault et al., 2006; Tull et al., 2009), though it is not thought that emotion dysregulation is specific to GAD, in keeping with the transdiagnostic formulation of the concept (Menin et al., 2009).

If this is the case, GAD emerges from an inability to regulate moods, and mood dysregulation is the driving force behind GAD. Indeed, it is not uncommon for people with mood dysregulation to present with "anxiety." When the anxiety is "about nearly all the things nearly all the time" and has few, if any, associated physical symptoms, it is likely associated with GAD. But if the "anxiety" has an irritable, agitated quality and is associated with a number of physical symptoms (increased heart rate, increased respiratory rate, sweating, etc.), it may actually be the dysphoric hypomania often associated with the activated state of a mood dysregulation, not simply GAD. Both types of anxiety can co-exist as well.

The "emotion dysregulation theory" approach to anxiety posits that people with symptoms of GAD experience emotions quickly and with high intensity. Emotion regulation therapy involves skills training in somatic awareness and experiential exposure exercises, and has been shown to be successful in treating people with GAD (Menin, 2004). Butler and colleagues (1991) found cognitive behavior therapy superior to behavior therapy in a randomized trial. Borkovec and Costello (1993) found applied relaxation and cognitive behavior therapy equal in efficacy in the treatment of GAD. Leischsenring and colleagues (2009), in a randomized study, demonstrated that psychoanalytic psychotherapy and cognitive behavior therapy were of equal efficacy in the treatment of GAD.

Among psychopharmacologists, there are fairly uniform guidelines for the treatment of GAD based on placebo-controlled studies. Benzodiazepines are found to be effective in reducing symptoms of anxiety but are not recommended as first-line treatment because of sedation, dependence, withdrawal symptoms, and risk of relapse to use. First-line treatment is generally the use of SSRIs. SNRIs and TCAs are also used (Baldwin & Polkinghorn, 2005; Davidson, 2001; Gould et al., 1997; Rickels & Rynn, 2002).

5.8 PANIC DISORDER

As with generalized anxiety disorder, panic disorder (PD) is a disturbance of anxiety and is associated with emotion dysregulation that encompasses similar problems as those seen in GAD (Campbell-Sills et al., 2006; Sloan & Telch, 2002).

Panic disorder is associated with a number of co-occurring diagnoses (Lecrubier, 1998; Marshall, 1996; Staecevic et al., 1992). Reports of generalized treatment have shown mixed outcomes (Brown et al., 1995; Keijsers et al., 1994; Pollack et al., 1994) but Tsao and others have found that treatment focused specifically on panic disorder has an ameliorative effect on comorbid conditions as well (Borkovec et al., 1995; Brown et al.,1995; Tsao et al., 1998).

While PD and GAD themselves co-occur, the predominant distinguishing feature between the two is that in PD, physical symptoms tend to be primary where in GAD, cognitive symptoms tend to be primary. Typically, a person with PD complains of sudden onset of shortness of breath, heart palpitations, sweating, nausea, dizziness, or other symptoms that may give a sense of impending doom. This may lead to anxiety and a snowballing of symptoms that includes a mental sense of anxiety, but the onset of symptoms is somatic. A person with GAD, on the other hand, may have physical restlessness but this results from a snowballing process that begins with mental worries.

The physically activated state that occurs in panic disorder may be indistinguishable from the activated state that occurs in the dysphoric hypomania of a mood dysregulation, often causing diagnostic and treatment confusion. Panic attacks are typically acute and time-limited where activated states of dysphoric hypomania are typically more protracted. When medication is indicated, the former may be appropriately treated with benzodiazepines while the latter should not.

5.9   SUBSTANCE USE DISORDERS

The interface between substance use disorders (SUD) and mood dysregulation is complex and cannot be discussed here in an adequate manner, in part because each substance of abuse is a different psychopharmacologic entity with different effects on mood and different pharmacologic action. Emotion dysregulation may be the single greatest factor leading to SUD. Drugs are often used to look for relief from negative emotional states (Jones et al., 2001; Le Moal, 2009) and affective lability and instability increase the likelihood of using drugs (Kessler & Merikangas, 2004). Negative affective states trigger drug craving, drug use, and drug relapse (Shiffman et al., 1996; Sinha & Li, 2007). Poor self-control in childhood, including low frustration tolerance and impulsivity, predict the onset of drug use and SUDs in adulthood (August et al., 2006; Ivanov et al., 2011; Moffitt et al., 2011).

An exquisitely complex reciprocal relationship exists between mood dysregulation and SUD inasmuch as mood dysregulation amplifies the behavioral reinforcement produced by substances. Repeated exposure to a psychoactive substance followed by a euphoric experience leads to further destabilization of mood. This process has effects on the underlying brain substrate, which is shared both by the affective and the addictive systems of the organ (Maremmani et al., 2006). "Substance-induced mood disorder" becomes indistinguishable from the natural mood cycles of an underlying affective disorder in such a system (Akiskal et al., 1977; Akiskal et al., 1979; Akiskal & Pinto, 1999). Notably, the neural circuitry implicated in affective reactivity and regulation is closely related to that proposed to underlie addictive behaviors (Cheetham et al., 2010; Koob, 2003; Koob, 2006; Li & Sinha, 2008).

Pursuit of the substance "high" predisposes individuals with mood dysregulation to substance abuse. Craving related urges closely resemble hypomanic/manic excitement: this excited, impulsive hyperactivity shares a common ground with addiction. Addiction is considered to be a conduct disorder developed through prolonged exposure to an intrinsically addictive substance, and it impairs one's capacity to achieve any ongoing sense of pleasant equilibrium (Maremmani et al., 2006). Ongoing use of the substance may result in repeated pleasure, but an acceptable level of regulation can never be sustained. The abuse/dependence cycle amplifies mood dysregulation in a snowballing process.

The affective temperament most closely associated with substance use disorders is cyclothymic temperament, regardless of the substance used (Rovai et al., 2017). Cyclothymic temperament was found to be an important negative predisposition to substance abuse in a survey of the prevalence of various temperament types in substance abusers (Erfurth et al., 2015). Recall that Akiskal and Perugi (2002) associated the cyclothymic temperament with mood dysregulation.

Cannabis is often the drug of choice for people with mood dysregulation. A cross-sectional study of medical students showed recreational cannabis use to be highest among those with irritable and cyclothymic temperaments (Infortuna et al., 2020). We all have endogenous cannabinoids in our bodies: these "endocannabinoids" are involved in the modulation of brain reward systems (Gardner, 1999). The two major cannabinoids in marijuana, delta-9-tetrahydrocannabinol (THC) and cannabidiol (CBD) both bind to endocannabinoid receptors in the body. THC and CBD have distinct and often opposing mechanisms of action, explaining some of the paradoxical effects of marijuana. For example, the sought-after "high" experienced by people smoking marijuana is due to the psychedelic effects of THC. But in vulnerable people, acute use of marijuana can precipitate symptoms of mania (Bally et al., 2014; Kahn & Akella, 2009), and chronic, heavy use can increase the risk of developing schizophrenia-like psychoses (Arsenault et al., 2004; Hall, 2006). These "psychotomimetic" properties of THC are antagonized by CBD, which weakens the psychoactive ("high") effect of THC (Dalton et al., 1975; Karniol & Carlini, 1973; Musty, 2004).

Marijuana's appeal to people with mood dysregulation is primarily because of its antianxiety properties. CBD has antianxiety, antipsychotic, and antiseizure effects. It is noteworthy that both the antiepileptic medications used as mood stabilizers and CBD, which has antiepileptic properties, are used to regulate mood. Many of the people I have treated for mood dysregulation say they self-medicate with marijuana. It appears that CBD is responsible for the intrinsic mood stabilizing effects of marijuana. For legal and political reasons, there is but a meager body of literature on this, but the mood stabilizing effects of marijuana have been documented (Grinspoon & Bakalar, 1998).

Along with cannabis use, alcohol use is also prevalent among people with mood dysregulation. Cyclothymia has been linked to an increase in alcohol abuse (Pombo et al., 2013) and bipolar disorder is the secondary diagnosis most associated with alcohol use disorders (Grant et al., 2004; Hermens et al., 2013). People with mood dysregulation self-medicate for a number of reasons (Khantzian, 1997), the first of which is anxiety (Swendsen et al., 2000), followed by depression. People also drink to mellow activated states of hypomania.

Alternating patterns of stimulant ("upper") and sedative ("downer") use occur in at least 50% of people with cyclothymia (Akiskal et al., 1979; Gawin & Ellinwood, 1988). People with bipolar disorder were found to be using higher doses of benzodiazepine sedatives than those without affective disorders (Clark et al., 2004) and those who were prescribed benzodiazepines were more likely to abuse them (Brunette et al., 2003). Despite abundant evidence against the use of benzodiazepines to treat symptoms of anxiety in chronic mood dysregulation, it can be difficult to convince anxious individuals that use of benzodiazepines is ill-advised and risky.

Psychostimulant medication has been used successfully to treat ADHD in those without an underlying bipolar diathesis. There has been concern that pediatric use of psychostimulants predisposes an individual to adult substance abuse. In fact, the data show that there is no increased risk of substance abuse in children treated with psychostimulants (Mannuzza et al., 2004; Wilens et al., 2003) and that it may, in fact, convey a protective effect against development of adult substance abuse.

The picture with mood dysregulation and opioids is more complicated than it is with other substances. Opioid dependence associated with cyclothymic and irritable temperaments (Maremmani et al., 2009) can have devastating consequences. But opioids can also have therapeutic effects as well. It appears that in a subgroup of the population, opioids are mood altering, in some behaving like antidepressants, in others precipitating hypomania/mania, at times even delivering a "high." I have treated people for opioid dependence with the partial opioid agonist buprenorphine and seen secondary antidepressant effects. Researchers have found opioids to have antidepressant effects in certain people (Emrich et al., 1982; Gold et al., 1982; Schaffer et al., 2007).

Schaffer and colleagues (2007) found that hydrocodone precipitated mania in a subpopulation of people with bipolar disorder and had an antidepressant effect in others. They postulate that opioids might unmask bipolar symptoms in an individual predisposed to such mood instability.

This data may explain why there appears to be a sub-group of individuals in the population who who are at elevated risk for narcotic dependence with the use of opioids prescribed for routine post-operative pain management. Many people who are prescribed hydrocodone after a post-operative procedure rather dislike the side effects, even if they benefit from the analgesia. But a subgroup of people taking hydrocodone experience a high from the medication, not unlike the subgroup of people with bipolar disorder developing mania with hydrocodone use. The high experienced by this group of opioid users is a treatment emergent response (e.g. akin to bipolar III disorder on the soft bipolar spectrum), and recognition of the presence of such a mood vulnerability might be useful in identifying a group at high risk for opioid dependence.

5.10   DISRUPTIVE MOOD DYSREGULATION DISORDER

Disruptive Mood Dysregulation Disorder (DMDD) is defined primarily by a baseline irritable/angry mood punctuated by temper outbursts. This mood disturbance is not altogether unlike the mood dysregulation we have been describing. It is curious, then, that we have been talking about mood dysregulation as related to the bipolar spectrum while the DSM-5 diagnosis DMDD is classified as a depressive disorder.

To address this discrepancy, it is important to know how the DMDD diagnosis was developed and came to be included in DSM-5 as a depressive disorder. Its inclusion is controversial for at least three reasons.

1. The very impetus for the development of the DMDD diagnosis was to curb the overdiagnosis of bipolar disorder among children who were irritable and explosive (Chen et al., 2016; Lochman et al., 2015; Meyers et al., 2017).

2. There is little statistical data supporting the validity of the DMDD diagnosis.
3. There is no consensus on treatment guidelines for the DMDD diagnosis (Axelson et al., 2011; Parens & Johnston, 2010; Stringaris, 2011).

To understand the DMDD diagnosis, one needs to understand the history of its development. In short, overdiagnosis of bipolar disorder and overuse of medication to treat bipolar disorder in children caused alarm and led to the reassessment of DSM diagnostic criteria.

When diagnoses of pediatric bipolar disorder increased 40-fold in less than a decade (Blader & Carlson, 2007; Moreno et al., 2007) and when the vast majority of these cases were being treated with mood stabilizers and atypical neuroleptics, a fierce debate emerged around the prevalence and presentation of pediatric bipolar disorder (Althoff, 2010; Biederman et al., 2004; Carlson & Glovinsky, 2009; Diler et al., 2009; Leibenluft, 2011). The need arose to determine whether or not a subgroup of irritable children without bipolar disorder could be distinguished by some phenotypic characteristic (such as medication response).

The issue was complicated by the fact that oppositional defiant disorder (ODD), pediatric attention-deficit/hyperactivity disorder (ADHD), and intermittent explosive disorder (IED) all shared overlapping symptoms of irritability and anger with pediatric bipolar disorder (BD). Adding to the confusion between these three diagnoses, a number of clinicians maintained that mania in pediatric bipolar disorder was not episodic as in adult bipolar disorder but that it was a more chronic, irritable manifestation of the disorder (Biederman et al., 1998; Faraone et al., 1997; Wozniak et al., 1995). Others argued that a syndrome of chronic irritability was distinct from pediatric bipolar disorder and that pediatric bipolar disorder cycled between elevated and depressed moods just as adult bipolar disorder (Geller et al., 1998).

Leibenluft and colleagues (2003) attempted to address this question by distinguishing two bipolar groups, studying a group of angry, explosive children. They separated those with narrowly defined bipolar disorder (experiencing manic episodes) from those with more broadly defined symptoms. They formulated a set of diagnostic criteria for the latter group which later came to define severe mood dysregulation (SMD).

Temper outbursts, negative mood, and hyperarousal were cardinal features of SMD. Children with SMD were found to have high rates of ADHD, ODD, anxiety disorders, and an elevated risk for unipolar depression but not for bipolar disorders (Brotman et al., 2006; Stringaris et al., 2009). Even so, children with patterns of episodic elation (not manic episodes) were 50 times more likely than other children with irritability to develop bipolar disorder later in life (Faraone et al., 1997; Mick et al., 2005; Stringaris et al., 2010). Clinical studies of children have shown that paroxysmal irritability is closely linked with manic episodes, while chronic irritability is closely related to depressive or anxiety disorders (Stringaris et al., 2012).

A significant amount of research was done on SMD, but the diagnosis was not included in DSM-5. In its stead, the new diagnosis that made its way into DSM-5 was disruptive mood dysregulation disorder or DMDD, despite universal objection that there was no research validation for it. Quite odd. There is little explanation as to how this came about.

The diagnostic criteria for SMD and DMDD are essentially the same with one significant difference. This difference is telling because it explains why mood dysregulation is tethered to—and beyond—the bipolar spectrum, unlike DSM-5 DMDD, in which mood dysregulation is categorized as a depressive disorder. The major difference between SMD and DMDD is that the diagnosis of SMD includes the **hyperarousal** criterion associated with manic states (and mood dysregulation) whereas DMDD does not.

Since the new diagnostic endeavor was driven by the impetus to reduce the diagnosis of childhood bipolar disorder, the SMD hyperarousal criterion seemed problematic as hyperarousal is an activated energy state seen in bipolar disorder. So it was dropped from the DMDD diagnosis because DMDD was intended to be distinguished from bipolar disorder. While still working within Leonhard's unitary classification, DMDD would then be classified as a depressive disorder. This classification was also in support of the longitudinal data suggesting high rates of depressive disorders associated with SMD (Rao, 2014).

Categorizing DMDD as a depressive disorder, I believe, is a mistake. The exclusion of the hyperarousal criterion was designed primarily to avoid the inclusion of individuals who exhibited periods of euphoric hypomania or mania and thus appeared bipolar. While such people may be excluded from the DMDD diagnosis, the DMDD diagnosis certainly DOES include individuals in states of hyperarousal who exhibit periods of dysphoric hypomania or mania, hallmark symptoms of the disorder. This reality would, at the very least, tether the DMDD diagnosis to the bipolar spectrum.

Disruptive Mood Dysregulation Disorder has two primary features: (1) severe and frequent emotional outbursts and (2) chronic and persistent anger or irritability that exists between temper tantrums. While not explicit criteria, these remain shared features of ADHD and ODD. Proponents of the RDoC system argue that though the DMDD diagnosis encompasses a heterogenous group of disorders, the impetus to separate out a new diagnostic entity from bipolar disorder was based on growing pathophysiologic evidence demonstrating a difference between mood dysregulation and bipolar disorder (Adelman et al., 2011; Brotman et al., 2010; Deveney et al., 2012; Thomas et al., 2012; Wiggins et al., 2016). Indeed, the goal of the RDoC endeavor is to classify symptoms on the basis of neuropathophysiological similarities, which is what the DMDD diagnosis seems to be doing. That DMDD encompasses the symptoms of people thought to be in subgroups with SDHD, ODD, and IED and BD indicates the transdiagnostic nature of the new DSM-5 entity. The somewhat counterintuitive classification of DMDD as a depressive disorder may be an apt illustration of the confusion between "depression" (not necessary for the diagnosis) and dysphoria (a negative mood phase often characterized by irritability, a cardinal feature of the DMDD diagnosis).

There is no clear, consistent recommended pharmacologic approach to the treatment of DMDD. Tourian and colleagues (2015) report that treatment options may include antidepressants, mood stabilizers, atypical neuroleptics, and psychostimulants. In placebo-controlled trials among children with DMDD, Risperdal, an atypical neuroleptic, was found to have some efficacy (Krieger et al., 2011) as did Ritalin, a psychostimulant (Waxmonsky et al., 2008) but no controlled trials have shown significant efficacy for any particular drug or class of drugs to treat DMDD.

The fact that DMDD responds to various classes of drugs reflects the heterogeneity of the group of disorders included within this diagnostic umbrella and is consistent with the transdiagnostic nature of DMDD, as seen with mood dysregulation. After all, the DMDD diagnosis emerged as clinicians sought to capture features of people diagnosed with ODD, ADHD, BD, and IED. One wonders whether the sub-group responsive to psychostimulants has co-occurring ADHD? Or whether the subgroup responsive to mood stabilizers has co-occurring bipolar (spectrum) disorder? Is there overlap between those with DMDD and those with mood dysregulation who, as adults, often live in chronic states of dysphoric hypomania? Perhaps DMDD is the DSM-5 footprint of our mood dysregulation.

5.11 RELATED CONDITIONS

There are several related conditions that, unlike those listed above, are not DSM-5 diagnoses but do intersect with mood dysregulation. They are related to mood dysregulation not so much in the transdiagnostic sense but in the sense that they are beyond the bipolar spectrum in their emergence from affective temperaments.

5.11.1 (SOME) TREATMENT-RESISTANT DEPRESSIONS

Some, but by no means all, cases of treatment resistant depression (TRD) seem to be cases of mood disturbance emerging from affective temperaments that behave in ways related to mood disorders on the bipolar spectrum. They are not bipolar disorders but they might be regarded to be beyond the bipolar spectrum. These depressions represent a subset of the larger universe of TRDs.

Definitions of TRD in the literature vary widely. Fava and Davidson (1996) define TRD as "failure to respond to standard doses (i.e., doses significantly superior to placebo in double-blind studies) of antidepressants administered continuously for at least 6 weeks." Fava (2003) later refers to TRD as inadequate response to one antidepressant trial of adequate dose and duration. Souery and colleagues (1999) define TRD as "failure to respond to two adequate trials of two different classes of antidepressants each given separately in adequate dosage for 6–8 weeks." Adequate treatment requires use of doses that are known to be superior to placebo in clinical controlled trials, and adequate duration requires sufficient time to allow for a robust clinical response, for example, 12 weeks (Quitkin et al., 1986).

One sub-group of people with TRD is that group whose "depression" masquerades as the negative mood of dysphoria. All too often, individuals seeking treatment for "depression" present with a negative mood which, on closer inspection, is an irritable dysphoria, a hidden manifestation of the activated state. Recognizing "depression" to be dysphoria unlocks a key to treatment that was previously not thought of or tried. Symptoms of TRD are often complex, involving depressed mood of the slowed/ depressive state at times and dysphoric mood of the activated state at others. But moods typically do not cycle. The nature of the dysphoric mood is missed by the practitioner, who prescribes antidepressant medication that fails to address the activated state. Dysphoria and activated negative moods persist despite repeated trials of antidepressant medications, and the individual is deemed to have treatment resistant depression. Depression that fails to respond to treatment with typical antidepressants is a red flag for mood dysregulation that should be approached with mood stabilizers.

Another group of people with TRD is that subset with a missed soft bipolar spectrum diagnosis (Akiskal & Mallya, 1987; Correa et al., 2010; Fava, 2003; Inoue et al., 2006; Sharma et al., 2005) whom Akiskal has described as having bipolar III disorder. These individuals have a depressive or hyperthymic affective temperament and/or bipolar family history, and a history of recurrent depression without spontaneous hypomania. Or they have recurrent depression with switches to hypomania provoked by the use of (typically stimulant) medication. These are people with "pseudo-unipolar" depression. As pseudo-unipolar depression is a mood disturbance emerging from an affective temperament that is not cyclical, I regard it to be a mood dysregulation. Once its association with the bipolar spectrum is recognized, treatment with mood stabilizers can generally resolve this type of TRD.

In these cases, TRD does not quite turn out to be "unipolar" depression: as mentioned before, it is thought that up to 50% of putative unipolar depressions are on the bipolar spectrum (Angst et al., 2005; Judd & Akiskal, 2003; Merikangas et al., 2007) and responsive to mood stabilizing agents. It is quite dramatic to see people with major depression who have failed numerous antidepressant treatments respond to lithium (Bowden, 1978; Mendels, 1976).

5.11.2   PSEUDO-UNIPOLAR DEPRESSION

Pseudo-unipolar depression (PUD) is a depression that appears to be unipolar (a la Leonhard) but at some point declares itself to be on the bipolar spectrum with the emergence of hypomanic or manic symptoms. Of course, in Kraepelin's scheme, the concept of pseudo-unipolar depression would not exist because affective disorders are not dichoto-mized but fall along a spectrum from depressed to manic. But we use the term pseudo-unipolar because the DSM continues to organize diagnoses a la Leonhard.

PUD can be a type of TRD or it can be a depression that "flips" to hypomania or mania, either spontaneously or with use of an antidepressant or psychostimulant.

Akiskal and colleagues (2000) regard PUD to be on the soft bipolar spectrum. As there is typically no evidence of mood cyclicity with PUD, we may think of it as falling on the ultraviolet end of the soft bipolar spectrum and as emerging from a depressive affective temperament, thus regarded to be a mood dysregulation. It shares the phenotypic charac-teristics of mood dysregulation beyond the bipolar spectrum such as family history of bipolar disorder and medication responsivity to mood stabilizers.

What makes PUD a mood dysregulation is the presence in the syndrome of both depressive/slowed states and activated states in a non-cyclical fashion, triggered by external events. People with PUD typically present appearing 'down' in a slowed state but switch to an activated state when triggered by medication.

How might we foresee whether or not a depression will develop these activated features? Looking at the data in hindsight, researchers have gathered a number of potential markers that might help identify PUD:

1) early onset of recurrent unipolar depression
2) post-partum depression
3) unipolar depression with a family history of bipolar disorder
4) depression with fatigue, hypersomnia, carbohydrate craving, diurnal craving, and seasonal mood-swing (Amsterdam & Hornig-Rohan, 1996)
5) young age of symptom onset
6) history of depressive mixed states
7) atypical features of depression (Akiskal, 2002; Baldessarini, 2000; Goodwin & Jamison, 1990)
8) sensitivity to treatment with lithium.

Below are some examples of PUD.

One group of people with PUD presented with symptoms of depression that switched spontaneously to hypomania or mania. In one study, 39% of people hospitalized for unipolar depression were subsequently diagnosed with bipolar disorder after spontaneous mood switches (Angst et al., 2005). Howland (1996) reported a series of depressed people with treatment emergent mania or hypomania (TEMH) who were subsequently stabilized on lithium: all people, it was later found, had family members with histories of bipolar disorder.

Another group of people with PUD presented with depression and became manic or hypomanic when treated with antidepressants or psycho-stimulants (TEMH) (Faedda et al., 2004; Ross, 2006). Prozac alone may precipitate a state of activation in individuals predisposed to such destabi-lization (Benazzi, 2007). Predisposing factors for TEMH appear to be (1) early age of onset of mood disorder, (2) family history of bipolar disorder, (3) low level of anxiety and physical complaints, (4) motor retardation, and (5) mood lability (Benvenuti et al., 2008).

Treatment of people with PUD is different from treatment of those with unipolar depression: it follows treatment for bipolar depression rather than that for unipolar depression. A missed diagnosis of PUD can result in improper treatment with antidepressant medication (Bowden, 2005). Akiskal and Pinto (1999) recognized that many PUDs were on the bipolar spectrum: their efforts to expand the diagnostic repertoire by defining the soft bipolar spectrum were, in part "to shield these patients from possible negative effects of antidepressants unprotected by mood stabilizers."

5.11.3 ATYPICAL DEPRESSION

Atypical depression has an interesting history. The term was first used by West and Dally (1959) to describe a subgroup of depressed people whose symptoms did not follow the typical pattern of decreased appetite and decreased sleep. These individuals were anxious, had "hysteria" and did not respond to classic treatment with tricyclic antidepressants or electroconvulsive therapy (ECT). They did, however, respond to treatment with monoamine oxidase inhibitors (MAOI's). Pollit and Young (1971) then observed a subgroup of depressed people who had "reverse vegetative symptoms," that is, instead of the typical symptoms of melancholic depression, they had the opposite symptoms (increased appetite/eating, increased sleep, and increased libido). These people also had symptoms of hysteria. Other people came to identify a similar subgroup of depressed individuals with the same atypical symptoms of depression. In the 1980s and 1990s, Columbia University researchers conducted landmark trials demonstrating the efficacy of MAOIs in people with these atypical symptoms of depression, operationalizing criteria for what became known as "atypical depression:" mood reactivity and two associated symptoms of overeating, oversleeping, leaden fatigue, and rejection sensitivity (McGrath et al., 1993; Leibowitz, 1988; Quitkin et al., 1989, 1991). A hallmark of atypical depression, mood reactivity, is also one of the cardinal symptoms of mood dysregulation in which individuals display over reactivity to personal slights and appear to have thin skin (the rejection sensitivity of atypical depression) that leads to interpersonal conflict. Atypical depression is thought to be on the soft bipolar spectrum (Perugi et al., 1998). Atypical depression appears to be a mood dysregulation as it shares a cluster of symptoms seen in the activated states associated with dysphoria (anxiety, reactivity, rejection sensitivity, and increased libido), and responsiveness to MAOIs.

Mood dysregulation is a transdiagnostic process and is seen with borderline personality disorder, ADHD, eating disorders, PTSD, generalized anxiety disorder, panic disorder, substance use disorders, and DMDD. When diagnoses overlap, it is tempting to focus on symptom detail, but in clinical practice, the quest for diagnostic certainty should never eclipse the endeavor to understand an individual and seek the most appropriate treatment, which may be symptom based and shared between co-occurring diagnoses.

The DMDD or disruptive mood dysregulation disorder diagnosis presents several interesting parallels to our discussion of mood dysregulation. First, just as the DMDD diagnosis was developed to separate a group of people with specific symptoms from those with bipolar disorder, so has the idea of mood dysregulation emerged from a need to distinguish a group of people with apparent mood fluctuations from those with a diagnosis of bipolar disorder. Second, both the DMDD diagnosis and the concept of mood dysregulation seek to distance themselves from the bipolar diagnosis and function to avoid some of the difficulties associated with the diagnosis of bipolar disorder. And finally, like the depressive disorder DMDD, the presenting symptoms of mood dysregulation typically masquerade as symptoms of "depression." When we line it up with DMDD, it just may be that the DSM has a template for our mood dysregulation after all.

KEYWORDS

- rapid cycling
- treatment-emergent hypomania
- developmental trauma disorder
- treatment resistant depression
- pseudo-unipolar depression
- atypical depression

CHAPTER 6

Biology

"Keep Calm and Carry On."
—British Ministry of Information, 1939

ABSTRACT

Under ordinary circumstances, one's moods are regulated by the body's elaborate system designed to maintain equilibrium or "homeostasis," orchestrated by the brain and its elaborate network of nerves which communicate via neurotransmitters and hormones. In a simplified scheme of the brain, there are three layers corresponding to evolutionary development. From innermost to outermost, the human brain has a core known as the reptilian brain, which mediates basic metabolic functions such as breathing and heart rate, the mammalian brain, which enables perception and expression of emotion and mediates attachment behaviors, and the new brain, which is responsible for reasoning, decision making, language, and higher cognitive function. Deep in the mammalian brain, also known as the limbic brain, is the amygdala, known as the mediator of the fight or flight reflex, mounted under times of acute stress. Acute stress mobilizes the sympathetic nervous system, a branch of the nervous system responsible for activation, arousal and motivation. Hormones, including the stress hormone cortisol, are also recruited in the stress response. Equilibrium is restored after the storm has passed by the counterpart to the sympathetic nervous system, the parasympathetic nervous system, which is designed to deactivate, to calm, and to preserve energy. Cortisol production and release is halted when no longer needed via negative feedback loops. In a healthy organism, these basic systems and numerous ancillary support systems regulate our somatic functions to maintain equilibrium. Mood regulation depends on the anatomic and functional integrity of these multiple systems.

Though it was little known until a copy was rediscovered in 2000, the motivational poster issued by the British government on the eve of World War II "Keep Calm and Carry On" was designed to galvanize the heart of the

Mood Dysregulation: Beyond the Bipolar Spectrum. Deborah A. Deliyannides, MD (Author)
© 2024 Apple Academic Press, Inc. Co-published with CRC Press (Taylor & Francis)

nation to assure its survival in the face of impending onslaught. In much the same way, species survival depends on our capacity to maintain equanimity in the face of stress. All living things are organized around the capacity to maintain equilibrium. The biologic process of achieving and sustaining equilibrium is called homeostasis. From amoebae to humans, survival depends on maintaining homeostasis amid the slings and arrows of life. Like the real estate mantra "location, location, location," the biologist's guiding lens is "homeostasis, homeostasis, homeostasis." Keep calm and carry on.

When we eat, we need to be able to process our food, utilize it, and return to a baseline level of metabolism. When we get the flu, we need to be able to fight it off and recover. When we see an assailant, we need to mount an acute stress response, then return to a state of calm when the threat has passed. And when we face interpersonal conflict, we need to be able to react with assertion while modulating our response if we want to live in a civilized world. The most trivial stresses we take for granted are under careful homeostatic control by complex systems in the body. Primitive organisms require simple mechanisms to maintain homeostasis, but humans require multiple, interconnected systems working in concert to maintain equilibrium. In our bodies, this process is orchestrated by the nervous system, which take marching orders from the brain.

The nervous system has a central division and a peripheral division. The central nervous system (CNS) is composed of the brain and spinal cord. The spinal cord contains "cables" of nerves carrying sensory information and motor directives between the rest of the body and the brain. The peripheral nervous **system** (PNS) has nerves plugged into the spinal cord from the far reaches of the body to monitor the environment and deliver information to the CNS for the processing and execution of its responses. It is composed of the somatic and the autonomic (with its sympathetic and parasympathetic branches) subsystems and their way stations. It connects the CNS to the body's organs and limbs. Communication between these anatomical structures is carried out by hormones and neurotransmitters which serve as the fundamental envoys of homeostasis, primarily by assuring healthy function of the body's (mostly negative) feedback loops. We will review each of these and their components briefly as well as their role in the homeostatic process.

6.1 BRAIN

As animals evolved, the brain became increasingly more elaborate to meet the needs of increasingly more diverse, complex, flexible, and adaptable organisms. Paul MacLean's (1982) oversimplified depiction of the "triune brain" captures this evolutionary process and is an accessible model we can use to begin understanding this daunting organ. It depicts the brain with three basic layers, one wrapped around the other as they developed in evolutionary order. These three layers, from oldest to newest, are the reptilian or instinctual brain, the mammalian (limbic) or emotional brain, and the new (neocortex) analytical brain (Figure 6.1).

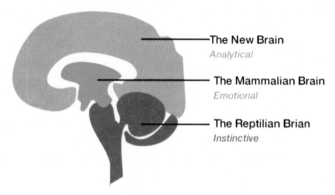

FIGURE 6.1 The triune brain

Deep inside the skull of every one of us there is something like a brain of a crocodile. Surrounding the R-complex is the limbic system or mammalian brain, which evolved tens of millions of years ago in our ancestors who were mammal but not yet primates. It is a major source of our moods and emotions, of our concern and care for the young. And finally, on the outside, living in uneasy truce with the more primitive brains beneath, is the cerebral cortex; civilization is a product of the cerebral cortex.

—Carl Sagan, Cosmos p.276–277

6.1.1 REPTILIAN BRAIN

The "R" complex, "reptilian brain," or brain stem (Figure 6.2) is something of a stump protruding from the lower back portion of the brain. It is the most primitive part of the brain but regulates the body's most vital functions. Somewhat like the boiler room in our homes in charge of maintaining a comfortable environment in which we can live, it does its work largely out of sight. The brain stem generates basic instinctual action, plans for primitive emotional processes such as exploration, feeding, aggressive/dominant displays, and sexuality. Many emotional processes in the brain are initiated in the brain stem (periaqueductal gray or PAG, reticular formation and nuclei of the vagus nerve). The brain stem is the regulatory center for such essential processes as cardiac and respiratory functions and the sleep-wake cycle. It mediates the activity of the cranial nerves and is the source of the motor and sensory tracts to the spinal cord, tracts through which nerves deliver signals to muscles to cause motor action, and tracts that deliver sensations from the environment to the CNS for processing. The brain stem is also called the "instinctual" or "motor" brain. The superior colliculi at the upper end of the brain stem and the inferior colliculi below them are responsible for processing visual and auditory information, respectively. Despite the primitive nature of the brain stem, these two systems help provide us a sense of "presence" within the world (Panksepp, 1998). Panksepp (1998) locates the primal neural representation of the self in the brain stem.

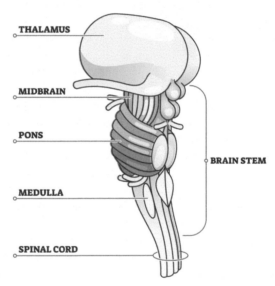

THALAMUS

MIDBRAIN

PONS

BRAIN STEM

MEDULLA

SPINAL CORD

FIGURE 6.2 Brainstem

6.1.2 LIMBIC BRAIN

If you've ever watched Animal Planet, you might have seen a mother sea turtle lay her eggs on the beach, swipe some sand over them, and leave, never to return. Reptiles don't bond with one another. Such is the limitation of the primitive, reptilian brain. Mammals are quite different, however. Mammals are affiliative. The evolution of animals living in social networks necessitated development of a more specialized brain. Thus, the second layer of the brain, limbic brain, to mediate expression of emotions necessary for attachment behavior. The limbic brain is called the "mammalian" brain because it is through limbic-mediated emotional communication that mammals develop the capacity to bond. Attachment behavior is unique to mammals, necessitating a brain system with more complex affect processing structures. Affiliative emotions are devoted to the mammal's survival.

While reptiles have reflexive mechanisms for fleeing danger, the limbic brain has a far more sophisticated fight or flight mechanism, enabling assessment before response to environmental threat. The limbic system regulates the internal state of the organism to prepare for necessary action. Affects are generated in the limbic system (primarily in the amygdala), and affect regulation, which enables social survival, depends on an intact limbic system.

The limbic brain orchestrates separation distress and social bonding functions as well as playfulness and maternal nurturing. We communicate with one another non-verbally via the limbic system. We are seldom aware of limbic communication: limbic communication occurs quickly and non-consciously in contrast to communication mediated through the neocortex, which is generally conscious and verbal. Four of the most important structures in the limbic system are highlighted here:

(1) Amygdala (2) Hippocampus (3) Hypothalamus/Pituitary Gland (4) Basal Ganglia

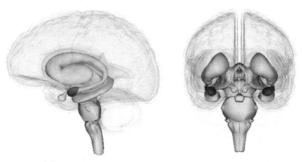

FIGURE 6.3 Amygdala

Source: Reproduced from "Amygdala" by Life Science Database (LSDB), 2009. Open Access, Creative Commons Attribution-Share Alike 2.1 Japan, CC-BY-SA-2.1-jp. Retrieved from: https://commons.wikimedia.org/wiki/File:Amygdala.png

1. **Amygdala:** (Figure 6.3) When animals became affiliative, it became a priority to determine "who is friend and who is foe?" A fast-acting escape mechanism was needed in the face of a foe, hence, the amygdala, the most important subcortical structure in the limbic system with its efficient fight or flight system. The amygdala is so named because of its almond shape. It is the most primitive part of the limbic system. It is actually not one structure but a cluster of nerve nuclei. The amygdala has connections to nearly 90% of the cortex, allowing it widespread affective control. The amygdala is the emergency command center of the limbic system. It functions to monitor the environment for danger and mediates fear and anger. Instinctual appraisals of the world are translated into bodily states within the amygdala: it is with the amygdala that we get our "gut instincts." With the amygdala we scan the environment for sensory input relevant to survival and extract its social significance (Freese & Amaral, 2009). When you hear a dog growl at something you can't see, he is probably sensing something menacing with his amygdala. The amygdala is the seat of "body memory:" traumatic experiences may be processed "somatically" in the amygdala and never reach conscious awareness but reverberate in the body, provoking symptoms of "panic attacks" when triggered non-consciously by stimuli processed in the amygdala that are evocative of the trauma.

    The fight or flight response is initiated in the amygdala: surgical removal of the amygdala results in docile, fearless behavior (Amaral, 2002). In the face of danger, the amygdala initiates a survival response before information has time to reach conscious awareness in the cortex. For example, if we see a snake in the grass, we recoil before even registering what we have seen or planning a response because the amygdala has instantaneously processed our fear on a non-conscious level, activating a behavioral reflex before we are aware of the danger. The fight or flight response emerges from an intrinsic system mediated by the amygdala through which an unconditioned fear circuit passes, ending in the periaqueductal gray (PAG) of the midbrain. Electrical stimulation of this circuit produces fear behavior in mammals (Panksepp, 1990).

Most of us have especially vivid memories of things that have been particularly important to us. The amygdala is responsible for this as it mediates the consolidation of emotionally salient experiences (Christianson, 1992; Pare, 2003). Under ordinary circumstances, these memories are then processed for long term storage in the hippocampus, another structure important for handling memory in the limbic system (Roozendaal et al., 2009).

The amygdala is one of the most crucial emotion mediating areas in the brain. Reduced activation in the amygdala is associated with increased emotion regulation (Ochsner & Gross, 2005) while increased activation of the amygdala is associated with emotion dysregulation. Thus, under ordinary circumstances (when we do not need to mount a stress response), "a quiet amygdala is a good amygdala."

The amygdala is one of the most excitable structures in the brain and is very sensitive to kindling. Kindling is a nerve excitation process by which repeated electrical stimulation comes to induce seizures. Animals that have been kindled show neuronal activation in widespread areas of the limbic system associated with a functional reorganization of the brain (Crain, 1982). Instability (i.e., increased activation) in the amygdala would be expected to have dysregulating behavioral effects because of its importance in mediating emotions. Kindled cats become temperamental and irritable (Adamec & Stark-Adamec, 1989) and kindled rats develop a form of nymphomania (Paredes et al., 1990). The altered behaviors exhibited by animals in whom seizures have been induced and who have undergone long-term brain changes parallel the changes we see in people with mood dysregulation. Noting these behavioral changes, Panksepp (1998, p. 95) hypothesizes that certain psychiatric disorders might be the result of kindling in the brain by exposure to chronic stress. Supporting this notion, he points out that many of these disorders (mood dysregulation?) are responsive to mood stabilizers, which are anti-kindling medications (Kalivas et al., 1993).

In short, the amygdala, central to the emotion-mediating limbic system, is crucial to affect regulation. Activation of the amygdala results in excitation and destabilization of emotional states. Seizure activity, kindling, and perhaps stress can precipitate this excitation and dysregulation.

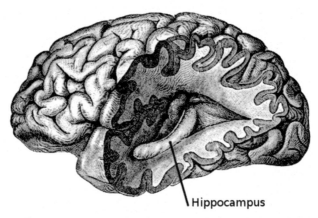

FIGURE 6.4   Hippocampus

Source: Reproduced from Henry Gray (1918) *Anatomy of the Human Body.* Public domain.

2. **Hippocampus:** (Figure 6.4) When systems are functioning properly, environmental information first processed in the amygdala is transferred to the hippocampus for storage and memory consolidation. The hippocampus is the second vital structure in the limbic system and is a (right and left) sea-horse shaped structure at the junction between the cortex and the limbic system. It has an important function in affect regulation and motivation in addition to its role in the processing of new information from short-term to long-term memory (Parkin, 1996). It is well connected to other important brain areas including multiple areas of the cortex, the amygdala, the hypothalamus and thalamus, and the endocrine system. Unfortunately, the hippocampus is particularly sensitive to damage produced by stress hormones (Diamond & Rose, 1994; Lupien et al., 2007; Sapolsky, 1996), explaining, in part, why stress compromises memory. Alzheimer's disease is characterized largely by degeneration of cells in the hippocampus.

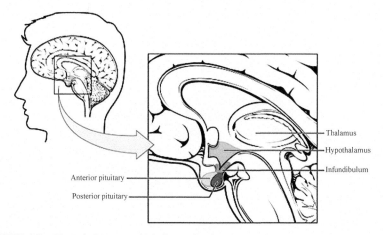

FIGURE 6.5 Hypothalamus and pituitary gland

Source: Image by Pharmattila. https://creativecommons.org/licenses/by-sa/4.0/

3. **Hypothalamus and pituitary gland:** (Figure 6.5) The hypothalamus and the pituitary gland are limbic structures as important as they are small. They are central to one of the body's two major stress response systems and together, serve as a major switchboard in regulating bodily functions. The hypothalamus connects the nervous system with the hormonal system *via* the pituitary gland and is important in mediating attachment as well as other emotion related behaviors. It has extensive connections with the frontal lobes and the brain stem. The hypothalamus regulates numerous metabolic processes in the body as well as the activities of the autonomic nervous system. We will talk further about its major role in hormone regulation.

4. **Basal ganglia:** (Figure 6.6) The swing of your bat, your graceful dance step, your one-handed catch, all of these are mediated by your basal ganglia. The basal ganglia, a lesser-known set of nuclei, are interesting and important to limbic function. They initiate and coordinate the smooth control of voluntary movements and are involved in procedural learning, cognition, and emotion. They come to attention more for their dysfunction than for their function: they are known primarily for their movement and behavioral disorders. Huntington's disease is characterized by random "choreiform" or dance-like movements caused by degeneration of

BASAL GANGLIA

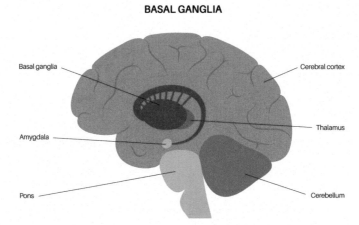

FIGURE 6.6   Basal ganglia

Source: Image by Leevanjackson. https://creativecommons.org/licenses/by-sa/4.0/deed.en

the neostriatum of the basal ganglia. Parkinson's disease, another movement disorder, involves degeneration of the dopamine producing cells of the substantia nigra of the basal ganglia.

The basal ganglia are also known for their role in mediating pleasure and reward. A cable of nerves crossing the basal ganglia called the *mesolimbic tract* is known as the "reward pathway." Nuclei along the reward pathway are involved in behavior reinforcement, reward learning, motivation, and pleasure. The most abundant neurotransmitter in this pathway is dopamine, which is associated with reward and pleasure. Dopamine disturbances in the basal ganglia have been associated with disorders such as Tourette's syndrome, obsessive-compulsive disorder, schizophrenia, movement disorders, and of course, addictions.

6.1.3   NEOCORTEX

As mammals evolved, the neocortex developed to process symbolic thinking and language. The *neocortex* or cortex is the seat of higher-order brain functions: language, cognition, generation of motor commands, spatial reasoning, sensory perception, and consciousness. Nearly all parts of the neocortex are involved in impulse control, which is central to affect regulation. Out of hundreds of areas in the cortex, we will look at just five of them that are the most relevant to mood regulation (Damasio, 1999).

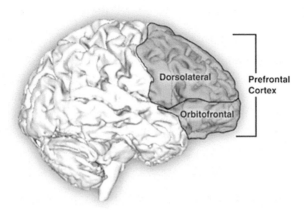

FIGURE 6.7 Prefrontal cortex

1. **Prefrontal Cortex (PFC):** (Figure 6.7) *Decision making, planning, impulse control: the heart of higher cognitive functioning.* The PFC is the anterior portion of the frontal lobe and sits above our eyes. With the PFC, we make decisions. We think about consequences. We exercise self-control and suppress behaviors. The PFC is regarded as the seat of executive control. It is also one of the vehicles for the expression of our personality. Successful mood regulation has been associated with increased activation of the PFC (Levesque et al., 2003).

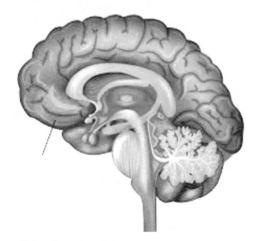

FIGURE 6.8 Ventromedial prefrontal cortex

2. **Ventromedial Prefrontal Cortex (vmPFC):** (Figure 6.8) *Processing fear and assessing risks.* The vmPFC is a part of the PFC. It sits center-front on the lower portion of the PFC and shares some overlapping functions with it. We exercise self-control with the vmPFC. We process risk and fear and make decisions with the vmPFC. The vmPFC is well connected to the amygdala and involved in its regulation.

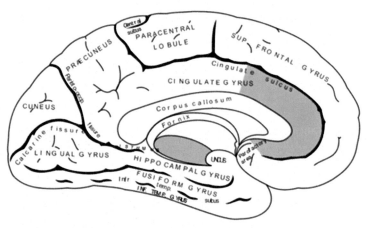

FIGURE 6.9 Anterior cingulate cortex
Source: Image by Mysid Brodmann. Public domain

3. **Anterior Cingulate Cortex (ACC):** (Figure 6.9) *Ability to regulate attention and emotion.* The ACC is wrapped within and behind the PFC. With the ACC, we can sustain attention, exercise self-control, regulate emotion, and make decisions. The ACC is sensitive to early life stress (Cohen et al., 2006; Danlowski et al., 2012; Korgaonkar et al., 2013), which we will see becomes important because of its role in emotional reactivity (Bush et al., 2000). A functionally related structure below the ACC, the dorsal ACC or dACC, shows activation on functional MRI (fMRI) in people experiencing the pain of social exclusion (Eisenberger & Lieberman, 2004; Wesselmann et al., 2013).

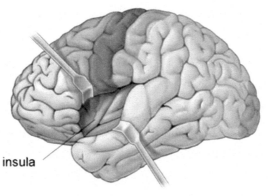

FIGURE 6.10   Insula

Source: Reprinted with permission from Harry Howard, Brain and Language. http://www. tulane.edu/~h0Ward/BrLg/index.html

4.   **Insular Cortex (Insula):** (Figure 6.10) *"I know who I am and how I feel."* The insula is buried deep within the cortex. Though relatively small, it is important to our social wellbeing. Conscious awareness of the self emerges from the insula as does empathy, compassion, and our understanding of interpersonal experience. The insula processes visceral feeling states, enabling us to experience emotions. It works in tandem with the ACC in the processing of emotion, the insula functioning as the primary sensory structure while the ACC functions as the motor structure, expressing emotion in bodily responses (Damasio, 2010).

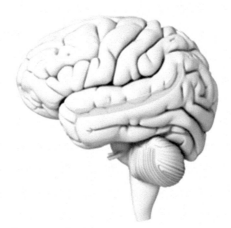

FIGURE 6.11   Superior temporal sulcus

5. **Superior Temporal Sulcus (STS):** (Figure 6.11) *"I understand what you are thinking."* The STS is the depression ("sulcus" or groove) separating the upper from the middle portion of the temporal lobe. It is thought to be important for the processing of words and the sounds of voices as well as the visual cues involved in the understanding of emotional communication. Neuroimaging studies have shown a correlation between activity in the posterior right STS and the ability of subjects to "mentalize" or imagine another's point of view (Dodell-Feder et al., 2011).

6.2 PERIPHERAL NERVOUS SYSTEM

The brain and spinal cord cannot maintain homeostasis by themselves. The central nervous system (CNS) needs the peripheral nervous system (PNS) (Figure 6.12), which gathers information from the environment and acts on what the brain instructs it to do. Unlike the CNS, the PNS does not have the protection of the skull and vertebral column, so it is more vulnerable to injuries.

The PNS consists of two parts, the somatic system, under conscious, voluntary control, and the autonomic nervous system (ANS), under unconscious, involuntary control.

The sensory portion of the somatic system transmits information such as taste and touch to the spinal cord and brain while the motor portion of the somatic system transmits action signals from the brain and spinal cord to targets such as muscles.

The ANS regulates bodily functions such as heart rate, respiration, temperature and digestion. It is intimately intertwined with the limbic system to maintain homeostasis and to regulate affect on an unconscious level. It has two opposing sub-systems: the *sympathetic* nervous system (SNS) and the *parasympathetic* nervous system (PPS).

The *sympathetic* nervous system is a system of *arousal and mobilization*. It prepares the body for emergencies and helps mediate the fight or flight response. Activating the sympathetic nervous system is akin to hitting the accelerator on a car: heart rate and respiratory rate increase and we are aroused and engaged for action. States of hyperarousal such as euphoria, irritability and agitation involve excessive sympathetic activation.

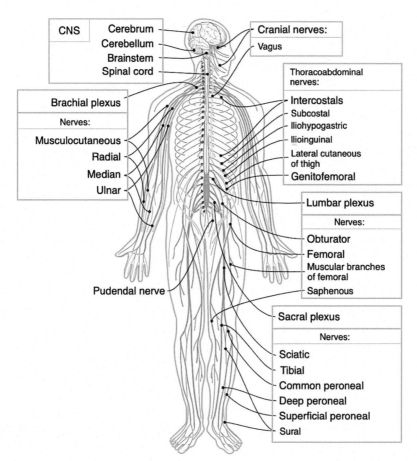

FIGURE 6.12 Peripheral Nervous System

The *parasympathetic* nervous system is responsible for *preservation* (digestion, sleep, healing, etc.). It exerts its modulatory effects by downregulating or counteracting sympathetic activation. Activating the parasympathetic nervous system is akin to hitting the brakes on a car: heart rate slows, respiratory rate decreases, gastric motility is reduced, and we withdraw. States of excessive hypoarousal such as depression involve too much parasympathetic activation.

. . .when the heart is affected it reacts on the brain; and the state of the brain again reacts through the pneumo-gastric [vagus] nerve on the heart; so that under any excitement there will be much mutual action and reaction between these, the two most important organs of the body (Darwin, 1872).

The vagus nerve is so named because it is a vagabond, wandering throughout the CNS with its varied functions, coordinating breathing, and gastrointestinal organs (hence, also called the "pneumogastric" nerve). Darwin described the crucial role of the vagus nerve in coordinating signals between the heart and the brain. It is the longest nerve in the autonomic nervous system and is the primary vehicle for parasympathetic control in the body. The vagus nerve (or the tenth cranial nerve) is the primary nerve associated with the limbic system. It has both motor and sensory fibers which form a rapid feedback loop for sympathetic and parasympathetic signals to help to regulate a number of visceral systems as well as facial, mouth, and throat muscles, and cognitive and emotional processing necessary to mediate social communication.

Neuroscientist Stephen Porges (2001) has hypothesized in his "polyvagal theory" that anatomical shifts in the structure of the vagus nerve have changed over the course of evolution, leading to our capacity for social engagement with one another. Over time, he argues, the vagus nerve came to control facial muscles and other functions necessary for successful social interaction. Embedded in his theory is the notion that social interactions function to regulate visceral states. Under normal circumstances, the nervous system processes sensory information from the environment and the internal organs, evaluating the milieu for safety. This process is called *neuroception* (as opposed to perception) because it is nonconscious. Emergence of the mammalian brain enabled limbic appraisal of the environment. When the environment is appraised as being safe, the defensive limbic structures are inhibited, enabling social engagement, and calm internal states to emerge. When vagal "tone" or function is regulated, people can engage comfortably with their environments. But in a dysregulated system, individuals may appraise the environment as being dangerous even when it is safe. This mismatch results in physiological states that support fight, flight, or freeze behaviors, but not social engagement behaviors. (Porges, 2009). We see this aberrant neuroception in much of the "anxiety" described in people with mood dysregulation. Porges' polyvagal theory of anxiety has been used to help people experiencing panic attacks learn to manage their physical symptoms by tuning in to their internal physiologic states and learning self-regulation through conscious awareness.

6.2.1 NEURAL COMPONENTS

Both the central and peripheral nervous systems are composed of nerve cells or neurons, which are unique to the body both in the appearance and function of the cell. They are not round, as typical cells are, and they communicate via electrochemical transmission rather than purely by chemical transmission. A brief review of neural components will help understand the physiology behind mood dysregulation.

1. **Neural Cells:** (Figure 6.13) Nerve cells or neurons are the building blocks of the nervous system. There are trillions of neurons in the nervous system, approximately 86 billion in the brain alone. Though neurons come in various shapes and sizes, in their most typical form they have a central cell body from which short spikes or "dendrites" protrude, connecting them to other neurons to form networks. A specialized long dendrite called the "axon" functions as a transmission cable and stretches out to neighboring neurons to stimulate responses in them via electrochemical signals called action potentials. Neurons communicate electrochemically when action potentials trigger release of neurotransmitters to neighboring neurons across a "synapse" or space between the neurons. Thoughts, behaviors, and emotions are all mediated by the simultaneous release and reuptake of neurotransmitters across millions of contiguous synapses in the brain. Neurotransmitters trigger reactions in nerve cells by activating receptors, protein complexes embedded in neuronal cell membranes. For example, the glucocorticoid cortisol triggers action in nerve cells by activating glucocorticoid receptors on neuronal membranes: this produces a cascade of other reactions within the neuron that carries on the communication process. When a neurotransmitter activates a membrane receptor, it is called a receptor *agonist:* when a neurotransmitter blocks a membrane receptor, it is called a receptor *antagonist.*

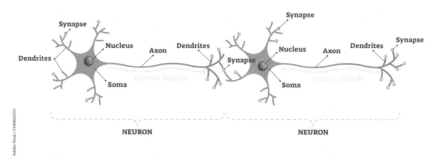

FIGURE 6.13   Two neurons with dendrites, axons, and a synapse

2. **Neural Networks:** Though a neuron may appear to be the smallest
    unit in the nervous system, the smallest *functional* unit is thought
    to be a *cortical unit* composed of tens to hundreds of neurons
    acting in concert to effect a response in a target area (Bush &
    Sejnowski, 1994) or a minicolumn composed of 100–200 neurons
    (Mountcastle, 1997; Szentagothai, 1975). Networks are generally
    groups of neurons that are not simply connected together physically
    but which *fire together to effect a particular function*. The most
    stimulating environments for brain development promote neural
    replication, foster dendritic growth, and facilitate firing of neurons
    in proximity to one another. Neurons develop networks when
    neural cells replicate and when new cells grow new dendrites
    that reach out to more neurons. The more dendrites a neuron has,
    the greater its capacity to network and the more effectively it can
    communicate with other neurons. On a macro level, this translates
    into more adaptive neural tissue. As neurons proliferate, they make
    connections with one another and begin to stimulate one another.
    The saying "neurons that fire together wire together" describes this
    very process by which the simultaneous firing of adjacent neurons
    increases the strength of the connections between them, creating
    a functional network. This principle was formulated by Donald
    Hebb and describes the "Hebbian learning" (Munakata & Pflaffly,
    2004) of a neural network. "Procedural behavior" or unconscious
    habits represent a form of Hebbian learning.

    The brain itself evolved by development of such networks that
    were selectively reinforced because of their adaptive properties.
    The Neuronal Group Selection theory (TNGS) explains how the

brain evolved to carry out its complex functions through this Hebbian process under genetic and epigenetic influence on the adaptive selection of neural networks (Edelman, 1987).

3. **Neural Nuclei:** A group of neurons that cluster together and have a specialized function is a nerve nucleus. The amygdala, for example, is a cluster of several nerve nuclei including the basolateral complex, the cortical nucleus, the medial nucleus, and the central nucleus. The basal ganglia are a group of nuclei important for motivation and movement and include the substantia nigra (implicated in Parkinson's disease) and the ventral pallidum (important in the reward system and relevant to addictive behavior).

4. **Neural Tracts:** Within the central nervous system, groups of neuronal axons traveling together like the wires that make up a cable are called a tract. Tracts of axons run from nucleus to nucleus in the brain. Tracts in the brain can either connect areas on one side of the cortex to each other (association fibers), areas on one side of the cortex to areas on the other side of the cortex (commissural fibers), areas in the cortex to areas in other parts of the brain (projection fibers), or other parts of the brain to one another. The reward pathway or "mesolimbic tract" in the basal ganglia is such an example. The most important tract in the brain is the *corpus callosum*, a large collection of association fibers that connects the right cerebral hemisphere to the left cerebral hemisphere, allowing communication between right and left cortices.

5. **Nerves:** A group of neuronal axons traveling together like the wires that make up a cable in the peripheral nervous system is called a nerve. The only difference between a tract and a nerve is that a tract is within the central nervous system and a nerve lies outside of the central nervous system.

6.3 COMMUNICATION SYSTEMS

Homeostasis. Clearly the "hardware" we have been describing developed to maintain equilibrium, but how does it do so? Maintenance of homeostasis in the body is dependent on feedback regulation. Feedback regulation requires a mechanism for terminating an initiated action once its goal has been achieved. A feedback regulating circuit in which the initiated action creates a shut-off mechanism to terminate a response once the goal has been achieved is called a negative feedback loop. Most regulatory systems in the body operate via such negative feedback loops. In a dysregulated negative feedback system, the shut-off mechanism fails.

There are two major vehicles of communication for regulatory feedback loops in the body linking the components of the nervous system with the body's major organs to maintain homeostasis. These are the hormone system and the neurotransmitter system.

Hormones are blood-borne chemicals that usually travel long distances between one organ and another. They are specialized chemicals released into the bloodstream with specific targets to effect specific actions. Their release is typically governed by the hypothalamus. As a rule, when their job is done, they signal the hypothalamus to stop requesting their presence, shutting off their production via a negative feedback loop. Neurotransmitters, on the other hand, are chemicals that travel miniscule distances in the synaptic space between one nerve cell and another along the axons that form tracts or nerves. They can also be governed in negative feedback loops by chemical receptors on neuronal cell membranes. Some specialized cells can communicate both through hormonal and neurotransmitter mechanisms (such as the neuroendocrine cells in the hypothalamus and pituitary gland).

6.3.1 HORMONES

Hormones are signaling molecules that circulate in the blood: their primary function is to maintain homeostasis. In a well-regulated system, hormones are secreted in response to chemicals that detect deficiencies in substances they are designed to restore homeostasis, and their secretion is suppressed when the level of those substances returns to normal.

For example, when we eat, our food provides fuel in the form of glucose, which is absorbed from the stomach into the bloodstream. Elevated levels of blood glucose trigger release of the hormone insulin from the pancreas to remove glucose from the bloodstream and store it for later use. Once blood levels of glucose have subsided, normalized levels of blood glucose "turn off" release of the hormone insulin. This is one of the body's typical negative feedback loops.

When insulin is not released from the pancreas in response to elevated levels of blood glucose, as in diabetes, the negative feedback loop fails and glucose levels can soar, creating potentially serious medical problems. In a similar way, when negative feedback loops responsible for mood regulation fail, mood instability results.

The hypothalamus and the pituitary gland have an unusually important role in the body's complex hormonal system because they regulate most of the major glands in the body: they control the thyroid, the adrenal glands, the testicles and the ovaries, regulating crucial physiological processes including body temperature, energy, metabolism, growth, reproduction, lactation, blood pressure, and pain. They are highly interconnected with other areas of the central nervous system including the amygdala and the autonomic nervous system, reflecting their crucial role in mood regulation.

We will pull out one strand of this system and look at it in detail since it is one of the two survival mechanisms that evolved to enable us to respond to stress: this is the Hypothalamo–Pituitary–Adrenal (HPA) axis. The HPA axis mediates communication from the hypothalamus to the pituitary gland to the adrenal gland and back and is central in our ability to mount a fight or flight response.

The hypothalamus sits above the small, pea-sized pituitary gland. It produces and secretes "releasing" hormones to the pituitary gland: releasing hormones from the hypothalamus signal the pituitary gland to send hormones to their target organs in the body, which then release corresponding hormones to their target tissues. For example, when the thyroid is the target, the hypothalamus releases thyrotropin releasing hormone to the pituitary gland, causing release of thyroid stimulating hormone to the thyroid, resulting in release of thyroid hormone. In a well-regulated system, when thyroid hormone has accomplished its task, normalized blood levels signal the hypothalamus to reduce release of thyrotropin releasing hormone to the pituitary gland, slowing signals to the thyroid gland to release more thyroid hormone. In a dysregulated system, the shut-off mechanism in this negative feedback loop fails.

In the HPA axis, the releasing hormone from the hypothalamus is called corticotropin-releasing hormone (CRH). This short molecule travels from the hypothalamus to the anterior pituitary gland where it stimulates the release of a much larger hormone, adrenocorticotropic hormone (ACTH). ACTH then travels through the bloodstream to the adrenal glands, perched on top of the kidneys, where it stimulates the release of glucocorticoid hormones from the outer layer (adrenal cortex) of these glands.

The body's primary glucocorticoid is the steroid cortisol, our major stress hormone. Cortisol is released from the adrenal cortex when blood sugar drops in times of stress to mobilize the fuel glucose from the body's "storage resources" in times of acute need. When cortisol levels in the body increase, glucose levels increase to provide fuel to activate the brain and muscles. Cortisol levels in the body naturally peak within an hour after we awaken and fall throughout the rest of the day.

In a well-regulated system, cortisol release from the adrenal gland helps us mount an acute stress response, giving us energy and enabling us to concentrate until the storm passes. When no longer necessary, glucocorticoids then signal the hypothalamus to stop sending CRH to the pituitary gland. The pituitary gland then stops sending ACTH to the adrenal glands and the surplus release of glucocorticoids from the adrenal glands stops.

In a dysregulated system, this negative feedback loop fails and glucocorticoids do not stop the hypothalamus from sending CRH to the pituitary gland. ACTH is continually released to the adrenal glands, which continue releasing glucocorticoids. Excessive and unnecessary cortisol then bathes neural tissues. As we will see, prolonged exposure to elevated levels of glucocorticoids is harmful to the body, especially to certain sensitive brain structures involved in mood regulation.

The hormonal HPA axis, as above, is designed to assure adequate levels of blood cortisol to mount a stress response. The neurotransmitter/hormonal SympathoAdrenoMedullary or SAM pathway is a component of the sympathetic division of the autonomic nervous system. Sympathetic (activating) nerves of the SAM system cause release of the hormone epinephrine (adrenaline) from the medulla (center) of the adrenal gland. Adrenaline increases heart rate and blood pressure, expands the air passages in the lungs, and dilates the pupils. The SAM system works in concert with the HPA axis to mount a stress response in the face of danger.

Suppose we encounter an unexpected threat. We need to make a snap decision: either to engage or to flee. The amygdala engages before we have even consciously registered the threat and begins activating both of the survival mechanisms we have online to mount a stress response: (1) the hormonal (HPA axis) and (2) the neurotransmitter/hormonal (SAM) system. Together these systems increase heart rate, respiration, and glucose levels to increase vigilance and arousal, narrow attention, and enable action in the face of danger. In a healthy person, these systems return the body to equilibrium once the threat has passed. Chronic stress disrupts this process and causes dysregulation to the system.

6.3.2 NEUROTRANSMITTERS

Though we may know much about the brain, we are really just beginning to learn about this immensely complex organ. So far, we know of over 200 neurotransmitters involved in the communication process between neurons and neuroendocrine cells. We can barely scratch the surface here in our discussion of this elaborate system. Some of the most important are:

1. **Monoamines:** monoamine neurotransmitters are derived from amino acids and are so named because they contain one amino group. Examples of monoamine neurotransmitters are *serotonin, dopamine, norepinephrine* (noradrenaline), and *epinephrine* (adrenaline):

    a. *Serotonin* plays an important role in regulating mood, sleep, anger, aggression, sexuality, appetite, vomiting, and body temperature. Serotonin pathways travel extensively throughout the cortex, brainstem, and spinal cord. Serotonin receptors are also widely distributed throughout the gastro-intestinal system. The serotonin system is very elaborate and involves over 10 different types of serotonin receptors.

    b. *Dopamine* is thought of as the "reward" or "pleasure" neurotransmitter. It is the primary neurotransmitter used in the "reward circuits" of the brain involved in the reinforce-ment of learned behaviors. Many addictive drugs increase the concentration of dopamine in these circuits. Dopamine distur-bances in the brain are associated with a number of diseases. Parkinson's disease results from reduction of dopamine in the midbrain. Schizophrenia is treated with dopamine blockers. Outside of the central nervous system, Dopamine has several functions as a hormone.

    c. *Norepinephrine* or *noradrenaline* functions as both a neuro-transmitter and a hormone. Within the brain, norepinephrine enhances cognitive function, but its stimulant properties can cause restlessness and anxiety. In the peripheral nervous system, it increases heart rate, blood pressure, glucose levels, blood flow to muscles, and prepares the body for the fight or flight response. There are relatively few norepinephrine neurons in the brain, but norepinephrine is the primary neurotransmitter in the sympathetic nervous system. Sympathetic nerves activate the medulla (center) of the adrenal gland, where norepinephrine is converted to epinephrine.

d. ***Epinephrine*** or *adrenaline,* like norepinephrine, functions as both a neurotransmitter and a hormone. While norepinephrine is manufactured in neurons, epinephrine is primarily manufactured in the endocrine cells of the adrenal medulla where it is released into the bloodstream. While norepinephrine is continuously released into the circulation at low levels and has effects on limited areas of the body, epinephrine is only released during times of stress and has effects on all body tissues.

2. **Acetylcholine** was the first neurotransmitter to be identified. It has important functions both in the central and in the peripheral nervous systems. Acetylcholine is important in mediating memory and learning in the brain. Deficiency of acetylcholine in the hippocampus is associated with Alzheimer's disease. CNS acetylcholine is associated with arousal, alertness, and maintenance of attention. In the peripheral nervous system, acetylcholine is important in both the somatic and autonomic nervous systems. Though it is found in the sympathetic nervous system, it is the primary neurotransmitter in the parasympathetic branch of the autonomic nervous system. In the somatic nervous system, it is the primary neurotransmitter at the junction between nerve and muscle cells. Many drugs we use block or mimic the effects of acetylcholine. Substances that block acetylcholine receptors are called anticholinergic agents. Many useful medications have anticholinergic side effects including dry mouth, constipation, and sedation. The sedating effects of the antihistamine diphenhydramine (Benadryl) are due to its anticholinergic properties.

3. **Amino Acids:** *Glutamate, GABA*, and *Glycine*

a. ***Glutamate,*** while necessary in regulated amounts, *will often prove to be implicated in mood dysregulation.* Glutamate is an amino acid and is the most abundant neuroexcitatory neurotransmitter in the CNS. Glutamate binds to the N-methyl-D-aspartate (NMDA) receptor and is important for the neuronal growth and plasticity required for memory and learning. This gives the glutamate system tremendous importance in the maintenance of normal biobehavioral function. The importance of glutamate in the pathogenesis of behavioral disturbances associated with early life adversity is fast becoming recognized because the stress reaction elevates glutamate levels in the brain, causing toxicity to neural tissues associated with relevant structures.

b. **GABA,** or γ-amino butyric acid, is a byproduct of glutamate and is the most abundant neuroinhibitory transmitter in the CNS, occurring in 30–40% of all cells. GABA counteracts the effects of glutamate. GABA has antiseizure and antianxiety properties. It is sold as a dietary supplement.

c. **Glycine,** like glutamate, is an amino acid, but like GABA, is inhibitory. Its primary function is as a building block in the manufacture of proteins. It is found primarily in the spinal cord and brain stem.

4. **BDNF,** or Brain-Derived Neurotrophic Factor, is a neurotransmitter modulator. It promotes cell growth and is necessary for development of brain plasticity, again, a process crucial for learning and memory. BDNF has been found to be deficient in people at risk for depression (Aydemir et al., 2005; Gonul et al., 2005; Karege et al., 2002). Lithium has been found to increase levels of BDNF in the brain (Einat et al., 2003). Maternal deprivation is associated with reduction in levels of BDNF in the PFC (Roceri et al., 2004).

6.4 BIOLOGY OF EMOTION REGULATION

Interdependent systems are involved in the regulation of emotions, which is more complex than we can adequately review here. In the most basic sense, mood regulation develops as limbic and HPA forces are brought under the inhibitory influence of the cortex and parasympathetic nervous system (Ochsner & Gross, 2007; Zelazo & Cunningham, 2007) as the organism matures in the context of the developmental caregiving relationship (Kring & Sloan, 2010).

6.4.1 BRAIN REGULATION

An oversimplified summary of mood regulation in the brain would depict two major areas of the brain competing with one another to find a balance: the impulsive/activating limbic system, driven primarily by the amygdala, in tension with the reflective/inhibitory cerebral cortex. It is not that simple, however. Emotions are not solely generated or activated in the limbic system, then inhibited or regulated by cortical structures. Imaging evidence indicates that emotions are activated by cortical structures as well as limbic structures (Barrett & Barr, 2009; Ochsner et al., 2009) and limbic structures exert influence on cognitive perceptual processes as well (Lewis & Todd, 2007). It is becoming increasingly clear that there is bidirectional influence between lower and higher brain structures in the processing of emotion (Ochsner & Gross, 2007; Thompson et al., 2008; Quirk, 2007).

The generation of emotions begins with our perception of a stimulus followed by our appraisal of it. Basic appraisal of a stimulus typically occurs on the limbic level. For example, the sight of a snake (perception) might evoke a response of fear (appraisal), generated in the amygdala. Basic appraisals may often lead to automatic actions that occur without conscious intent, leading to reactive impulses (LeDoux, 2000; Rolls, 1999). But the sensory information leading to basic appraisals reaches higher level brain areas for further processing. Circuits connecting the amygdala with the ACC are responsible for emotional appraisals and emotion regulatory processes Cardinal et al., 2002; Quirk, 2007). Conceptual appraisals (e.g., the verbalization "I am afraid of snakes") are made in cortical areas, primarily the rostromedial PFC and dorsomedial PFC (Cato et al., 2004; Lindquist & Barrett, 2008).

Most of human neuroscience research on emotion regulation has been devoted to the study of "reappraisal" (Ochsner & Gross, 2008). Reappraisal is the reinterpretation of the meaning of a stimulus in ways that alter its emotional impact. One can see how this is an important strategy for mood regulation. Reappraisal is highly effective for enhancing positive and reducing negative emotion and promoting interpersonal relationships (Gross, 1998; Gross & John, 2003). Reappraisal in our model might result in the thought, "It's just a garter snake: no need to be afraid."

Neuroimaging studies have consistently shown activation of the lateral PFC during reappraisal challenges (Goldin et al., 2008; Ochsner et al.,

2002). The medial PFC has been strongly implicated in reappraisal by its association with the processes of making judgments about internal mental states (Lieberman, 2007; Ochsner et al., 2004) and its increased activity when making self-referential judgments (Kelley et al., 2002; Ochsner et al., 2004). The dorsal ACC is associated with monitoring conflict between competing appraisals (Botvinik et al., 1999), correlating with affective states that involve conflict such as the distress elicited by rejection (Eisenberger et al., 2003). And several studies have implicated the lateral OFC in downregulating negative emotion (Goldin et al., 2008; Levesque at al., 2003; Phan et al., 2005).

But more important in mood regulation than the individual structures involved in cognitive reappraisal are the circuits between these structures. Because the amygdala has such an important role in the perception and appraisal of stimuli, it has connections to 90% of the cortex, with many circuits dedicated to impulse control. Normal affective functioning depends not only on intact structures in the brain but on the functional integrity of the circuits created by these structures. Three cortical structures, the lateral prefrontal cortex (LPFC), the anterior cingulate cortex (ACC), and the orbitofrontal cortex (OFC) are components of a circuit with the amygdala which enables conscious modulation of emotionally evocative stimuli, a process at the heart of mood regulation (Banks et al., 2007; Frank et al., 2014; Kim & Hammon, 2007; Ochsner et al., 2004). Disruption of this circuit is seen in mood dysregulation.

The language networks of the insula and the ACC are necessary for healthy mood regulation. The insula allows us to reflect on and name our emotional experiences. The ACC is involved in impulse control and decision making. The insula/ACC network is important because language gives us access to a rich base of self-regulatory strategies through which to modulate negative emotions (Botting & Conti-Ramsden, 2008; Silva et al., 1984). Secure children learn a complex vocabulary to describe their emotions, to understand how they feel and communicate their feelings to others (Ciccetti & White, 1990). Children who suffer abuse or neglect and are deprived of interaction or whose language is impoverished lack these tools and are vulnerable to mood dysregulation.

The right neocortex is more extensively connected to the limbic system than the left, for which reason Schore (2003) contends, in his affect regulation theory, that the right neocortex is more important than the left neocortex for affective regulatory processes (Nebes, 1971; Schore,

2016). The left hemisphere is thought to predominate in the mediation of most linguistic behaviors while the right hemisphere is regarded to be important for broader emotional aspects of communication (Van Lancker & Cummings, 1999). Witting and Roschmann (1993) demonstrated that the right brain is more responsive to emotional stimuli than the left in keeping with Schore's characterization of the right brain as the seat of non-verbal, non-conscious communication.

6.4.2   VAGAL REGULATION

Researchers studying mood regulation are particularly interested in how the vagus nerve regulates the heart under conditions of stress (Thompson et al., 2008). Recall that the vagus nerve is the primary vehicle for para-sympathetic control in the body and the primary nerve associated with the limbic system. One of its major functions is to slow the heart. Vagus nerve function allows us to regulate distress and arousal, mediating social communication (Kovacs et al., 2008). Differences in vagal "tone" or adapt-ability are associated with differences in stress reactivity/ soothability and attention (Calkins, 1997). Well-regulated vagal function or "tone" allows for flexible mood regulation. With good vagal tone, we can become upset, anxious, or angry with people without becoming aggressive. Low vagal tone is an indicator of poor mood regulation (Forbes et al., 2006).

Parasympathetic (vagal) regulation develops rapidly in the early years of life during which the quality of infant care affects an individual's capacity for self-regulation (Propper & Moore, 2006). Greater parasym-pathetic regulation has been shown to be associated with fewer behavioral problems and greater emotion regulation in preschool and to be correlated with increased parental support (Calkins & Keane, 2004). Some have hypothesized that problematic early attachment relationships can lead to inadequate vagal system development and subsequent problems modu-lating vagal tone. This, in turn, can lead to inappropriately aggressive behavior (Cozolino, 2010, p. 234).

Mood regulation depends on sympathetic and parasympathetic nervous system balance and responsivity. An index called Heart Rate Variability (HRV) measures the fluctuation of heart rate in response to breathing and is a gauge of the balance between the sympathetic and parasympathetic nervous systems. A robust HRV reflects effective arousal capacity and the ability to control impulses and emotions (Mather & Thayer, 2018). A poor

HRV is associated with poor impulse control (Allen et al., 2000; Porges, 1996) and mood dysregulation (Cohen, 2003; Latalova, 2010; Moon, 2013). Activities such as exercise, yoga, and biofeedback that increase HRV improve mood regulation. Increased HRV has been associated with improved self-regulation as well as a sense of emotional wellbeing (Geisler et al., 2010; Segerstrom & Solberg Nes, 2007).

6.4.3 HPA REGULATION

The HPA axis is involved in maintaining mood regulation in the normal course of development in the absence of chronic stress. Neuroendocrine arousal systems mature rapidly in the infant, leading to declining emotional lability and increased self-control (Gunnar & Vasquez, 2006). Developmental studies indicate that these changes are affected by the quality of the individual's early caregiving experience and lead to the ability for self regulation and adaptive response to stress (Boyce & Ellis, 2005; Gunnar & Donzella, 2002).

In healthy adults, cortisol has been found to foster the emotion regulatory process (Denson et al., 2014; Jentsch et al., 2019; Lam et al., 2009) and to reduce stress-induced negative affect (Het et al., 2012; Hat & Wolf, 2007). Not so in adults subjected to chronic stress, as we will see.

Our survival as a species is organized around the capacity to maintain an internally balanced environment or homeostasis. All physiologic functions in the body serve this regulatory process under the control of the brain. Affect regulation is central to this homeostasis and is integral to physiologic equilibrium.

As animals evolved, the brain became more complex to meet the needs of the organism. The primitive reptile brain regulates our instinctual needs and controls the basic bodily organs such as the heart, lungs, and gastrointestinal tract. With the appearance of mammals, the brain evolved a second layer, the limbic system or "emotional brain" to mediate attachment and social behavior. The most important structures in the limbic system are the amygdala, which initiates the fight or flight response, and the hippocampus, in which memory is processed and consolidated. The most recent layer of the brain, the neocortex, enabled the emergence of symbolic thought, language, and impulse control. The neocortex is vastly connected with the limbic system allowing for cortical control of limbic generated emotions, a process necessary for mood regulation.

The nervous system is made up of nerve cells, nerve networks, tracts, and nerves. Communication between these structures and other structures in the body occurs via blood-borne hormone messengers and via neurotransmitters, chemical messengers that enable communication between one nerve cell and another across the synaptic space between neurons. Nerve growth factors facilitate the elaboration of dendrites on nerve cells to multiply and create nerve networks that enhance these pathways and increase our capacity for emotional adaptability.

These same hormones and neurotransmitters are the primary vehicles of communication employed to achieve and maintain homeostasis within the body. Hormones are chemicals that typically travel long distances within the body through the bloodstream to carry their signals while neurotransmitters travel miniscule distances between neurons to effect their action. When normal life stressors disturb our homeostasis, we have two systems with which to mount a stress response, one using the HPA axis and one using the neurotransmitter/hormonal sympathetic adrenal medullary (SAM) circuit. In a healthy system, once the stress has passed, negative feedback loops "turn off" these stress responses.

The anatomical and functional integrity of these systems as they mature in concert is necessary for healthy mood regulation. As we will see in the next chapter, environmental stressors can disturb this process, predisposing a vulnerable individual to mood dysregulation.

KEYWORDS

- **homeostasis**
- **triune brain**
- **sympathetic/parasympathetic nervous system**
- **neurotransmitter**
- **feedback loop**
- **HPA axis**

CHAPTER 7

Etiology

The true character of a society is revealed in how it treats its children.
—Nelson Mandela

ABSTRACT

If an affective temperament is the underpinning of a mood dysregulation, how does the environment determine whether or not such a genetic predisposition will ever develop into visible symptomatology? Clues to this are found in the field of epigenetics, the study of how the environment affects gene expression. There may be no environmental influence so impactful to the brain and mind as early life attachments. Attachment is regarded to be a homeostatic necessity. Attachment theorists have observed that infants with secure attachments to their caregivers develop the capacity to regulate their physical arousal while those without secure attachments do not. Epigeneticists are beginning to explain the molecular basis for how caregiver behavior affects an individual's capacity for self-regulation. Studies have shown that the more mother rats lick their pups, the more of the pups' genes become activated to produce receptors for cortisol, the stress hormone that enables the organism to regulate itself. Humans subjected to chronic stress such as insecure attachments and developmental trauma mount a chronic stress response. In contrast to the acute stress response, in which cortisol production regulates the organism and is time-limited, the chronic stress response is a runaway process. Under chronic stress, cortisol production is ongoing and toxic to the system. Long term cortisol exposure destroys the integrity of crucial structures in the brain associated with mood regulation and stimulates production of the neurotransmitter glutamate, another agent which damages nerve cells throughout the brain. Together, cortisol and glutamate create physiologic and functional disruption of crucial elements in the brain that results in mood dysregulation. The degree to which this process unfolds is a function both of the individual's biological diathesis and of the environs for its genetic expression.

Mood Dysregulation: Beyond the Bipolar Spectrum. Deborah A. Deliyannides, MD (Author)
© 2024 Apple Academic Press, Inc. Co-published with CRC Press (Taylor & Francis)

If a mood dysregulation emerges from an affective temperament, is it genetic? The determinants of our behavior are dynamic and depend on the ongoing interaction of genetic and environmental factors throughout our lives. Not every gene we are born with is activated or "expressed" in the person we are, and the genes we inherit are altered in our interactions with the world we encounter. The burgeoning science of epigenetics is uncovering the influence of the environment on gene expression (Fagiolini et al., 2009; Tsankova, 2007). In the most material sense, we are truly products of our environments.

We have posited that an affective temperament is a predisposition or "diathesis" to mood dysregulation. According to the epigenetic model, mood dysregulation is partially genetic, the result of this diathesis, and partially the result of the environment. The diathesis-stress model of pathophysiology captures the essence of the interaction between environmental influence and genetic predisposition. According to this model, symptoms of a disorder may emerge if a genetically predisposed organism is subjected to sufficient environmental stress. Thus, individuals vulnerable to mood dysregulation, either by an affective temperament or by genetic relation to others with affective disorders, under sufficient conditions of stress, may become symptomatic. Likewise, that same genetically predisposed individual, if maintained in a protected environment that is sufficiently stress-free, may not display symptoms at all. In one example, constitutionally shy monkeys (i.e., those with a diathesis to withdraw,) when reared by confident mothers, became more confident (Mastropieri & Carroll, 1998; Suomi, 1997). Of course, a genetically vulnerable individual maintained in a stress-free environment may still exhibit symptoms of a disorder if the genetic loading or diathesis is too great.

7.1 ENVIRONMENTAL STRESS

Because of the complex interplay between our genes and the environment, stress can have an impact on us before we are born. Studies have shown that maternal stress during pregnancy leads to emotional and cognitive deficits early in life (Talge et al., 2007) as a result of fetal overexposure to maternal stress hormones (Edwards et al., 1993; Seckl, 2004). Overexposure to stress hormones is a unifying theme for the genesis of mood dysregulation.

Overexposure to stress hormones occurs under conditions of chronic stress to the organism. Any number of conditions contribute to chronic stress in childhood resulting in mood disturbances. In adults, maternal anxiety, depression (Kovacs et al., 2008), and family disturbances, including inadequate parenting or parental threat to the infant (Repetti et al., 2002), punishment, shaming, serious quarreling and fighting (Flinn & England, 1995), sibling abuse (an underrecognized source of chronic stress) (Caffaro, 2014), and peer rejection (Coie et al., 1990) can all be sources of chronic stress, leading to poor mood regulation. Animal and human studies consistently find that early life adversity is associated with the later development of mood disorders (Carr et al., 2013; Faravelli et al., 2012; Nemeroff, 2004; Maccari et al., 2014; Oskis et al., 201; Shea et al., 2005; Van den Bergh et al., 2008).

Not all stressors are created equal. Relationship or social stressors are actually more deleterious to physiologic function than nonsocial stressors such as physical exertion or cognitive stress. Your body is more likely to be stressed by an argument with someone than by a strenuous workout at the gym. Rats show greater changes in blood pressure, heart rate, and hormone secretion when subjected to social stressors than to other environmental stressors (Sgoifo, 1999).

Social stressors are deleterious because attachment is a homeostatic necessity. From an evolutionary perspective, attachment is a preservative reflex. We do not have a choice in the matter: we are programmed to seek attachments (MacDonald & Leary, 2005; Panksepp, 2005). One of the motivational circuits Panksepp describes as being biologically hard-wired in all mammals is the seeking circuit, which is in service of our biological need for attachment. He poses the question "Why does depression hurt?" His answer: when attachments are lost, brain separation–distress systems create psychological pain. Separation from an attachment figure activates the seeking system. If the attachment figure cannot be found, the panic/grief system is activated. Prolonged panic leads to system shut down and a depressive despair phase begins. Depression is posited to have evolutionary advantages such as conservation of resources following unalleviated separation distress (Panksepp & Watt, 2011).

The panic circuit is a social circuit as it is activated in relation to attachments. Its circuit arises from the midbrain periaqueductal gray (PAG) and passes through the amygdala and other parts of the limbic system, then the ACC (involved in expression of emotion) before ending in the thalamus.

The panic circuit is activated when young mammals are separated from their support systems and are placed in social isolation. Separation leads to vocalizations ("isolation calls") to arouse the attention of a caregiver. The proximity of a caregiver is generally sufficient to inhibit isolation calls in both humans and animals (Pettijohn, 1977). Oxytocin, prolactin, testosterone, and opioids all alleviate separation distress. Maternal separation distress leads to enhanced emotional reactivity in mice (Uhleski & Fuchs, 2010). Interestingly, the corresponding behavior in humans was interpreted as "borderline" (personality disorder) symptomatology in children and adolescents (Bradley, 1979), a manifestation of which appears to be mood dysregulation.

This same process appears to be true in humans; disruption in attachments appears to cause arousal of the separation distress system, activating the panic circuit. It appears that people who suffer from panic attacks often have childhood histories characterized by separation-anxiety problems (Torgersen, 1986). Chronic arousal of the panic system by persistent social isolation leads to over-responsiveness of the HPA axis resulting in excess levels of circulating cortisol. This has long-term psychiatric consequences (Mendoza et al., 1978) including lifelong difficulties with social adjustment (Mason, 1968).

Separation distress is operative in mood disorders as well. The primary developmental cause of depression in humans is social loss (Wilner, 1985). The most effective treatment for monkeys suffering from social isolation was found to be exposure to younger monkeys whose playful interaction drew them out (Mason, 1982). Dogs make excellent surrogate mothers to isolated monkeys. By the same mechanism, pets can be important to humans in promoting mental health and emotional equilibrium (Panksepp, 1998).

Infant maternal attachment is not only a preservative reflex but has significant regulatory effects on the child from the time of birth. Rats have been studied extensively in research focusing on the impact of early experience in shaping the neurobiological systems responsible for stress reactivity. In a widely replicated study, rat pups separated from their mothers for 15 minutes daily were compared with rat pups separated from their mothers for 3 hours daily. The rat pups separated from their mothers for 3 hours daily were more vulnerable to stress. They startled more easily, secreted more stress hormones, showed mild cognitive impairment, were more anxious, more reactive, more listless, and consumed more alcohol than rat pups separated from their mothers for only 15 minutes daily (Cirulli & Alleva, 2003; Sanchez, 2001). Rat pups given maternal handling, on the other hand, showed decreased stress reactivity and decreased fearfulness throughout adulthood (Ader & Grota, 1969; Levine et al., 1967; Viau et al., 1993; Zarrow et al., 1993).

Maternal attention to rat pups even programmed the pups' future maternal behavior (Szyf et al., 2008). In a much-cited epigenetic study, Michael Meaney (2001) showed that newborn rat pups intensively licked by their mothers in the first 12 hours after birth themselves gave birth to rat pups who were braver and showed far lower levels of stress hormones than the offspring of rat pups with inattentive mothers who did not lick or groom them after birth.

Have you ever felt you just needed a hug? Tactile stimulation is important to the development and maintenance of physiological and psychological regulation (Fosshage, 2000) and has calming effects on the organism. Animals stop crying rapidly when gently touched. The brain's endogenous opioid system can achieve this effect as well (Panksepp et al., 1980). All mammalian stress studies demonstrate that physical touch is an important early modulating influence of the caregiver on the infant (Barnard & Brazelton, 1990; Fields, 1993; Spitz, 1965) and show that maternal touch and contact have a lifelong organizing effect on the stress management systems of the young, functioning to regulate the stress response (Champagne, 2008; Hofer, 1995; Meaney, 2001). Feldman and colleagues (2010) replicated the results of animal studies in humans, demonstrating regulation of the stress response to touch between mothers and infants.

The hormone cortisol is called up when we are subjected to demands or placed under acute stress. Cortisol in short, infrequent bursts is adaptive and helpful. But too much cortisol too frequently can be damaging to neural tissues. Perhaps because of the enormous developmental demands placed on infants, nature provided humans with a physiologic buffer from the effects of stress on neural tissue. During the first 12 months of life, the body has a window of developmentally programmed reduced sensitivity to the effects of stress hormones such as cortisol (Lashansky et al., 1991; Gunnar, 2003), which buffers the infant from environmental insult. Thus, infants are less responsive to cortisol in the first year of life. This period of hyporesponsivity to cortisol continues in some form until adolescence when adult patterns of stress response begin. At this point, individuals become more vulnerable to the effects of stress hormones, which is why we see an increased risk of emotional difficulties among teens (Spear, 2000).

Despite reduced sensitivity to cortisol during early life, infants and young children are not immune to the effects of stress, particularly when it is chronic. Depressed mothers are less verbally interactive with their infants, show less quality stimulation, and make less affectionate contact with them (Fleming et al., 1988; Righetti-Veltema et al., 2003). Their infants, in turn, have more difficulty with mood regulation. They fuss more, smile less, and are less able to contain their distress (Field et al., 2007; Weinberg et al., 2006). Distress and emotion regulation remain problematic in 3 and 4-year-old children of depressed mothers as well (Hoffman et al., 2006; Maughan et al., 2007). Depressed mothers are "absent" from their infants in some measure; though not fully abandoning their infants, their depression is still perceived as a social loss. Social loss has been posited to be the most significant epidemiological stressor and the precursor of depression (Bowlby, 1980, Heim & Nemeroff, 1999). Depressed mothers predispose their young to mood dysregulation.

7.2   ATTACHMENT THEORY

Long before Panksepp's work, the British psychiatrist/psychoanalyst John Bowlby recognized the evolutionary importance of attachment for survival related affect regulation. Observing infants' crying, clinging, and frantic efforts to reunite with caregivers, Bowlby postulated that these were adaptive responses designed to restore security and protection from the caregiver, realizing that such efforts allowed the infant to return to equilibrium and, ultimately, to survive.

Bowlby (1969) believed the key to the attachment bond was in the mother's soothing, or regulating, the infant's fear states. He understood that healthy attachments were vital to an individual's survival and capacity to deal with stress, largely through the mechanism of mood regulation. As Bowlby describes it, the infant's need for attachment is so intense only maternal touch can terminate the seeking behavior:

> Every mother knows that a child who is cold, hungry, tired, in ill health, or in pain is likely to be "mummyish." Not only is he reluctant for his mother to be out of sight but often he demands to sit on her knee or to

be carried by her. At this intensity, attachment behavior is terminated only by bodily contact, and any breaking of contact produced by mother's movements evokes intense attachment behavior afresh–crying, following, clinging–until such time as the two of them are in contact again. (Bowlby, 1969).

Infants are incapable of regulating themselves on their own in the early days of life (Hostinar & Gunnar, 2013) and depend on caregivers for their maintenance, necessitating secure attachments. The intersubjective research of contemporary attachment theorists furthered our understanding of how early mother-infant bonds impact affect regulation (Beebe & Lachmann, 1994; Diamond & Hicks, 2005; Gianino & Tronick, 1988; Lyons-Ruth, 2008; Lyons-Ruth et al., 1987; Schore, 2014) in the context of interpersonal relationships. We now know that infants who have secure attachments with early caregivers have lower rates of affective disorders (Murray et al., 2011).

Secure attachments develop only with affectively attuned responsiveness on the part of the caregiver toward the infant. Such affective attunement has pervasive effects on the infant's capacity to organize affective life, (Stolorow et al., 1987) to develop a sense of self, (Stern, 1985) and to establish self-regulation of affect (Beebe & Lachmann, 1998, 2002; Sander, 1983).

Working with Bowlby in the 1960s and 1970s, developmental psychologist Mary Ainsworth explored the nature of attachments, observing infants in a test called the "Strange Situation Procedure." First, mothers and infants were introduced to a pleasant room with toys for the infant and left alone together. Once the mother and infant had settled in, a stranger entered the room. After some time, the mother departed, leaving the infant in the room with the stranger. Then, the mother returned to the room and the stranger departed. After spending time with the infant again, the mother departed once more and left the infant completely alone. Three minutes later, the mother returned to reunite with the infant (Ainsworth et al., 1978). The most interesting behavior was observed when the infant was reunited with the mother after being left alone. Three primary attachment patterns were observed. A fourth was noted and added later by Ainsworth's student Mary Main (Main & Solomon, 1990).

1. Infants with secure attachments settled into the room easily with their mothers and were fully engaged in play. When the stranger

entered, they stopped playing momentarily and focused on their mother but were reassured and returned to play. When their mother left, they became upset but were able to engage with the stranger and return to play, though not quite as engaged as before. When their mother returned, they reached for her to be picked up and were soothed. When mother left once more, they cried until she returned, but upon her return they reached for her again with joy, expecting to be soothed. They engaged in play with their mother and once re-regulated, they returned to the task of playing by themselves.

Children whose parents are generally sensitive and attentive to their needs develop secure attachments. Secure attachment promotes the development of effective mood regulation (Cassidy, 1994). Under optimal circumstances, the development in the first few years of life involves transitioning between caregiver-initiated emotional regulation and self-initiated emotional regulation (Cole et al., 2004; Fox, 1994). Adults with secure attachment styles tend to be self-confident, trusting, open to, and interested in close relationships with romantic partners, and are likely to form stable and long-lasting bonds in which they are good partners. They are comfortable asserting their needs and do not linger in relationships with people who do not meet their needs.

2. Infants with avoidant attachments also engaged readily in play when first introduced to toys with their mothers. The entrance of a stranger was barely noticed, and they continued playing with toys when their mother left them with the stranger, expressing little upset or fear. When their mother returned, they looked for her but made no eye contact, showed no excitement, and were not fully engaged with her, exploring other toys or stiffening when she held them. When their mother left them completely alone, they showed little distress and were able to engage in some level of play while she was out of the room. When she returned again, they showed little emotion and again did not fully engage with her, turning away to play with toys and only partially engaged in play with little exploration. When the mother lifted them, they became limp and made no eye contact.

Ainsworth postulated that these infants whose neediness was consistently met with parental rebuff or rebuke regulated their emotions by deactivating the attachment system. These children learned to cope with negative emotions by not experiencing them. Infants with avoidant attachments detach from their caregivers and do not look to them for soothing; they autoregulate. In stressful situations, their affect is down-regulated by avoidant and withdrawn behavior.

Children who have learned to avoid experiencing negative emotions often avoid emotionally charged social situations in the wider world as they mature. Intimacy and close personal relationships involve vulnerability so they are often avoided. People with avoidant attachments are often uncomfortable with self-disclosure and may be socially inhibited (Cooper et al., 1998).

3. Infants with anxious or anxious/ambivalent attachments did not readily leave their mother to play with toys when they first entered the room. When they did begin to play, they kept one eye on their mother at all times. They stopped playing and had a strong negative reaction when the stranger entered the room, returning to mother for her attention. They cried when mother left the room. When left alone with the stranger, they could not be soothed and would not return to play. When mother returned, they both clung to her and hit her, and could not be soothed. When the mother offered them a toy, they pushed it away. When left alone, they were so intensely distressed that mother returned before three minutes had elapsed: they were then almost inconsolable. Once settled, they could not play again unless mother was nearby.

Children whose parents are unpredictable develop an anxious/ambivalent style of attachment. At times, their parents may be available and warm, at other times rejecting and cold. They may be well-meaning but misattuned and insensitive, intruding on their infants' needs. Infants with inconsistent parents learn they cannot take parental security for granted and become hypervigilant for rejection cues (Bartholomew & Horowitz, 1991). Where infants with avoidant attachments are detached and hypoaroused, infants with anxious attachments are hypervigilant and hyperaroused. They become stuck in chronic cycles of unfulfilled attachment-seeking/attachment-anxiety, fixated on their caregivers. In adulthood, this manifests as preoccupation with relationships that can compromise self-worth and autonomy. With a focus on the other at the expense of the self and reactivity to the external environment, people with anxious attachments have a diminished capacity for autoregulation.

Adults with this style of attachment lack self-confidence, are chronically worried about rejection and abandonment and are prone to bouts of jealousy and rage at partners who are perceived to be untrustworthy. Their constant need for reassurance in the context of their emotional reactivity often leads others to feel overwhelmed by them, setting them up for the very rejection they fear (Bartholomew & Horowitz, 1991). They are eager to become involved in romantic relationships despite the difficulties they pose and are likely to engage in inappropriate self-disclosure. They often fall in love quickly and indiscriminately and experience frequent breakups and reunions.

4. Infants with disorganized attachments were unable to play spontaneously. They played happily with the stranger. When reunited with their mother after she left the room they responded inconsistently; at times, they approached her, at other times, they appeared to freeze, and at other times, they appeared to avoid her. The overall pattern of ambivalent, avoidant, and ignoring behavior suggested emotions of fear. The behavior pattern was disorganized and chaotic.

Children develop this type of attachment when caregivers are unpredictable in their responses and do not know how to soothe them or meet their needs. This generally occurs when a parent is disturbed or has been traumatized. Adults with disorganized attachments have difficulty self-soothing and typically have trouble regulating their emotions. This interferes with trusting and developing healthy social relationships and often with seeking help. There is a strong association between a history of abuse or neglect and disorganized attachments (Carlson et al., 1989; Liotti, 2004).

An illustrative description of how attachment impact affect regulation and subsequent behavior is given by van der Kolk and Fisler (1993):

Secure attachments with caregivers play a critical role in helping children develop a capacity to regulate physical arousal. Loss of the ability to regulate the intensity of feelings and impulses is possibly the most far-reaching effect of trauma and neglect. It has been

> shown that most abused and neglected children develop disorganized attachment patterns. The inability to modulate emotions gives rise to a range of behaviors that are best understood as attempts at self-regulation. These include aggression against others, self-destructive behavior, eating disorders, and substance abuse...Affective dysregulation can be mitigated by safe attachments, secure meaning schemes, and pharmacologic interventions...

Insecure or disorganized attachments do not necessarily develop because of parental abuse or neglect. The infant's temperament plays a role in the bonding process (Fox et al., 1991). Attachment theory has maintained that the developing relationship between the infant and the caregiver is co-constructed, emerging from the caregiver's responses to the signals the infant produces (Zeanah & Fox, 2004). Attachment theorists believe that temperament affects the way infants call to their caregivers in times of stress (Sroufe, 1985), which has an effect on the caregiver's response.

Ainsworth understood that the parents' sensitivity to the infant was at the heart of a secure attachment and that this was dependent on the caregiver's capacity to understand the infant's mental state (Ainsworth et al., 1978), carrying out a "reflective function" (Meinz & Main, 2011) or "mentalization" (Fonagy & Target, 2005). Mentalization is the imaginative psychological activity that enables us to understand the mental state of ourselves and others. Human bonds rely on the mentalizing process to create a full sense of togetherness (Frith & Frith, 2006; Van Overwalle, 2009).

The mentalization capacity to imagine another's mind is important for affect regulation. Fonagy (2003) proposes that the way temperament unfolds into personality is not directly via attachment style but by the mentalization process, the infant's creation of internal working models and expectations of the caregiver's mental world, derived from his attachment experience, "transforming genotype into phenotype." In other words, the infant brings his affective temperament to the table, but the caregiver brings her capacity to "see" or "mentalize" the infant to the table; it is the interplay between the two—not the affective temperament alone—that determines the infant's capacity to regulate his mood. It is within the mother-infant dyad that the infant's genetic potential becomes expressed. How does attachment relate to mentalization? And how do we get from mentalization to affect regulation?

Secure attachments emerge from consistent, empathic care. Development of such security requires the caregiver's attentiveness and capacity to understand and see the infant. Implicit in this process is the need for the caregiver to mirror back to the infant what he sees, to articulate and label the infant's feelings. The caregiver must be able to acknowledge, hold, and tolerate the wide range of the infant's emotions and feed them back to him in a modulated form. This is the beginning of affect modulation. This capacity of the caregiver to "mentalize" or reflect on the infant's feelings before feeding them back to the infant creates the opportunity for the infant to learn to mentalize himself.

A caregiver who has the reflective capacity to understand the state of the infant's mind understands his feelings and treats him as an agent. The infant perceives this and develops a sense of himself as an autonomous being who is understood and cared for. As the infant comes to know himself in the face of the mentalizing caregiver, he comes to regulate his affect without the caregiver. As language comes online, mood regulation becomes more refined with verbal representation of affect. In the naming of his affect, the infant gains the capacity to "work with," "modulate" and "regulate" his emotion. Doubtless, we have all had the experience of having an immediate angry reaction to something, then pulling back to reflect and say to ourselves "Once I thought about it, it made more sense and I wasn't so upset." Mentalization involves the capacity to "name" affect, which, once named, can more easily be regulated.

Failures in caregiver mentalization create insecure attachments and consequent difficulty with mood regulation in the infant. At times a parent cannot mirror an infant's distress, perhaps because doing so evokes painful memories in the parent. The parent may take a defensive/ avoidant posture toward the infant (Fonagy & Target, 1997), essentially dismissing the infant and leaving him without an experience of being "mentalized" or recognized. When this type of interaction becomes chronic, the infant's attachment to the caregiver becomes insecure. The infant then develops a mental representation of the parent as unreliable, and his expectation is that he will not be soothed. His mental representation of the caregiver will affect his expectations of other caregivers, creating anxiety. This, in turn, affects his caregiver and a cycle of mood dysregulation ensues.

Attachment related stress has neurological effects on brain circuits associated with social cognition and the capacity for mentalization (Nolte et al., 2013), including, as we have noted, the superior temporal sulcus

(STS). Structures associated with the capacity to mentalize have been found to be associated with reduced fMRI (functional magnetic resonance imaging) activity in individuals subjected to chronic stress in comparison with normal controls.

Failures in mentalization may not appear to result from violent or egregious childhood abuse or neglect, but to the helpless infant who is biologically programmed to seek attachment for survival, recognition itself is a literal lifeline, and its absence becomes a chronic physiological stressor.

Bowlby's attachment theory became the basis for affect regulation theory, elaborated by Allan Schore (2016), who describes the neurobiology of emotional development as it unfolds in the context of mother-infant attachment. He posits that the infant learns to regulate affect within the setting of the infant-caregiver interaction, and that the caregiver's regulatory function permanently affects not only psychological development but the development of brain structures themselves. Affect theorists have come to describe the environmental stress of infant/childhood abuse or neglect as "relational trauma."

The biological alterations in the brain caused by relational trauma have been increasingly recognized to be an important source of mood dysregulation (Schore, 2016). Secure attachments between mother and infant have hidden regulatory effects (Hofer, 1995) on physiology (sensorimotor, thermal, metabolic, and nutritional) which become disturbed with disruptions in attachment. An infant's earliest connections protect its vulnerable brain during development from the harmful effects of hormones and chemicals generated in the face of stress. Unfortunately, insecure attachments play a crucial role in the genesis of physiological chronic stress reactions and the affective fallout that results from them, as we will discuss. Caregivers with whom infants form their first relationships can either be the most powerful defense against harmful stressors or the most powerful source of stress in their lives.

7.3 EPIGENETICS

In the epigenetic dance, we regard attachment styles to play a crucial environmental role in the gene expression that affects one's capacity to regulate affect. Environmental influences on genetic makeup that determines the expression of the organism's traits are at the heart of epigenetics. But what is the biological mechanism by which attachment has an impact on our genes, thereby on affect regulation? Mammalian studies have helped us begin to understand how maternal behavior affects gene expression on a molecular level.

Deoxyribonucleic acid (DNA) contains anywhere from 50,000 to 140,000 genes which code for proteins and ribonucleic acid (RNA). Genes are simply templates. The information encoded on genes needs to be "unlocked" by "transcription" before proteins or RNA can be produced. Transcription is a molecular translational process by which the DNA code is "read." It determines which amino acids, proteins, nerves, networks, organs, and neurotransmitters will be created from genes on the DNA. This transcription process can be blocked, however, by "silencing" genes carried on the DNA. Genes can be silenced or blocked if special molecules are attached to them. Methyl groups are the most common molecules that attach to DNA to block gene transcription, silencing gene expression. This gene silencing process is called "methylation."

The gene silencing process can be affected by environmental influences, as demonstrated in animal models.

Genes that are blocked or silenced are not transcribed, so "would be" traits are dormant and never find expression. In this way, the environment alters the molecular mechanisms that regulate gene activity (Siegal, 2010) determining which genes will be available for transcription (Szyf et al., 2007).

An example of this process is illustrated by a rat model in which a rat pup's resilience to stress depends on its experience of maternal behavior. As we have discussed, stress is mediated by glucocorticoid stress hormones such as cortisol. The capacity to mount a robust stress response depends on cortisol's target organ having a wealth of glucocorticoid receptors to respond to this hormone. With an inadequate number of glucocorticoid receptors, the organism cannot adequately respond to stress, even when cortisol is present. Thus, a gene that codes for glucocorticoid receptors becomes important in ensuring the capacity to mount a strong stress response.

Would it be possible to influence how many glucocorticoid receptors an organism has? Glucocorticoid receptors are proteins created after transcription of the glucocorticoid receptor gene. If the glucocorticoid receptor gene is silenced, it will not be available for transcription, so there will be fewer glucocorticoid receptors.

Recall that the addition of methyl molecules to DNA is a mechanism by which genes are silenced (Meaney & Szyf, 2005). Thus, methylation of glucocorticoid receptor genes in the rat hippocampus silences them or blocks their transcription so that glucocorticoid receptors will not be produced.

Maternal rat behavior, it turns out, affects the methylation process. The experience of maternal licking and grooming of rat pups in the first week of life determines the extent to which glucocorticoid receptor genes in the hippocampus will become methylated or silenced (Weaver et al., 2001).

In an environment enhanced by maternal licking and grooming, methylation of glucocorticoid receptor genes is reduced. Reducing the methylation of glucocorticoid receptor genes frees them to produce more glucocorticoid receptors. The more glucocorticoid receptors in the rat pup's hippocampus, the greater its capacity to handle glucocorticoids and to mount a healthy stress response.

On the most molecular level then, maternal licking and grooming translates into structural changes in the brain that affect the rat pup's capacity to adapt to stress and to self-regulate: more maternal licking and grooming results in more rat pup stress resilience. This is one epigenetic example of how early caregiver-infant interaction can provide for the organism's resilience in the face of stress. Dynamic gene-experience interactions such as this occur in all systems of our bodies throughout our lives.

Though environmental influence on gene expression is ongoing, researchers do believe that epigenetic influence is greatest in anatomic structures that are most rapidly developing such as those in the fetus and neonate (Dobbing, 1981), so our earliest experiences have the greatest impact on neural development. Neural growth (via BDNF) and plasticity of neurons result from gene expression stimulated by early caregiver attention.

"Neuroplasticity" refers to the capacity of the nervous system to adapt its structural and functional organization to altered circumstances to meet developmental or environmental changes (Leuner & Gould, 2010). In a healthy system capable of such adaptation, environmental stimulation can cause neurons to grow longer and more abundant dendrites, creating larger networks and more dendritic connections with other neurons. Neural plasticity enables nerve networks to physically rewire and nerve cells to functionally retool, giving us flexibility and adaptability for varied emotional and behavioral responses in stressful situations.

Early experiences that lead to mood regulation are mediated by this very cellular adaptation, which is the neuronal basis for self-regulation (Fonagy, 2002), enabling us to "maintain flexibly organized behavior in the face of high levels of arousal or anxiety" (Srouf, 1996). The more adept we are at mood regulation, the wider the range and the higher the intensity of emotional experience we can have while maintaining adaptive behavior. Neural integration is crucial for this regulatory process to occur (Siegal, 2000). It is what allows us to respond to that irritating guy while keeping our cool rather than reacting to him and blowing a fuse.

7.4   CHRONIC STRESS RESPONSE

Acute stress functions differently in the body from chronic stress. The body is designed to respond effectively to acute stress and then to recover. Under ordinary circumstances, our homeostatic mechanisms can handle stressors such as an injury, a brief illness, or an environmental threat and return to equilibrium. But we were not designed to handle chronic stress. Early life adversity, child abuse and neglect, maternal deprivation, relational trauma of all kinds result in chronic stress for which the body was not designed. Under the burden of chronic stress, the negative feedback loops that enable the acute stress response to switch on, then off, do not switch off, and neural tissues are bathed in excess levels of toxic stress hormones.

A healthy stress response occurring in the context of a secure attachment (Gunnar et al., 1996; Spangler & Schieche, 1998; Sroufe & Waters, 1979) is achieved through maternal responsiveness. This leads to lower cortisol levels, which reflect a properly functioning negative feedback loop and correspond to a well-regulated HPA axis (Gunnar et al., 1996). An infant with a secure attachment will exhibit an appropriate level of distress (brief separation from mother) and maintain well-regulated cortisol levels; as the acute stressor passes, a transient elevation in circulating cortisol will signal the hypothalamus to stop releasing CRH, which turns off the stress response and the production of further cortisol.

Insecure, anxious, and disorganized attachments can lead to interferences in the HPA axis (Hertsgaard et al., 1995; Spangler & Grossman, 1997; Spangler & Schieche, 1998) and pose a risk for the development of behavioral and emotional problems (van Izjendoorn et al., 1999), creating circumstances of chronic stress for an infant (and child). One of the most toxic chronic stressors is maternal neglect, the "silent abuse." Maternal neglect has been associated with several neurodevelopmental disturbances (Chugani et al., 2001; Graham et al., 1999; Rule et al., 2002), compromising the body's capacity to modulate its stress response and leading to mood dysregulation by causing chronic HPA axis overactivity (Plotsky & Meany, 1993; Heim et al., 2001).

How does chronic stress lead to HPA axis and autonomic system disturbance? In short, chronic stress causes the normal homeostatic negative feedback loops to fail. The stress hormones, norepinephrine, and glucocorticoids, instead of putting the brakes on the system, actually accelerate the system in a positive feedback loop. When this occurs, stress hormones remain chronically elevated.

An example of this occurs when ACTH from the pituitary gland drives the adrenal glands to overproduce the "rescue" hormone cortisol, which fails to turn off an inured hypothalamus, so more CRH is secreted. The HPA axis becomes chronically overactive resulting in ongoing elevation of cortisol levels.

Under normal circumstances, the hippocampus plays an important role in shutting down the HPA axis stress response. However, in cases of chronic stress, persistently elevated levels of cortisol override the normal hippocampal response, bathing it in cortisol. The hippocampus is one of the most vulnerable neural structures to cortisol damage. Cortisol-mediated damage to or atrophy of the hippocampus further impairs the hippocampal shutoff mechanism and leads to a prolonged stress response (Herman & Cullinam, 1997; Jacobson & Sapolsky, 1991; Sapolsky et al., 1986). Despite this positive feedback loop, the body never becomes irretrievably out of control because homeostatic forces are so effective and the checks and balances so numerous that the system finds ways to restore some manner of equilibrium, even if it is at the expense of other functions.

In a dysregulated system, the stress response is amplified in a positive feedback loop such that cortisol is on duty at all times. A cortisol surge is necessary for an acute stress response but chronically elevated levels of cortisol in the body are present in conditions of chronic stress. Chronically elevated cortisol levels can lead to several pathogenic processes and disease states including major depression, insulin resistance and diabetes, hypertension and atherosclerosis, bone loss, and disorders related to impaired immune function (Gold & Chrousos, 1999; McEwan, 1998). It can also lead to psychopathology because of cortisol's toxicity to the mood regulating structures in the brain.

7.4.1   NEURAL EFFECTS OF CHRONIC STRESS

Relational trauma creates chronic stress that we know precipitates epigenetic effects resulting in mood dysregulation. As we discussed, this epigenetic process begins with caregiver behavior that affects the way DNA is transcribed. This, in turn, has implications for the organism's capacity to handle stress. Early life adversity is associated with mood disorders via epigenetic changes leading to a cascade of events including DNA methylation of glucocorticoid receptor genes, glucocorticoid receptor dysfunction, hyperactivity of the HPA axis, and compromise to brain structures important to affect regulation (Na et al., 2014; Tata & Anderson, 2010).

Chronic stress leads to HPA axis hyperactivity and elevated levels of cortisol. HPA axis hyperactivity results in chronically elevated levels of circulating glucocorticoids, including cortisol. Cortisol exposure is particularly toxic to neural tissues, resulting in neuroinflammation and neuronal injury (Ganguly & Brenhouse, 2015).

Chronic stress causes inhibition of neurogenesis, disruption of neuronal plasticity and synaptic connectivity, neural toxicity, and, in a positive feedback loop, activation of the stress response; this further exacerbates the effects of the stress response by exposing target tissues to elevated levels of cortisol (Gunnar & Quevedo, 2007; Pagliaccio et al., 2013; Wolf et al., 2005).

Most of the research done on the toxic effects of glucocorticoids on the brain has focused on three anatomic areas, each of which is important to mood regulation—the hippocampus, the PFC, and the amygdala. Significantly, cortisol's effects on the hippocampus and PFC are different from those it has on the amygdala.

In the hippocampus and the PFC, cortisol erodes nerve cells, causing dendritic and neuronal shrinkage. This creates functional impairment in two crucial anatomic structures necessary for mood regulation.

In the amygdala, however, a structure which we would like to keep quiet, cortisol causes dendritic and neuronal elaboration, activating the function of an area which needs silencing for mood regulation ("A quiet amygdala is a good amygdala").

The PFC is most notable for the executive functions that support the various cognitive processes involved in self-regulation (Curtis & D'Esposito, 2003; Goldberg, 2001; Miller & Cohen, 2001), thus, damage to this area is associated with mood dysregulation. High doses of glucocorticoids impair PFC-dependent working memory (Arnsten & Goldman-Racik, 1998; Lupien et al., 1999; Roozendaal et al., 2004). Child abuse or neglect can result in stress that activates the HPA and SAM in a positive feedback loop leading to elevated levels of cortisol and increased levels of norepinephrine, causing dysfunction of the PFC (Arnsten, 1999) and symptoms of disrupted attention (Glayser, 2000).

The hippocampus, as we know from our epigenetic example, has a high concentration of glucocorticoid receptors, making it particularly vulnerable to the toxic effects of chronic cortisol exposure (Sapolsky, 1985; Virgin Jr. et al., 1991; Zhang et al., 2017). Excess levels of cortisol in the hippocampus and PFC are associated with neuronal shrinkage (Roozendaal et al., 2009). Because the hippocampus is involved in learning, memory, and emotions, elevated levels of cortisol associated with chronic stress create memory impairment (Ivy et al., 2010; Kim & Diamond, 2002; Park et al., 2008) and have an impact on cognitive function as well as mood regulation.

In contrast to its effects on the hippocampus and PFC, excess cortisol in the basolateral complex of the amygdala caused by chronic stress results in neuronal remodeling of synapses and dendritic branching (Roozendaal et al., 2009. This is manifest behaviorally in increased anxiety (Conrad et al., 1999; Mitra & Sapolsky, 2008; Vayas & Chattarji, 2004; Vyas et al., 2006). Strengthening the connections between neurons in the amygdala may be the mechanism for the development of anxiety (Chattarji, 2008; McEwen & Chattarji, 2007) as it enhances amygdala activity.

However, cortisol is not altogether directly responsible for this neuronal damage; it is a mediator. Cortisol actually brings about its effects through the excitatory neurotransmitter glutamate.

Under ordinary circumstances, glutamate, like cortisol, is useful in acutely stressful circumstances. Stress-induced increases in the level of cortisol can mediate the learning of fear by increasing levels of the excitatory neurotransmitter glutamate in the limbic system (Eckersdorf et al., 1996; Popoli et al., 2011).

The excitatory neurotransmitter glutamate thus becomes the agent through which cortisol becomes toxic to neural tissue. Excess cortisol has deleterious effects on neuronal architecture by stimulating the release of glutamate (Karst, 2005; Popoli et al., 2012). Glutamate causes damage including inflammation and retraction of neuronal dendrites (Gorman & Docherty, 2010; McEwan, 2005; Pittenger & Duman, 2008). This leads to shrinkage of nerve networks and brain structures and impairs tissue function. Stress-induced increase in extracellular glutamate in the hippocampus is secondary to increase in glucocorticoids (Lowy et al., 1993).

Glial cells are specialized cells that support neurons. One of their functions is to clear glutamate from neuronal sites. Failure of glial cells to clear glutamate results in neuronal damage (Rothstein et al., 1993; Tanaka et al., 1997), a process now thought to be linked with neuropsychiatric disorders (Pitt et al., 2003).

7.4.2   STRUCTURAL EFFECTS OF CHRONIC STRESS

What structural changes occur in the brains of organisms subjected to chronic stress and what are their implications? Few studies have been done in humans to determine whether or not chronic stress is associated with any structural changes in the brain, but Ansell and colleagues (2012) performed magnetic resonance imaging (MRI) scans on a cohort of over 100 people with "cumulative adversity." They found reduced gray matter in the PFC, ACC, and insula, areas we know are pivotal for symbolic thought and emotional processing. Drawing inferences from the putative functions of these circuits, they concluded that these changes interfered with emotional, social, and self-regulatory behaviors. This evidence of abnormalities in brain structures responsible for mood regulation in humans subjected to chronic stress who developed behavioral/emotional disturbances lends strong support to our model of the role of chronic stress in the development of mood dysregulation.

Structural damage leads to functional impairment from neuronal damage to neural circuits. The amygdala/ventromedial prefrontal cortex (vmPFC) pathway is responsible for the extinction of fear. By its connections with the amygdala, the vmPFC allows us to process risk and fear and to make decisions. Disruption in this pathway impairs one's capacity to modulate responses to threats, leaving one vulnerable to situational overreactivity. Social defeat stress has been found to impair neuronal activity in the vmPFC of rodents (Abe et al., 2019) and infant stress has been found to lead to decreased amygdala/vmPFC activity, leading to such disturbance. Child neglect has been associated with a disruption in connections between the amygdala and the PFC (Nooner et al., 2013; Herringa et al., 2013; Pagliaccio et al., 2015), and between the amygdala and the vmPFC (Burghy et al., 2012), predisposing the individual to mood dysregulation.

Other stress-induced structural disruptions of circuits in the limbic system important for mood regulation are those responsible for attention control. Early life stress has been shown to impair these circuits (Correll et al., 2005; Mueller et al., 2010; Tottenham & Galvan, 2016). Attention control is associated with the executive attention networks including the ACC, the lateral frontal cortex (LFC) and PFC, and the basal ganglia (Posner & Fan, 2008). The ACC, located behind the LFC and PFC, acts as a switchboard, mediating between those two aspects of the cortex and lower behavioral control regions such as the basal limbic and structures in the autonomic nervous system. Emotional control and self-regulatory behaviors are found to depend on attention control. Shifting attention away from negative stimuli can improve mood and reduce overall stress (Bardeen et al., 2015; Gross & Thompson, 2007; McRae et al., 2010). When these structures are compromised, mood dysregulation can result.

Stress-induced damage to neural circuits affects the PFC and limbic system, two crucial areas necessary for the cognitive reappraisal upon which emotion regulation is dependent. Damage to circuits linking the hippocampus (memory processing) to the PFC (cognitive processing) and the ACC (attention, decision, and emotion regulation) are tightly linked to the affective and cognitive abnormalities seen in mood disorders (Mayberg, 1997). Emotion dysregulation is increasingly seen neurologically to reflect a cortico-limbic disturbance. Sicorello and Schmahl (2021), approaching emotion dysregulation from a transdiagnostic perspective, address it as a fronto-limbic imbalance. Kebets and colleagues (2020), also using a transdiagnostic model, demonstrated increased activation in fronto–limbic areas in individuals with ADHD, bipolar disorder, and borderline personality disorder with emotion dysregulation. There is ample evidence that chronic stress has deleterious effects on neural structures that lead to mood dysregulation.

7.4.3 IMMUNOLOGIC EFFECTS OF CHRONIC STRESS

Psychoneuroimmunology has revealed the effects of chronic stress on the brain as mediated by the immune system. Immune cells in the body are activated by tissue injury and psychosocial stress, releasing proinflammatory substances called cytokines (such as interferon and interleukin). Cytokines mount an inflammatory response in the body that affects neurotransmitter metabolism, neuroendocrine function, synaptic plasticity, and neural circuits that regulate mood, motivation, anxiety, cognitive function, and sleep (Capuron & Miller, 2011).

Proinflammatory cytokines produce stereotypical behavioral changes called the "sickness response" characterized by depressive-like symptoms such as anhedonia, fatigue, psychomotor slowing, decreased appetite, sleep alterations, and increased sensitivity to pain (Hart, 1988; Kent et al., 1992). Chronic treatment with antidepressants has been shown to abolish depressive-like symptoms induced by certain proinflammatory cytokines (Castanon et al., 2001; Yirmiya, 1996).

One putative mechanism for the production of the sickness response is the disruption caused by proinflammatory cytokines on mood-enhancing neurotransmitters such as serotonin and dopamine. Interferon-α reduces the number of available serotonin 1A receptors in the brain (Cai et al., 2005). Other cytokines mediate the breakdown of tryptophan, an amino acid building block of serotonin, leading to reduced levels of this neurotransmitter in the brain (Dantzer et al., 2008; Schwarcz & Pellicciari, 2002). Similarly, cytokines have been shown to impair both the synthesis and reuptake of dopamine (Kitagami et al., 2003).

Chronic exposure to cytokines alters the function of the HPA axis by flattening the diurnal (morning) spike of cortisol (Matthews et al., 2006; Raison et al., 2010). While cortisol levels are not chronically elevated by cytokines, they are elevated by chronic stress, which feeds back in deleterious ways to the immune system (Pariante & Lightman, 2008) in a positive (runaway) feedback loop, causing more cytokine release. Cytokines also stimulate the release of glutamate, which mediates cytotoxicity which, in turn, decreases the production of trophic factors in the brain including BDNF (Hardigham et al., 2002; Haydon & Carmignoto, 2006) leading, among other things, to a reduction in neurogenesis, a hallmark of chronic stress (Duman & Monteggia, 2006). This has important implications for mood regulation as neurogenesis in the hippocampus has been positively correlated with improved mood states and increased cognitive performance (Brachman et al., 2015; Wolf et al., 2009).

Certain neural structures may be preferentially targeted by the cyto-toxic effects of proinflammatory cytokines. The basal ganglia, playing an important role in motor activity and motivation, appear particularly sensitive to the effects of cytokines (Brydon et al., 2008; Capuron et al., 2007) as does the ACC (Eisenberger & Lieberman, 2004; Lozano et al., 2008), implicated in depression, arousal, anxiety, and alarm. In addition, memory impairment has been seen with cytokine-associated loss of hippocampal neurons (Wan et al., 2007).

In this chapter we discussed the diathesis–stress model of illness to understand the epigenetic nature of mood dysregulation. Mood dysregulation, after all, is not simply determined by one's inherited affective temperament but by the interaction of that genetic predisposition with environmental influences.

An infant might be born with a certain affective temperament, rendering her vulnerable to development of a mood disorder. Raised in a nurturing environment, she might never have mood symptoms, but under chronic physiological strain (e.g., relational stress), might develop a mood dysregulation.

Chronic stress leads to hyperactivity of the HPA axis and subsequent excess levels of cortisol. Cortisol causes neural tissue damage by stimulating the activity of glutamate, the brain's major excitatory neurotransmitter. Excess glutamate destroys nerve cells, particularly in the sensitive area of the hippocampus (the seat of memory) and the PFC (responsible for executive functions including self-control), and enhances nerve growth in the amygdala (resulting in anxiety and irritability). Cells in the ACC responsible for attention, decision making, and emotional regulation, are also damaged. Likewise, chronic stress causes parallel (and cascading) damage via the immune system where cytokines effect similar damage to many of the same structures. The net effect of these changes is disruption in cognitive and affective processing circuits and ultimately, mood dysregulation.

It would be a gross oversimplification to suggest that mood dysregulation is all the result of attachment disruptions and relational stress with its associated HPA axis/cortisol/glutamate impact on the emotion-mediating circuits we have outlined here. This hypothesis is put forward because there is current data to substantiate it—not because it presumes to be comprehensive. Our views will be expanded and modified as our understanding of these complex systems deepens with ongoing research.

KEYWORDS

- epigenetics
- diathesis-stress model
- attachment theory
- glucocorticoid receptor
- chronic stress response

CHAPTER 8

Treatment

Each player must accept the cards life deals him.
But once they are in hand, he alone must decide how to play
the cards in order to win the game.

—Voltaire

ABSTRACT

Important aims of the DSM diagnostic classification system are to guide research and treatment. As "mood dysregulation" is not a diagnosis, there is no research or treatment specific to it that we can discuss here with any authority, but as it is a transdiagnostic phenomenon, we can certainly discuss treatment of its symptoms as they are associated with various diagnoses. Environmental interventions that promote regulation such as sleep interventions, meditation and neurofeedback are useful for people with mood dysregulation. When mood dysregulation develops over time in individuals subjected to chronic relational stress, psychotherapy has unique potential to help them learn to regulate themselves in the context of stable attachments. Medication treatment of mood dysregulation can be complex for a number of reasons. First, many people with mood dysregulation have been diagnosed and treated for depression with untold medications for many years and are reluctant to try yet another agent. When subsequently told their symptoms are likely to respond to mood stabilizers, many refuse the recommendation because they assume they are being told they have a diagnosis of bipolar disorder (which they associate with mood stabilizers): DSM nomenclature has a problematic grip even on the lay person. Mood stabilizers are the most successful pharmacologic intervention with mood dysregulation because they target dysphoria, a state of activation related to the bipolar symptom of hypomania. Antidepressants, antianxiety agents, and other medications used to treat symptoms characteristic of other diagnoses are often helpful as well because of the transdiagnostic nature of mood dysregulation. As we learn more about the biologic underpinnings of mood disorders, new pharmacologic agents are being developed that promise to be useful in treating mood dysregulation.

Mood Dysregulation: Beyond the Bipolar Spectrum. Deborah A. Deliyannides, MD (Author)
© 2024 Apple Academic Press, Inc. Co-published with CRC Press (Taylor & Francis)

There is no literature on the treatment of "mood dysregulation" as I have described it: The literature that does exist on the treatment of mood dysregulation pertains to DSM-5 DMDD, which has partial relevance inasmuch as mood dysregulation as discussed here is a transdiagnostic process that overlaps with DMDD. Treatment of the transdiagnostic process of emotion (dys)-regulation as written about in the literature, however, is relevant to the treatment of mood dysregulation.

When dealing with transdiagnostic processes, there are two fundamental approaches to treatments: diagnosis focused treatments and symptom focused treatments. Thus, some treatments may be geared toward people with the mood dysregulation seen in borderline personality disorder while other treatments may focus on treating mood dysregulation with predominant symptoms of anxiety, regardless of diagnosis. Symptom based treatments usually cross diagnostic boundaries and are typical of mood dysregulation.

The choice of treatment for mood dysregulation must take into account the underpinnings of the problem, the desires of the individual, and the availability of social and economic resources. People who come for help after years of misdiagnosis or failed treatments often lack confidence in recommendations and refuse treatment options that might prove to be very successful. Many come seeking a magic pill, too impatient to invest in more reflective or time-consuming modalities of treatment. It is typical for treatment to be multimodal and to include psychotherapy. When medication is indicated, it is not unusual for two or more medications to be necessary for a satisfactory response.

As a rule, if an individual's symptoms are creating problems for them (or others they care about), it is time for intervention with a mental health professional. Psychotherapy is typically practiced by psychiatrists (MD), psychologists, (PhD, PsyD), social workers (MSW, LCSW), marriage and family therapists (MFT), and other professionals. Medication can be prescribed by physicians, physician's assistants, and nurse practitioners. Psychiatrists and psychopharmacologists are physicians typically certified in psychiatry. Most psychopharmacologists limit their practices to prescribing medication, but some take on the broader role of practicing psychotherapy as well. If there is a question of diagnosis, a psychiatrist should be consulted. If there is a question of mood dysregulation, a psychiatrist should be consulted, not a primary care MD. Many primary care physicians can comfortably treat depression but few, if any, can properly treat mood dysregulation.

A description of symptoms including their onset and course is inadequate information for any clinician to begin to determine the proper treatment. Given what we know about the etiology of mood disorders, genetic and environmental information must be gathered. It is important to know the psychological profiles of as many relatives on the family tree as possible when embarking on treatment. Who in the family has had any mood disorder? Bipolar disorder? Depression? Temper problem? Substance abuse? Schizophrenia? Suicide? It is also important to know as much as possible about the relational background of the person being treated including their early attachments. What is the narrative of the individual's life? Who were their caregivers? What was the nature of their relationship with them? Did they experience any major conflict or trauma growing up? Overt trauma or abuse is generally obtainable information but sexual trauma may not be, especially if an individual has no access to relevant memories. Most difficult to uncover may be a history of neglect, a trauma that can leave no awareness in a subject asked to report its presence; this is a history we may only be able to infer.

An individual with a family member who has bipolar disorder may have a diathesis or predisposition to a mood dysregulation that epigeneticists would say becomes expressed under the environmental stressor of an attachment disturbance. Some individuals with mood dysregulation have symptoms that reflect more of a biological diathesis than any environmental determinant. They may have clear mood fluctuations and no overt history of attachment failure or relational trauma. Medication is often the primary treatment modality for them. Other individuals appear to have less of a biological predisposition and more of an environmental vulnerability to mood dysregulation. They may have more chronic dysphoric states and clear histories of maternal deprivation or overt abuse. Psychotherapy may be the primary modality of treatment for them. Treatment needs to be individualized depending on a person's history, needs, resources, and desires, and the most appropriate approach may not always be immediately obvious.

8.1 ENVIRONMENTAL TREATMENTS

Emotional stability depends in large part on the most routine activities in our lives, given the body's drive to maintain equilibrium and the association of mood regulation with overall homeostatic control. Simple behavioral changes can have an enormous impact on mood by affecting the body's biological milieu.

Sleep: One of the ways in which the body seeks homeostasis is by maintaining its circadian rhythm. Processes that occur naturally in the body every 24 hours regardless of fluctuations in light, are called circadian ("circa" = about; "diem" = day). Light-sensitive photoreceptor cells in the retina send signals to the "master circadian clock," the suprachiasmatic nucleus (SCN) of the anterior hypothalamus. The SCN controls many circadian functions such as the natural sleep–wake cycle, body temperature, cortisol secretion, and mood. The circadian master clock is synchronized when the light-sensitive nerve cells in the retina, exposed to natural light, send signals to the SCN that cause suppression of melatonin. Melatonin is a hormone associated with sleep onset that is crucial to maintaining the circadian clock.

Dysregulations in circadian rhythm can affect mood (Bechtel, 2015; Viaterna et al., 2001). There is increasing evidence for the hypothesis that the primary mechanism for bipolar spectrum disorders is a disturbance in the regulation of circadian rhythm (Abreu & Braganca, 2015; Alloy et al., 2017; McClung, 2007; Murray & Harvey, 2015), whether by lifestyle/behavioral desynchronization of the circadian clock or by a genetic defect in the circadian pacemaker (Goodwin & Jamison, 1990).

Irregularities in bedtime, irregularities in mealtime, variations in normal social routines, exercise, and significant life stressors can all disrupt circadian rhythm (Caliyurt, 2017), causing mood disturbances. Shen and colleagues (2008) found that people diagnosed with cyclothymia or bipolar II disorder had quicker onset of major affective episodes than those whose lives were more regulated. The well-established association between mood and the routine behaviors that regulate our circadian rhythm has led to "social rhythm therapy," which aims to help people with mood disturbances achieve mood stability by structuring their daily routines including the sleep–wake cycle (Frank et al., 2000; Haynes et al., 2016).

Co-occurrence of sleep disturbances is seen with nearly all mood disorders. Whether or not this cause or result, or simply an epiphenomenon (an unrelated association) is unclear because mood disturbances can be induced by sleep deprivation. Sleep deprivation has been associated with irritability and affective volatility (Horne, 1985). Functional magnetic resonance imaging (fMRI) scanning of individuals subjected to sleep deprivation showed a 60% increase in amygdala reactivity compared with controls and a significant reduction of functional connectivity between the amygdala and the medial prefrontal cortex (mPFC) (Sotres-Bayton et al., 2004). In other words, cortical regulation over limbic activation diminishes with sleep deprivation.

In addition, sleep deprivation interferes with the initial formation of emotional memories (encoding) (Harrison & Horne, 2000; Morris et al., 1960) and with their long-term consolidation (Beaulieu & Godbout, 2000; Hennevin & Hars, 1987). Emotional memories are crucial for the reappraisal tasks that are so important for healthy emotion regulation.

There is no single activity an individual can control that has more impact on mood regulation than their sleep schedule (Bauer et al., 2006; Umlauf & Shattell, 2009). Sleep hygiene takes on a new level of urgency when mood dysregulation is present. Brain imaging studies and laboratory, clinical, and home documentation of material gathered after spontaneous recall from sleepers suggest that dream (rapid eye movement [REM]) sleep in particular helps regulate mood (Cartwright, 1989). The challenge for a person with mood dysregulation is not just assuring an adequate amount of sleep each night but also maintaining a regular sleep schedule.

Artificially imposed shifts in our exposure to natural sunlight can lead to mood disturbances. People who work the night shift can develop sleep phase delay (Shirayama et al., 2003) and shift work sleep disorder (Drake et al., 2004) which can cause depression. Similarly, jet lag in people flying east to west can precipitate depression while in people flying west to east it can precipitate mania (Jauhar & Weller, 1982).

Several treatments for mood disorders take advantage of the effect of light exposure on mood. Bright light therapy (BLT) is used to treat depression and involves exposure to high-intensity artificial light (a "light box") to reset one's circadian clock. Use of a light box for 30 minutes in the early morning often relieves symptoms of depression within 3–5 days. BLT is an effective treatment not only for seasonal affective disorder (winter depression) but also for unipolar and bipolar depression (Even et al., 2008; Kupeli et al., 2018; Lam & Levitt, 1999). Interestingly, predictors of response to BLT are excessive eating, excessive sleeping, and lethargy, all features of atypical depression (Terman et al., 1996), which we have described as a mood dysregulation beyond the bipolar spectrum.

Conversely, the creation of "virtual darkness" by blocking exposure to blue light (which stimulates retinal photoreceptors and results in melatonin suppression) is a treatment for several conditions including mania. We are exposed to blue light throughout the day (and night), particularly when we look at computer screens, iPhones, or TV. Virtual darkness can be effectively achieved by wearing amber glasses to block blue light, simulating conditions of darkness. Colorless "blue light glasses" marketed for use with computers and other blue light emitting devices do NOT block blue light. Blue light blocking glasses must have amber lenses. Amber glasses have been used successfully to treat insomnia (Fargason et al., 2013; Shechter et al., 2018), to improve sleep quality and mood (Burkhart & Phelps, 2009), to treat sleep phase delay (Esaki et al., 2016), and to shorten manic episodes (Wirz-Justice & Terman, 2016). Like the use of bright light therapy, the use of amber glasses to create virtual darkness is thought to affect mood by resetting the circadian clock.

Diet: Another crucial behavioral measure one can engage in to help regulate mood is that of diet. While eating meals at regular intervals is one aspect of regulating mood (Lopresti & Jacka, 2015), choosing healthy foods is another. Limiting processed sugars reduces mood instability. There is significant evidence that a ketogenic diet has mod stabilizing properties by inducing a metabolic state of ketosis (El-Malakh & Paskitti, 2001; Phelps et al., 2013; Brietzke et al., 2018). The ketogenic diet limits carbohydrates and is high in fats. It includes foods such as red meat, fish, eggs, and cheese. Dipsticks to measure the concentration of urinary ketones are readily available over the counter to help monitor whether or not a state of ketosis is achieved. Eating disorders can co-occur with mood dysregulation, causing one problem to exacerbate the other (Dingemans et al., 2017; Spence & Courbasson, 2012). In these situations, other interventions are generally necessary to help regulate both diet and mood.

Mindfulness: Myriad studies demonstrate the benefits of meditation and mindfulness in regulating mood (Watford & Stafford, 2015; Tang et al., 2016; Hill & Updegraff, 2012).

> Mindfulness involves nonjudgmental attention to present-moment experience. In its therapeutic forms, mindfulness interventions promote increased tolerance of negative affect and improved wellbeing. However, the neural mechanisms underlying mindful mood regulation are poorly understood. Mindfulness training appears to enhance focused attention,

supported by the anterior cingulate cortex and the lateral prefrontal cortex (PFC). In emotion regulation, these PFC changes promote the stable recruitment of a nonconceptual sensory pathway, an alternative to conventional attempts to cognitively reappraise negative emotion. In neural terms, the transition to nonconceptual awareness involves reducing evaluative processing, supported by midline structures of the PFC. Instead, attentional resources are directed toward a limbic pathway for present-moment sensory awareness, involving the thalamus, insula, and primary sensory regions. In patients with affective disorders, mindfulness training provides an alternative to cognitive efforts to control negative emotion, instead directing attention toward the transitory nature of momentary experience. Limiting cognitive elaboration in favour of momentary awareness appears to reduce automatic negative selfevaluation, increase tolerance for negative affect and pain, and help to engender self-compassion and empathy in people with chronic dysphoria (Farb et al., 2012).

Perhaps nobody has better articulated the principles and practice of mindfulness and its many therapeutic applications than Jon Kabat–Zinn in *Full Catastrophe Living* (Kabat-Zinn, 2013). Kabat–Zinn is the founding director of the Stress Reduction Clinic and the Center for Mindfulness in Medicine, Health Care, and Society at the University of Massachusetts Medical School. As our emotions are embedded in our bodies, mindfulness begins with a focus on breathing, to which we all have access, all the time. From there, we learn to pay closer attention to the larger world around us including our emotions and the interactions we experience with others. By learning acceptance of our moods and suspending judgment of ourselves, we can gain conscious control over our feelings. Most forms of yoga involve mindfulness; yoga is effective at improving heart rate variability (HRV) (Ross & Thomas, 2010), which, in turn, regulates mood.

As yoga improves HRV, so does exercise, and as such, it is a natural mood regulator (Stein et al.,1999; Routledge et al., 2010). Exercise in individuals with bipolar disorder and depression has been shown to improve mood (Wright et al., 2009). Vigorous exercise has even caused manic symptoms (Melo et al., 2016). There are several proposed candidate biomarkers for the role of exercise in mood modulation with BDNF and oxidative stress as two main promising components of exercise's antidepressant effect (Hearing et al., 2016; Gomez-Pinilla et al., 2010).

8.2 NEUROFEEDBACK

Neurofeedback is a therapeutic intervention using feedback generated by a computer program that assesses brain wave activity as measured by an electroencephalogram (EEG). The program creates sound or visual signals from the EEG to "feed back" a visual graph of one's brain waves so the individual can then learn to respond in ways that regulate their brain waves to alleviate target symptoms. The goal of neurofeedback treatment is to achieve brain self-regulation.

Neurofeedback, a type of biofeedback, has been used to treat seizure conditions, behavior disorders, ADHD, autism, acquired brain injury, anxiety, depression, PTSD, and emotion dysregulation. When combined with real-time functional magnetic resonance imaging (rt-fMRI) as a feedback tool, neurofeedback appears most effective in mitigating symptoms of depression and anxiety. It has also been effective in treating symptoms of PTSD, borderline personality disorder (in which mood dysregulation features prominently), and schizophrenia (Linhartova et al., 2019).

To oversimplify, neurofeedback involves training individuals with the use of EEG-guided biofeedback to gain "top–down" cognitive (PFC) control over "bottom-up" emotional (limbic) processes. Most neurofeedback studies target the amygdala for therapeutic action (Bruhl et al., 2014; Paret et al., 2014; Zotev et al., 2011) because of its importance for emotion regulation. Neural connections between the amygdala and the PFC are regarded to be a major pathway responsible for this process (Diekhof et al., 2011; Hartley & Phelps, 2010; Vivani et al., 2014). Fronto–limbic imbalance in these networks (reduced inhibition from the PFC and ACC on the amygdala leading to hyperactivation of the limbic system) is thought to be responsible for emotion dysregulation in PTSD (Admon et al., 2013; Aupperle et al., 2012; Ronzoni et al., 2016). Neurofeedback has been successful in reversing this process, reducing amygdala activation and activating PFC activity (Nicholson et al., 2016; Zotev et al., 2013).

Numerous other rt-fMRI neurofeedback studies have demonstrated improved emotion regulation with reduced amygdala activity and increased activity in various cortical structures. Paret and colleagues (2016) demonstrated reduced amygdala activity and improved emotion regulation in people with borderline personality disorder after four ft-fMRI neurofeedback sessions. Johnston and colleagues (2010) demonstrated the use of rt-fMRI neurofeedback to help individuals up-regulate very specific emotion networks to achieve emotion regulation, networks involving the precuneus, the mPFC, and the ventral striatum.

While neurofeedback is a very effective treatment for mood dysregulation, two hurdles limit its use. Few clinicians are trained in the complexities of reading EEGs and the technicalities of administering neurofeedback. In addition, insurance companies typically do not cover the cost of the treatment.

8.3 PSYCHOTHERAPY

Psychotherapy is one of the most fundamental and flexible tools we have to achieve mood regulation. Daniel Hill and others have posited that the cultivation of mood regulation is a common mechanism of all forms of psychotherapy (Bradley, 2000; Hill, 2015).

Cognitive behavior therapy (CBT), dialectical behavior therapy (DBT), behavior activation (BA), and interpersonal therapy (IPT) are all short-term psychotherapies that help individuals learn to regulate their moods by teaching them cognitive tools and behavioral skills. Intensive short-term dynamic psychotherapy (ISTDP) is a short-term psychotherapy that is dynamically oriented. Eye movement desensitization and reprocessing (EMDR) is more physiologically based and is uniquely suited for people who have had trauma. Psychoanalytic psychotherapy is a long-term therapy in which an individual learns mood regulation in the context of a relationship with the therapist and has a unique place for people with mood dysregulation emerging from relational trauma.

- **Cognitive Behavior Therapy** seeks to challenge and change cognitive distortions that create disturbances in moods and relationships. The focus of CBT is on current thought patterns and on how to replace them with more adaptive ideas. CBT has been shown to improve medication compliance and reduce relapse rates in people with bipolar disorder (Reilly-Harrington & Knauz, 2005). In another study, CBT was found to be beneficial in improving social functioning and reducing relapse rates (Jones, 2004). People often ask for referrals to "a CBT therapist." In reality, there is no "pure" CBT therapist While some therapists may focus more on CBT than on other techniques, most therapists are eclectic and will incorporate CBT techniques into their treatment approach.

- **Dialectical Behavior Therapy** was developed by Marsha Linehan at the University of Washington in 1984 as a modified form of CBT, originally to treat people with borderline personality disorder. Where CBT might help people focus on understanding how maladaptive thoughts can lead to maladaptive behaviors, DBT helps people understand how those maladaptive thoughts and behaviors affect their relationships; DBT is designed to help people increase their emotional and cognitive regulation by understanding their "relationship triggers" and to develop coping skills to avoid excessive reactivity. DBT has now found its way into broader applications such as the larger umbrella of mood dysregulation (Linehan et al., 2007; Neacsiu et al., 2014a). An important component of DBT is education, and it is often offered in a group setting. DBT is increasingly used for treating emotion dysregulation in contexts other than borderline personality disorder such as depression, anxiety, and substance abuse (Caviccioli et al., 2019; McMain et al., 2001; Neacsiu et al., 2014b).
- **Behavior Activation** is derived from CBT. In BA, "cognitions" are treated as behaviors. Also, BA helps increase motivation by enabling people to "connect the dots" between their goal-oriented behaviors and the positive feelings that follow. It stimulates movement toward personal goals by positive reinforcement in a self-perpetuating cycle of improving mood. People keep a record of their activities and their associated moods; when they learn to associate certain behaviors with positive moods, those behaviors are positively reinforced. It is a variant of CBT and has been demonstrated by fMRI to result in brain changes that reflect greater emotion regulation (Ochsner et al., 2001; Ritchey et al., 2011).
- **Interpersonal Therapy** was first developed in 1969 at Yale University as an adjunctive treatment to medication for people with depression. It is designed to be highly structured and to be limited to 3–4 months, focusing on helping people consider ways in which their thinking patterns (and distortions in them) affect their relationships, and how to modify them to improve those relationships. Homework and assignments are a part of the treatment. A meta-analysis showed IPT to be effective in treating depression both alone and in combination with medication (Cuijpers et al., 2011).

- **Intensive Short-term Dynamic Psychotherapy** emerged in the 1960s out of the need to reach treatment resistant patients, in particular, those with histories of attachment trauma. In 40 or so treatment sessions, the goal is to access unconscious factors that perpetuate problems (such as depression or anxiety) that have been warded off because they are too frightening or painful. This is achieved through paying attention to physical and mental feelings in the context of a relationship with a therapist.
- **Eye Movement Desensitization and Reprocessing** is a treatment generally used for post-trauma reactivity and helps with mood dysregulation. It has classically been used to treat symptoms of acute trauma, but many practitioners are extending its use into wider applications including the treatment of anxiety and stress (Shapiro, 2018). While its mechanism is not clear, the treatment is thought to reprocess emotional/body memories (stored in the amygdala/ limbic system, being too traumatic for words), to semantic/narrative memories (now stored in the frontal lobe where they can be processed in a conscious, verbal manner), a process similar to that which occurs during REM sleep (Stickgold, 2002).

  REM or dream sleep serves an important role in helping us integrate our emotional experiences into narrative memory (Stickgold, 1999) and is thought to be associated with mood regulation (Perlis & Nielsen, 1993). When something painful happens, we often feel better once we talk about it. Initial exposure to affective stimuli is associated with encoding primarily in the amygdala and hippocampus, creating "emotional memories" (Dolcos et al., 2004; Kilpatrick & Cahill, 2003). Emotional memories are not conscious or verbal. It is hypothesized that during REM sleep, emotional memories are consolidated and integrated with pre-existing semantic memories in the cortex. where we experience things on a linguistic/ symbolic level (Walker & van der Helm, 2009). This transfer/ consolidation process detoxifies the emotional nature of the memory by disengaging it from its physiologic "panic" (limbic) response and making it available, instead, to verbal (cortical) reflection.

  People benefitting from EMDR report after treatment that they no longer feel physically gripped by inchoate feelings associated with trauma but can talk about the emotions associated with a traumatic experience in a calmer, rational, narrative manner. van der Kolk (2007) demonstrated in a study comparing EMDR with Prozac and placebo in the treatment of PTSD that EMDR was more effective in reducing symptoms of reactivity than Prozac or placebo.

- **Psychoanalytic Psychotherapy:** It is always refreshing to hear psychoanalysts talk about emotions. Psychiatrists often describe emotions in a somewhat disembodied manner whereas psycho-analysts speak about emotions in the context of relationships, which, after all, are where we live and feel and falter. We regulate our moods through our relationships (Hofer, 1984). Regardless of the temperament with which we were born, our relationships, throughout our lives, affect our emotions. Psychoanalytic psycho-therapy is a long-term therapy that helps people understand and regulate their emotions and better their existing relationships in the context of a new relationship.

Given the unique role that relationships play in the genesis of mood dysregulation, long-term psychotherapy has a crucial role in helping people learn to regulate their moods. Medication therapy should never be seen as an alternative to psychotherapy—the two are different modalities of treatment that go hand-in-hand. Many people seek medication because they believe it (passively!) brings quick results, but there are aspects of depression and anxiety that will always remain outside the reach of what medication could be expected to relieve but that long-term psychotherapy is uniquely suited to address.

Psychoanalysts are psychotherapists with special training in the interplay between the conscious and unconscious forces within one's mind and how these forces affect their relationships. Contemporary psychoanalysis encompasses both conscious cogni-tions (like short-term therapies) and dynamic, unconscious aspects of behavior. Contemporary psychoanalytic thinking is permeated by attachment theory, making psychoanalytic psychotherapy espe-cially suitable for people with mood dysregulation emerging in the context of relational trauma as the psychoanalytic setting fosters relational repair. Not all long-term psychotherapies last a lifetime. Many people successfully achieve their goals in psychotherapy in a year or so.

Attachments are not only important in infancy but remain so throughout our lives. Healing relationships can result in biological and physiological changes in the substrate of the brain that have lasting, beneficial effects on mood regulation. Having a new, positive relational experience with a therapist can buffer the physiological stress response and lay the foundation for the development of mood regulation (Stansbury & Gunnar, 1994), which is ultimately effected through neuronal restructuring.

Corrective emotional experiences can reverse the negative consequences of early adversity (Gunnar & Quevedo, 2007), in part because neuronal restructuring is an ongoing process that is dependent upon ongoing experience. The emotional impact of physiological changes resulting from relational trauma need not be permanent because the brain is plastic.

For those who are naysayers, imaging studies have shown functional brain changes in anatomic circuits associated with increased mood regulation in treatment cohorts after psychotherapy (Malejko et al., 2017; Schmitt et al., 2016). In a study of long-term psychotherapy, Buchheim and colleagues (2012) found fMRI evidence of changes in mood regulating circuits in the brain. In particular, they noted decreased reactivity in the anterior hippocampus/amygdala complex (a limbic structure that is overactive in mood dysregulation ("a quiet amygdala is a good amygdala") and increased activation in the mPFC (a part of the neocortex involved in behavioral inhibition) after 15 months of psychotherapy.

John Bowlby (1969, 1973) believed the organism's search for attachment was an evolutionarily driven biological program. The work of subsequent attachment theorists rests on this biologic premise. Arietta Slade (2008) studied the implications of attachment theory as applied to adult "mentalization based therapy" (Bateman & Fonagy, 2004) for those with mood dysregulation. She explained the way in which insecure attachments lead to mood dysregulation in terms of mentalization theory. When a parent fails to recognize and mirror the infant's feelings back to her, she may not develop a sense of her own self or a capacity to become conscious of her own emotional state, limiting her capacity for emotion regulation. Mentalization therapy aims to reverse this process by restoring a reliable attachment, which enables a person to begin to articulate how she feels. In the naming of feelings, she can begin to manipulate, regulate, and master them.

According to affect regulation theory, we learn to regulate our moods during infancy in relationships with our caregivers. Responsive caregivers model behavior that leads us to develop expectancies of others' responses to us. These patterned expectancies become imprinted in the right hemisphere of our brains as "procedural memories" or "scripts."

Right brain procedural memories are nonverbal and nonconscious. Based on these early experiences, we develop expectations of our caregivers that we carry unconsciously into adulthood as a repository of procedural memories that govern our reflexive responses to others in various situations. Have you ever caught yourself making an assumption about another person you later realized was totally wrong, but when you thought about it, realized it was based on your experience with a family member or someone familiar to you? Most likely your mistaken assumption arose from an unconscious procedural memory.

Soothing caregivers teach us to calm ourselves when agitated, shamed, or frightened, leaving us with procedural memories from infancy for adaptive affect regulation throughout life. Distressed or neglectful caregivers leave us feeling abandoned and in a state of upset. We then fail to develop procedural memories for how to calm or soothe ourselves. Instead, we are left with procedural memories of runaway anxiety, shame, or fear. This translates into mood dysregulation until corrective procedural memories can replace these problematic scripts.

Becoming aware of and changing procedural memories that lead to (maladaptive) "reflex behaviors" and mood dysregulation is the work of affect regulation psychotherapy; this treatment affords a relationship in which one can reflect and develop a verbal narrative (in the left brain), becoming aware of what one already "knows" on an otherwise visceral (right brain) level. It also affords an opportunity to relearn new patterns of interacting in a constructive relationship so that new procedural memories can be created, leading to improved mood regulation.

Many have said that achieving this process requires "right-brain-to-right-brain" communication between the individual and his therapist in which affect regulation is achieved by nonverbal (right brain) communication between both members of the dyad

(Dales & Jerry, 2008; Schore & Schore, 2014). It is difficult to capture the nonverbal dynamics of what transpires in the consultation room between a person and his analyst, so crucial to the therapeutic process is the unconscious right brain communication, however, Lachman and Beebe (1996a) illustrate this affect regulation process, then describe the therapeutic action of self and mutual regulation in the dyad of psychoanalytic psychotherapy in a live clinical setting (Lachmann & Beebe, 1996b).

Other types of psychoanalytic psychotherapy involve helping people identify inaccurate or distorted interpretations of others' behaviors that might lead them to overreact, precipitating a cycle of problematic behaviors in relationships. This process occurs even as the individual and her therapist are aware that she has been subjected to a traumatic past and that the very brain with which she makes judgments is the substrate compromised by that same traumatic past. Such situations can be complex since impulse control in addition to judgment is physiologically disturbed by the stress reaction associated with the relational trauma.

Unfortunately, early life adversity can leave people with a vulnerability to treatment resistant depression (Plakun, 2012; Stevenson et al., 2016) through mechanisms that may be outside of conscious awareness. Psychoanalysis, however, can help bring these vulnerabilities to light and improve treatment outcomes immeasurably. Adverse early life experiences may:

1. Predispose people to negative rather than positive effects of treatment because of a predisposition to expect harm rather than help from a caregiver (Mintz & Belknap, 2011). This is the "nocebo" effect
2. Undermine a person's ability to trust an authority figure like a doctor because of failures of past authority figures
3. Interfere with one's capacity to develop a sense of agency and overcome treatment resistance
4. Lead to communication disruption and distress so that messages are conveyed by behavior (impulsive/self-destructive) rather than verbally

Plakun (2012) reminds us of an important sub-group of people with treatment resistant depression who are held back for reasons about which they may not be aware but which are embedded in their traumatic histories. This is the predisposition to expect harm rather than help from something designed to help: the nocebo effect. Such predisposition leads to the fulfillment of negative expectations.

Encompassed within the nocebo effect is the difficulty of a person (often mistreated) to trust an authority figure such as a physician or therapist. The nocebo effect is not inconsequential to the discussion of treatment resistant depression, be it with medication or other types of therapy. People with negative or traumatic experiences with caregivers often come with expectations that they will not or cannot be helped (or that they are going to be harmed), which become self-fulfilling prophecies. The nocebo effect is particularly sad when it is clear that a person is a good candidate for medication. Most of the psychoanalytic literature on the nocebo effect refers to negative expectations toward treatment with medication (Alfonso, 2009; Mallo & Mintz, 2013; Stratigos, 2009; Tutter, 2006; Wing Li, 2010). Very little of it refers to negative expectation toward treatment with psychotherapy.

There are different schools of psychoanalysis including self-psychological, relational, interpersonal, intersubjective, ego-psychological, Kleinian, and Freudian to name a few. Many therapists, while they may have a primary orientation, are often eclectic, employing concepts from other approaches as they find them useful. More important than the orientation of the psychoanalyst is the chemistry between the individuals in the treatment dyad. For a therapeutic process to be effective, the person seeking help needs to have respect for and confidence in the analyst as well as a sense that the analyst is genuinely concerned about their wellbeing.

Psychoanalysts increasingly regard psychotherapy to be a somatic treatment in and of itself. Cozolino (2002) describes psychotherapy as a biological treatment as follows:

> With the advance of neuropsychology, psychotherapy is increasingly becoming recognized as a somatic intervention. Attachment and trauma studies show that the complex neural and regulatory networks criss-crossing the brain are environmentally programmed and experience-dependent. These are the same circuits that therapists attempt to influence in reshaping the brain in ways which lead to more positive adaptation later in life. The idea that psychotherapy is a kind of reparenting may be more than a metaphor; it may be precisely what we are attempting to

accomplish at the level of the epigenome ... and places psychotherapy at the heart of biological interventions.

It is well established that when combined with medication, psychotherapy for the treatment of depression is more effective than medication alone (Arnow &Constantino, 2003; Cuijpers et al., 2009; de Jonghe et al., 2001; Manber et al., 2008; Pampallona et al., 2004). Fonagy and colleagues (2015) conducted an important study at the Tavistock Clinic demonstrating that individuals with treatment resistant major depression receiving psychoanalytic psychotherapy had a 40% better chance of partial remission during every 6 months over the course of 4 years than those receiving other treatments. People often come to us seeking a quick fix with medication but it is misguided to expect medication to be the sole solution to mood dysregulation.

8.4 MEDICATION TREATMENTS

Perhaps because the majority of the literature on emotion regulation has been written by psychologists, there is virtually no discussion of the treatment of emotion dysregulation with medication. However, this need not be so. Medication is an important and complex piece of its treatment picture. The use of medication in the transdiagnostic model of mood dysregulation is focused primarily on the particular symptom(s) of concern.

The first question to be looked at in each situation is whether or not medication is indicated. The expected limitations of what medication can do and the functional impairment caused by the symptoms to be treated form the parentheses around this decision. How urgent is it to manage the symptoms at hand? Paranoia and psychosis per se do not require medication as these symptoms can be benign. Psychotic symptoms need to be assessed individually for their level of danger. Is the individual I am seeing acutely suicidal? Is this person actively self-destructive? Is an important relationship in her life being threatened by her behavior? Is his job in jeopardy? Is her academic career at risk? Is the psychotherapy at an impasse because of rumination, depression, or anxious impulsivity? When these questions are answered in the affirmative, it may be the appropriate time for a medication evaluation.

The second question, if the situation is not acute, is whether or not the person has tried other treatment modalities, and if so, what has been the outcome. Medication as a first treatment is often reserved for situations in which reflective awareness is very limited or fails to control or modulate behavior sufficiently. Depending on whether and how much mood regulation is achieved through psychotherapy, the use of medication might be considered. When symptom resolution at the outset is more urgent, concurrent initiation of medication and psychotherapy can be the most appropriate approach.

When therapists are considering when to refer someone to a prescriber for medication, several things should be considered. What will be the impact of any changes medication might have on the psychotherapy? Will the prescriber's presence in the clinical picture enhance or detract from the therapeutic process? Therapists should **never** refer someone for a medication consultation having given that person the therapist's own diagnostic opinion or medication recommendations, in particular, a diagnostic assessment of ADHD or medication recommendations for psychostimulants or benzodiazepines. These pre-emptive communications undermine a consultative endeavor and have the potential to create uncomfortable or antagonistic situations if the prescriber has a different viewpoint from the therapist, whom the referred individual is more likely to trust. If the therapist has diagnostic or treatment impressions regarding the person being referred for a consultation, it is crucial to communicate these directly to the prescriber before the consultation appointment. Never through the person being referred.

Though medication management of mood dysregulation typically focuses on symptoms rather than diagnosis, it is important to recognize the difference between dysphoria and depression so as not to miss a mood dysregulation tethered to the bipolar spectrum. This distinction is crucial to determine the proper treatment. Target symptoms may be associated with treatment resistant depression, suicidal ideation, self-harming behavior, aggression, excess irritability, mood reactivity, and, at times, self-medication with substances. Symptoms of mood dysregulation are a feature of depressive disorder with mixed features, borderline personality disorder, ADHD, DMDD, and other diagnoses, all of which must be assessed in the context of associated symptoms without undue focus on diagnosis.

People with mood dysregulation are often impatient, irritable, and impulsive, and look to medication to work miraculously and efficiently to regulate their moods. Not only will they bypass potentially more effective treatment modalities but they may also have idealistic expectations of medication. It is not unusual to try several medications before finding an effective and tolerable agent, and medication management of mood dysregulation often requires use of more than one drug. Some medications have side effects, and no medication is a miracle cure. Having a realistic view of what to expect from medication increases the chance of its successful use.

On the other hand, some people are more reluctant. Some will not take medication until they have trawled the internet for every possible side effect. Any side effect one may fear can be found for the looking. If one worries about it, one will find it. Those who proceed with hidden fear will find their fear realized, even if they are not aware of feeling anxious about taking medication. People who come for help armed with a list of potential side effects have a greater likelihood of treatment failure than those who do not.

Anxiety has a strange way of hijacking the body. When anxious people take medication, they often experience physical symptoms of anxiety of which they are unaware. Symptoms of anxiety can manifest in just about any way imaginable but those associated with taking a new medication are typically misinterpreted to be "side effects." Anxious people starting medication on as low as one-hundredth of a usual starting dose, a dose that could not physiologically produce true side effects, have reported "side effects." These are not true side effects of medications but are symptoms of anxiety causing a stress reaction in the body that results in physical symptoms. While typical symptoms may be headache, nausea, diarrhea, dizziness, or sedation, virtually any physical symptom has been mistakenly reported as a "side effect" of medication. These symptoms are not imaginary. They are real. But they are not medication side effects. They are symptoms of anxiety. They are physiologic responses to stress, the body's manifestation of non-conscious anxiety. Stress, even when one is not aware of it, can cause HPA axis overactivity and a cascade of physiologic responses that can mimic virtually any known true medication side effect. The distinction between symptoms (of anxiety) and side effects is crucial.

The phenomenon of feeling worse from unexpected "side effects" of a medication expected to have beneficial effects is often due to the "nocebo response," an unconscious expectation of being harmed by someone or something otherwise expected to be helpful. The nocebo effect is typically related to anxiety.

Mitsikostas and colleagues (2014), in a meta-analysis studying the nocebo effect in clinical trials of depression, found that "nearly half of the people had 'side effects' to placebo," and 4.5% of people on placebo dropped out of studies because of "side effect" intolerance. What this means is that in a series of placebo-controlled drug trials, nearly half of the people who were taking placebo or inert pills, which should not have been associated with any side effects, experienced "side effects." This is dramatic data in support of the notion that *perceived "side effects" are symptoms of anxiety – nearly 50% of the time!*

Symptoms are associated with a medical condition. Side effects are associated with a medication. The internet is a precarious source of information and should be used with caution by a discerning person. It does not reflect what physicians know to be typical side effects statistically associated with the various pharmaceutical agents. The importance of consulting a medical professional as opposed to the internet is, in part, to obtain the opinion of a person uncontaminated by the noise of false "side effect" reports, a professional who has seen hundreds of people on various medications, and who understands a realistic range of what is to be expected—and what not—with a given medication.

It is particularly difficult for those of us with extensive clinical experience and confidence in our recommendations to watch people struggle with the nocebo effect, which is common in the treatment of depression (Dodd et al., 2015) and anxiety. Having said this, any medication can behave in an idiosyncratic manner in any individual. However, these reactions are rare and should never prevent a person from starting a recommended medication.

8.4.1 ANXIETY

People routinely present with anxiety, assuming they should be treated with "something like Klonopin." Yet the treatment of anxiety is anything but straightforward. Is the anxiety intermittent or chronic? Is it associated with physical symptoms? Is there an associated mood disorder? Is the mood disorder depression, dysphoria, or both? Panic attacks? Most important, does it occur in the context of irritability and/ or agitation or activation, suggestive of the dysphoric hypomania of a mood dysregulation? Answers to these questions will guide treatment. In general, chronic anxiety should not be treated with benzodiazepines.

Benzodiazepines can be used intermittently for anxiety in the context of various mood disorders. The agitation of dysphoric hypomania should always be considered in anyone presenting with "anxiety."

8.4.1.1 BENZODIAZEPINES

Benzodiazepine medications are the most popular medications used to treat anxiety, but they are grossly overprescribed. They are never recommended as a first-line medication for chronic anxiety because of their potential for tolerance, dependence, and withdrawal. As a result, nonhabit-forming drugs are the treatment of choice when medication is indicated for chronic anxiety.

The mechanism of action of benzodiazepines appears to be the facilitation of (inhibitory) GABA neurotransmission in the brain (Costa et al., 1975; Haefely, 1980). Benzodiazepines, like alcohol, belong to the sedative-hypnotic class of substances. As such, regular use of these agents leads to tolerance (the body's adaptation such that the same dose has diminishing effects over time) and withdrawal (the body's physiologic reactivity when the benzodiazepine receptors are no longer being stimulated by use of the medication). Symptoms of benzodiazepine withdrawal are elevated heart rate, increased blood pressure, shaking, shortness of breath, tremors, sweating, anxiety, and panic attacks. Benzodiazepine withdrawal symptoms are indistinguishable from anxiety and panic attacks (Higgit et al., 1990; Mackinnon, 1982; Petursson, 1994). This makes benzodiazepines particularly unsuitable to treat chronic anxiety as ongoing benzodiazepine use becomes necessary to prevent withdrawal symptoms rather than to treat the original/primary/underlying anxiety. Daily benzodiazepine use for chronic anxiety creates and perpetuates medication dependence.

Benzodiazepine use produces a vicious cycle once taking the medication for extended periods. It can be difficult to help a person break this cycle for several reasons. First, reduction or cessation of the drug causes withdrawal, which is (erroneously) interpreted as "anxiety," justifying ongoing use. Many people argue that since a doctor has prescribed benzodiazepines, the medication must be good for them (while refusing other safer drugs prescribed for anxiety). Furthermore, many people do not understand the difference between medication dependence and drug addiction (Miller & Gold, 1991). Medication dependence is a physiological phenomenon that can occur to anyone taking a pharmacologic agent that can cause tolerance. Medication addiction is a behavioral phenomenon that occurs in people for whom physiological dependence has created psychological dependence leading to maladaptive behavior. Some of the people we see taking benzodiazepines are simply dependent on them but others are addicted. However, both groups are scared to discontinue taking them because of the "anxiety" they anticipate feeling while discontinuing the medication. Some of this "anxiety" is psychological and can be treated in other ways. Potential physiological withdrawal effects that can mimic anxiety can be avoided in several ways to enable comfortable and successful withdrawal from benzodiazepines. People with simple physiological dependence on benzodiazepines can generally discontinue them more easily than people with psychological addiction as the former group can easily find an incentive to discontinue the medication while the latter cannot do so. For these reasons, recommendations for first-line medication treatment of chronic anxiety uncomplicated by mood dysregulation is not benzodiazepine medication (Bandelow et al., 2017; Reinhold et al., 2011).

8.4.1.2   PROPRANOLOL

Propranolol (brand Inderal) is not marketed as an antianxiety medication. It is approved by the FDA primarily as an antihypertensive medication and is also used to treat migraine headaches and heart arrhythmia. However, it is widely useful in the treatment of short-term anxiety and panic attacks (Easton & Sherman, 1976; Laverdure & Boulenger, 1991; Vasantkumar et al., 1977).

Propranolol is one of a class of β-blocker drugs. Epinephrine and norepinephrine bind to β receptors in the heart and lungs and in the blood vessels to mount a fight or flight response, increasing heart rate and respiratory rate, and mobilizing the system for action. β-blocker medications prevent this response.

People will often report that anxiety begins with physiological symptoms that develop into psychological symptoms. This is because the nonconscious sympathetic fight or flight response precedes any conscious sense of anxiety. Once conscious, anxiety feeds back signals to the body, which then amplifies sympathetic overactivity in a runaway positive feedback loop. Anxiety triggers palpitations, shaking, and shortness of breath which, in turn, trigger more anxiety, triggering more palpitations, shaking and shortness of breath, and so on. For this reason, propranolol is a well-known trade secret of performers with stage fright wishing to block the physical symptoms of anxiety (Brantigan et al., 1982; Packer & Packer, 2005). People whose symptoms of anxiety are largely physiological (palpitations, tremors, hyperventilation, etc.) benefit the most from the use of propranolol to treat their anxiety (Kelly, 1980).

Propranolol, interestingly, has other potential psychiatric applications that may not be related to its action as a β-blocker. It has been useful in treating anger outbursts in several settings. Yudofsky and colleagues (1981) found propranolol effective in treating people with chronic brain syndromes manifesting rage episodes and violent behavior. Silver and colleagues (1999) used propranolol successfully to treat chronically hospitalized aggressive people. Williams and colleagues (1982) found propranolol useful in treating outbursts in children with organic brain dysfunction. Perhaps more relevant to our discussion, propranolol has well known antiseizure properties (Fischer et al., 1985 Fischer, 2002; Jaeger et al., 1979).

While the exact mechanism of action is not known, this puts propranolol into the antiepileptic (AED) class of drugs that, as we will see below, is one of the most important classes of medications we use to treat people with mood dysregulation. It might be a researchable question to determine whether or not propranolol has applications in treating any of the symptoms associated with mood dysregulation such as irritability, agitation, or impulsivity. And if propranolol has significant AED properties, one has to wonder whether its usefulness in treating those with somatic symptoms of "anxiety" is a reflection of activated dysphoria (i.e., a mood dysregulation) mislabeled as "anxiety."

8.4.2 DEPRESSION

Because we know that antidepressant medication has the potential of effecting a "switch" in individuals with bipolar spectrum disorders (Altschuler et al., 1995; Chun & Dunner, 2004; Goldberg & Truman, 2003; Henry et al., 2001; Kupfer et al., 1988), one must always be on guard for a switch in energy states when using antidepressant medication in people with mood dysregulation because of its association with the bipolar spectrum. Indeed, this is the basis on which many people with so-called pseudo-unipolar depression were thought to have bipolar disorder (Akiskal, 1996a; Gil et al., 2020), i.e., those with bipolar III disorder on the soft bipolar spectrum, which we have described as having mood dysregulation.

Before embarking on treatment for depression, it is imperative to assess whether or not the symptom at hand is the depressed mood of the depressive state (sad, hopeless, pessimistic, down, blue) or the dysphoric mood of the activated hypomanic/manic state. I never assume a person is in a depressive state simply because they say they are depressed. Sadness is the hallmark mood of the depressive state. Anxiety can accompany sadness without necessarily indicating dysphoric hypomania or a mixed state. If I see no clear evidence of dysphoria, activation, or even some degree of anxiety in a "depressed" individual presenting with sadness, even in the face of a family history positive for bipolar disorder, I tend to assume the individual has "simple" depression and to proceed with antidepressant treatment, albeit cautiously. One never knows when pseudo-unipolar depression will emerge.

8.4.2.1   MONOAMINE REUPTAKE INHIBITORS

The antituberculosis drug iproniazid, a monoamine oxidase inhibitor (MAOI), was found serendipitously in the 1950s to have antidepressant effects. Shortly thereafter, an antihistamine derivative imipramine was also found to have efficacy in treating depression. Both medications increased levels of norepinephrine in the brain, so it was hypothesized that depression was the result of norepinephrine deficiency (Schildkraut, 1965). Around that same time, Coppen and colleagues (1963) found that the amino acid tryptophan, a precursor of serotonin, potentiated the antidepressant effects of MAOIs in people with depression. Serotonin deficiency was then also thought to be associated with depression (Coppen, 1967). The discovery that norepinephrine and serotonin, both monoamine neurotransmitters, were deficient in the brains of depressed people, led to the monoamine hypothesis of depression, which has been regnant until quite recently. Most of the first-line agents we use today to treat depression still have their primary impact on the monoamine system. To call them "antidepressants" is something of a misnomer because most of them are also first-line treatment agents for symptoms of anxiety, and many medications in this class have several other applications.

Monoamine based medications are typically the first-line agents used in treating people with depressed moods. As mood elevators, they have stimulant (albeit mild) properties. Thus, they must be used carefully in people with mood dysregulation whose affective temperaments make them vulnerable to their stimulant effects, which can cause switches from the depressive to the hypomanic/manic state in people with bipolar disorder (Ghaemi et al., 2003).

Frequently a person comes to treatment already on a monoamine medication. It is important to assess what effect that medication has had on their symptoms, primarily to establish that it has not precipitated or worsened irritable dysphoria. One must assure that the monoamine 1) has not worsened any symptoms and 2) has had some beneficial effect before maintaining the individual on such medication. A history of the progression of symptoms can help in assessing whether or not the medication has helped or worsened the clinical picture. If there is a suspicion that the medication may be exacerbating irritability, a medication washout is indicated before proceeding with further treatment.

1. **Tricyclic antidepressants (TCAs):** The TCAs were introduced in the 1950s as first-generation antidepressants and are primarily norepinephrine reuptake inhibitors. TCAs were the primary medications used to treat depression until Prozac reached the market in 1986. They are still used to treat depression and anxiety, especially when insomnia is a symptom feature, as they tend to be sedating. Common TCAs are imipramine (Tofranil), amitriptyline (Elavil), and desipramine (Norpramin).

2. **Selective Serotonin Reuptake Inhibitors (SSRIs):** Prozac was the first SSRI to be introduced in 1986. SSRIs eclipsed the TCAs largely because the TCAs caused so much sedation, dry mouth, and weight gain. SSRIs are also first line "broad-spectrum" medications to treat anxiety (Tsapakis & Travis, 2002), depression, and other conditions. They are known for their sexual side effects and can also cause sleep disturbances. Common SSRIs are fluoxetine (Prozac), sertraline (Zoloft), paroxetine (Paxil), citalopram (Celexa), and escitalopram (Lexapro).

3. **Serotonin Norepinephrine Reuptake Inhibitors (SNRIs):** The first SNRI, venlafaxine (Effexor), was introduced in 1993. Both SNRIs and SSRIs have fewer side effects than TCAs, but both are still plagued with sexual side effects. Like TCAs and SSRIs, SNRIs are also first-line medications for the treatment of anxiety. SNRIs can be particularly difficult to discontinue because of their

withdrawal effects (nausea and dizziness in particular) so they are not always the first medications of choice. Some common SNRIs are venlafaxine (Effexor), duloxetine (Cymbalta), desvenlafaxine (Pristiq), milnacipran (Savella), and levomilnacipran (Fetzima).

4. **Norepinephrine Dopamine Reuptake Inhibitors (NDRIs):** Bupropion (Wellbutrin) is the only formal antidepressant in this class and has been on the market since 1985. Methylphenidate and its derivatives (e.g., Ritalin, Concerta and Focalin) are stimulant medications used to treat ADHD and are also NDRIs. Possibly, for this reason, bupropion is more likely than any of the other monoamine based medications to cause irritable dysphoria or hypomanic/manic switching in individuals predisposed to bipolar mood disorders (Akiskal's bipolar III or mood dysregulation). Bupropion is often a preferred medication for use in treating depression because it does not have sexual side effects.

5. **Norepinephrine reuptake inhibitors (NRIs):** The NRIs have been explored for various uses including narcolepsy and neuropathic pain. Only one of them is now used to treat depression, reboxetine (Erdonax). Another NRI, atomoxetine (Strattera) is used to treat ADHD.

6. **Monoamine oxidase inhibitors:** The MAOIs are first-generation antidepressants, also introduced in the 1950s. They have been used successfully for some forms of treatment resistant depression. When the presenting symptoms are depression (sadness) with mood reactivity and rejection sensitivity, a particular mood dysregulation of atypical depression may be present and MAOIs might be a good first-line treatment. Care must be taken to avoid precipitation of a hypomanic (euphoric or dysphoric) episode as monoamine-based medication use in individuals with mood dysregulation related to the bipolar spectrum can be destabilizing (Ghaemi et al., 2003). At the same time, MAOIs are the one class of antidepressant which may be less likely to cause mood instability (Fiedorowicz & Swartz, 2004). MAOIs may be especially suited to treating bipolar spectrum depression (Agosti & Stewart, 2007; Mallinger et al., 2009), In fact, Quitkin and colleagues (1993) believe MAOIs to be uniquely effective in treating atypical depression, which they see as a bipolar spectrum disorder. The first two MAOIs introduced, phenelzine (Nardil) and tranylcypromine (Parnate) are still the

most frequently used medications in this class, even though strict dietary precautions must be observed to avoid ingestion of too much tyramine. Excess amounts of this amine in the blood can lead to the release of norepinephrine, causing an acute spike in blood pressure (a hypertensive crisis), which can be fatal. To avoid this danger, a new class of MAOIs called "reversible MAOIs" was developed which did not have dietary requirements and were thus safer than the MAOIs. They were less effective, however, and never gained the traction enjoyed by the original MAOIs. MAOIs fell out of favor with the advent of Prozac in 1986, the first in a plethora of safer antidepressant medications. Unfortunately, dietary restrictions and the risk of hypertensive crisis associated with MAOIs have caused unjustified avoidance of this class of medication, which is often effective where many other first-line antidepressants are not.

8.4.2.2   GLUTAMATERGIC AGENTS

The high incidence of treatment resistant depression (Nierenberg et al., 2006; Rush et al., 2006) highlighted the inadequacies of the monoamine hypothesis (Goldberg et al., 2014; Hirschfeld, 2000; Sanacora et al., 2012) and led to the exploration of other mechanisms that might underlie depression and other mood disorders. Research from many different fields including gene expression, neuronal circuitry, and intracellular signaling mechanisms over the past several decades turned the focus away from the monoamine hypothesis and led to the "neuroplasticity hypothesis" (Pittenger & Duman, 2008; Racagni & Popoli, 2008).

Central to the neuroplasticity hypothesis is the notion that impaired neuronal resilience, reduced neuroplasticity, and neurodegenerative changes are all mediated by glutamate in response to a stress-induced increase in cortisol levels (Zarate, 2008). Well-regulated glutamate levels are increasingly recognized to be necessary for healthy neural growth and adaptation. Interest has shifted to the study of glutamate as research has shown glutamate transmission to be abnormal in several limbic and cortical areas in the brains of individuals with mood disorders (Sanacora et al., 2012). In the brains of people with unipolar depression, levels of glutamate are consistently found to be reduced (Auer et al., 2000; Block et al., 2009; Yuskel & Ongur, 2010), whereas in the brains of people with bipolar disorder, levels of glutamate are found to be elevated during manic, mixed and depressed states (Dager et al., 2004). Furthermore, glutamate receptor antagonists such as lithium, lamotrigine, and ketamine are found to have antidepressant action (Dutta et al., 2015; Trullas & Skolnick, 1990; Zarate et al., 2006).

Over 50% of the neurons in the brain use glutamate as their primary neurotransmitter (McGreer et al., 1987), making it central in mediating cognitive and affective processes. The medications we use today to treat mood disorders have many complex mechanisms elaborated in the neuroplasticity hypothesis. While in many ways the glutamate hypothesis has come to eclipse the monoamine hypothesis, it is still in its infancy. Below are just three glutamate-based medications used to treat depression. These medications all act to lower levels of the excitatory neurotransmitter glutamate.

1. **Riluzole** (Rilutek) is an inhibitor of glutamate release. It has neuroprotective properties, making it useful in neurodegenerative disorders. In 1995, it was approved by the US Food and Drug Administration (FDA) for the treatment of Amyotrophic Lateral Sclerosis (ALS or Lou Gehrig's Disease). It was originally developed in France as an antiepileptic agent. As we will discuss, antiepileptic drugs (AEDs) have found special use as mood stabilizers. Riluzole is useful for the treatment of major depressive disorder (Sanacora et al., 2007), and bipolar depression (Zarate et al., 2005). There is no evidence that Riluzole can induce hypomania/mania.

2. **Memantine** (Namenda) is an NMDA (glutamate) receptor blocker with neuroprotective properties. It is marketed to treat Alzheimer's disease but it has been used off-label for its putative antidepressant effects. Some studies show its efficacy as an antidepressant (Amidfar et al., 2018), but a meta-analysis of its use as a sole agent in treating depression showed no benefit over placebo (Kishi et al., 2017). Ferguson and Shingleton (2007), however, argue that the use of high doses of memantine can achieve an adequate antidepressant response. It was found to have a synergistic effect when used in combination with imipramine and fluoxetine (Zarate et al., 2003). There is some evidence that memantine reduces mania-like symptoms in animal models (Gao et al., 2011), suggesting it might not cause destabilization of mood in those with mood dysregulation.

3. **Ketamine** (Ketalar), among other things, is an NMDA (glutamate) receptor blocker. It was first approved for medicinal use in the USA in 1970 as an anesthetic. Its dissociative and hallucinogenic effects led to its use as a recreational drug ("Special K"). It is now approved by the FDA for use in treatment resistant depression.

Like lithium, ketamine blocks the NMDA glutamate receptor. Though most of the studies on ketamine have focused on unipolar depression, the successful use of ketamine for the treatment of bipolar depression has been demonstrated as well (DiazGranados et al., 2010a; Lara et al., 2013; Zarate et al., 2012). Ketamine has also been found to treat the anxious, irritable, and agitated features of people with bipolar disorder (McIntyre et al., 2020).

Intravenous ketamine infusions in people with major depression resolved symptoms of depression in an average of 40 minutes (Carlson et al., 2013; Ibrahim et al., 2011; Machado-Viera et al., 2009; Phelps et al., 2009), but the median relapse time after treatment is within 2 weeks (DiazGranados et al., 2010b; Matthew et al., 2010; Thakura et al., 2012). Repeated/ maintenance infusions have resulted in rapid relapse unless used in conjunction with adjunctive medication to extend ketamine treatment response (Diamond et al., 2014; Murrough et al., 2013). Ketamine has been used intranasally and orally to treat depression. Sublingual ketamine has also been used with some success to treat depression (Lara et al., 2013).

Lithium and lamotrigine have both been shown to have efficacy in resolving symptoms of TRD when used as adjunctive agents with ketamine (Barbee & Jamhour, 2002; Gabriel, 2006; Ivkovik et al., 2009; Schindler & Anghelescu, 2007; Thomas et al., 2010), but it does not appear that either of them has been studied as a sole agent for TRD as has ketamine.

Lithium has been found to potentiate the antidepressant effects of ketamine (Chiu et al., 2015).

Thirty percent of depressed people treated with IV ketamine do not respond to the treatment (Scheuing et al., 2015). Side effects of concern associated with ketamine related to its dissociative and hallucinogenic effects are schizophrenia-like behaviors (Krystal et al., 2003). Ketamine has been associated with oxidative stress in the body (Zou et al., 2007), which causes inflammatory damage to organs. In addition, there have been reports of ketamine-induced mania (Allen et al., 2019; Banwari et al., 2015; Lu et al., 2016), suggesting caution in its use for those with mood dysregulation and bipolar spectrum disorders.

The "S" enantiomer of ketamine, esketamine (Spravato), is marketed in the USA for treatment resistant depression in nasal spray form. In contrast to IV ketamine, esketamine is self-administered and must be used in conjunction with oral antidepressant medication. Sublingual ketamine has also been used with some success to treat depression (Lara et al., 2013) in research studies but is not authorized for general use at present. Oral ketamine is available but antidepressant effects are slower than those with IV ketamine (Rosenblat et al., 2019).

8.4.2.3 OTHER AGENTS

In addition to treating depression with monoamine reuptake inhibitors and glutamatergic agents, there are other scattered agents used such as agomelatine (Valdoxan), a melatonin agonist, opipramol (Insidon), a sigma agonist, mirtazapine (Remeron), an α-blocker, and mianserin (Tolvon), a serotonin and α-blocker.

8.4.3 MOOD STABILIZERS

Mood disorders on the bipolar spectrum (and on the soft bipolar spectrum, such as pseudo-unipolar depression) typically respond to mood stabilizers (Akiskal & Perugi, 2002). It would stand to reason, then, that mood disorders tethered to the bipolar spectrum neurophysiologically would respond to this class of medication as well, which is what we see clinically. People with mood dysregulation appear to respond preferentially to mood stabilizers over antidepressants (see Appendix B).

1. **Lithium:** When I say "lithium," people hear "Thorazine." The stigma attached to this substance is matched only by the ignorance surrounding it. Lithium is a resoundingly successful medication that labors under the stigma associated with bipolar disorder, for which reason it is, unfortunately, underutilized for the treatment of mood dysregulation. However, the response of people with mood dysregulation once treated with lithium is virtually a biomarker, locating mood dysregulation on its well-founded position adjacent to but beyond the bipolar spectrum (Malhi et al., 2013).

Lithium is the prototypic mood stabilizer and the only medication in its pharmacologic class. It is a natural element (No. 3 on the periodic table) so it is not available to Big Pharma for marketing. This adds to its lack of popularity; to a large degree, lithium has become eclipsed by the "new and improved" agents that pharmaceutical companies peddle to capture the imaginations of psychiatrists. It remains a mainstay treatment for bipolar disorder but there is little in the literature about the use of lithium for other purposes.

Lithium promotes the growth of new nerve cells (Chen et al., 2000; Fiorentini et al., 2010) in the hippocampus in particular, and has been shown to increase the volume of gray matter in the brain (Monkul et al., 2007; Moore et al., 2000; Phillips et al., 2008). Lithium increases the volume of brain structures involved in emotion regulation (Bauer et al., 2003; Moore et al., 2000; Sassi et al., 2002). Its neuroprotective effects are well documented (Manji et al., 1999; Rowe & Chuang, 2004). The incidence of dementia was found to be decreased in people with bipolar disorder taking lithium (Gerhard et al., 2015; Kessing et al., 2010; Terao et al., 2006); since then, interest has been developing in using low-dose lithium as prophylaxis for dementia (Chiu & Chuang, 2011; Forlenza et al., 2014; Zhong & Lee, 2007). Few of our medicines both protect and regenerate nerve cells as lithium does.

Lithium's mechanism of action is very complex and not altogether understood but one major aspect of it appears to be a reduction in levels of the excitatory neurotransmitter glutamate in the brain (Jope, 1999; Kendell et al., 2005; Shaldubina et al., 2001) by blocking the NMDA glutamate receptor (Hashimoto et al., 2003) and by increasing levels of the inhibitory neurotransmitter GABA (Malhi et al., 2013). Prolonged exposure to glutamate can be neurotoxic to brain tissue just as prolonged exposure to cortisol, so lithium's disruption of glutamatergic activity is part of its therapeutic action.

Lithium is singularly the most effective and rapidly acting of all mood stabilizing medications (Davis et al., 1999; Licht, 1998; Goodwin & Zis, 1979; Coryell, 2009). It is not simply known for its antimanic properties but for its antidepressant properties as well (Bauer et al., 2000; Keck & McElroy, 2003; Neubauer &

Bermingham, 1976). Lithium's antidepressant effects are thought to relate to its NMDA (glutamate) antagonist properties (Mohseni et al., 2017; Souza & Goodwin, 1991). Lithium is also one of two agents (ketamine being the other) known to have antisuicidal efficacy (Rybakowski, 2020; Hafeman et al., 2019).

Lithium is still regarded to be the first-line treatment for people with bipolar disorder (Bauer & Mitchener, 2004). One of the reasons people stigmatize lithium is because of its association with use in people with bipolar disorder who are overmedicated, particularly those who are oversedated. Doses of lithium used in people with mood dysregulation are quite different from those used in people with bipolar disorder. People with mood dysregulation often require one-sixth the dose of people with bipolar I disorder. At these doses, there are few to no side effects (About-Saleh & Coppen, 1989) and little risk of long-term complications. In contrast to many other psychotropic medications, lithium has no sexual side effects.

Lithium works within 2–3 days in contrast to 2–3 weeks for many other psychoactive medications, and lithium can be stopped immediately without withdrawal symptoms. Routine blood level monitoring is generally not necessary when using low dose lithium for the treatment of mood dysregulation. When low dose lithium proves inadequate for full symptom control, it can readily be used in combination with other medication to minimize potential side effects associated with higher doses of either agent. In low doses used for the types of mood dysregulation described here, lithium is the safest and most side effect free of all mood stabilizing medications.

Lithium has been used successfully to treat people with cyclothymia (Akiskal, 1995; Baldessarini et al., 2011; Bowden, 2005; Katzow et al., 2003). Treating a group of people diagnosed with cyclothymia and subsyndromal mood swings, (mood dysregulation?) Shou (1968) reports:

...the stabilization brought about by continuous lithium administration very often led to a such complete change in the patients' lives that their children suddenly realized that they had never known their parents as entirely normal before, and the parents almost monotonously declared that for the first time in many years they felt they were once again themselves.

Interestingly, in its typical application for people with bipolar disorder who frequently experience euphoric hypomania/mania, lithium is not the most welcome medication because it tempers the "high," whereas, in people with dysphoric hypomania, lithium is a most welcome medication for the same reason: it attenuates the unwanted feelings of irritability, agitation, and anxiety.

Nearly everyone I have treated with lithium for mood dysregulation, including those who were most reluctant to try it because of the associated stigma, has been impressed with the mood regulating capacity of the medication. The primary obstacle to the more widespread use of lithium is the unfortunate stigma associated with it. Once people become more educated about the true nature, action, and therapeutic effects of the medication there is bound to be a revival of lithium.

2. **Antiepileptic drugs:** Antiepileptic drugs (AEDs) are the primary class of mood stabilizers used by psychiatrists. They were first used by neurologists for the treatment of seizure disorders. AEDs do not all work in the same way, though the primary mechanism of action for most of them appears to be reduction in levels of the excitatory neurotransmitter glutamate and elevation in levels of the inhibitory neurotransmitter GABA (Davies, 1995; White et al., 2007), resulting in reduced neuronal excitability in the brain (Kwan, 2001). Other complex actions such as methylation of DNA are mechanisms of action for AEDs (Mitchell et al., 2005; Phiel et al., 2001).

The three AEDs approved for the treatment of bipolar disorder are divalproex (Depakote), carbamazepine (Tegretol), and lamotrigine (Lamictal). All are known as effective medications for acute and maintenance treatment of bipolar disorder (DeLeon, 2001).

Among these, lamotrigine uniquely has antidepressant effects (Bowden et al., 1999; Calabrese et al., 1999), which are thought to be related both to monoamine and glutamatergic mechanisms. (1) lamotrigine behaves as a serotonin and norepinephrine reuptake inhibitor (Southam et al., 1998), and (2) lamotrigine blocks the release of the excitatory neurotransmitter glutamate (Ostadhadi et al., 2016). People prone to stimulant switching may be vulnerable to increased irritability, agitation, and problematic activation on lamotrigine because of its antidepressant activity. Confounding this complication, the side effect of akathisia, an agitated restlessness, can also be associated with lamotrigine. Having said this, lamotrigine is generally quite well tolerated and is often used as a first-line mood stabilizer because of its efficacy and antidepressant activity.

Another AED, topiramate (Topamax), while not approved for the treatment of bipolar disorder, has been found effective in mood regulation, particularly when co-occurring with obesity or eating disorders as it has appetite suppressant properties (Chengappa et al., 2002; Guille & Sachs, 2002; McElroy et al., 2000). However, because it can induce depression in some individuals, it must be used with care (Klufas & Thompson, 2002; Phabphal & Udomratn, 2009). Other AEDs used to treat mood dysregulation are gabapentin (Neurontin), oxcarbazepine (Trileptal), tiagabine (Gabitril), and zonisamide (Zonegran). Like lithium, these medications can be used alone or in combination with medications from other classes.

3. **Neuroleptics:** Neuroleptic or antipsychotic medications for mood dysregulation are often misunderstood because of their association with schizophrenia. Unfortunately, though most medications in this class are approved and routinely used for bipolar disorder, they may still be referred to as antipsychotic medications because of their indication for schizophrenia.

Neuroleptics are subdivided into two basic categories. The older category, first generation (or "typical") neuroleptics, includes medications such as thiothixene (Thorazine), thioridazine (Mellaril), and haloperidol (Haldol). These medications work primarily by blocking dopamine receptors. Because they caused troublesome side effects including tardive dyskinesia, a potentially irreversible movement disorder, they were replaced by second generation (or "atypical") neuroleptics. Medications in the second category of neuroleptics work primarily by reducing dopamine transmission, blocking serotonin 2A receptors (Yatnam et al., 2005; Brugue & Vieta, 2007), and stimulating serotonin 1A receptors (Meltzer & Massey, 2011). Some of the medications in this class are olanzapine (Zyprexa), risperidone (Risperdal), ziprasidone (Geodon), aripiprazole (Abilify), quetiapine (Seroquel), and asenapine (Saphris). Second-generation neuroleptics have mood-stabilizing, antianxiety, and (mild) antidepressant effects. Nevertheless, they are used with caution because atypical neuroleptics are also associated with tardive dyskinesia (Woerner et al., 2011).

Phelps (2005) describes an interesting case in which an individual presented with "agitated dysphoria" following a dose increase of sertraline after seven years of successful treatment with the medication for depression. Her symptoms remitted with the use of quetiapine, a second generation neuroleptic which functioned, in this case, as a mood stabilizer. I regard this to be an illustration of pseudo-unipolar depression, "agitated dysphoria" indicating a state of activation precipitated by sertraline as seen in bipolar III disorder, a mood dysregulation.

Newer atypical neuroleptics are being developed as it becomes clear that these agents can be useful as mood stabilizers. Two recent additions, marketed specifically for bipolar depression, are lurasidone (Latuda) and lumateperone (Caplyta). Both are dopamine/serotonin blockers like other atypical neuroleptics but have unique antidepressant properties.

8.4.4 PSYCHOSTIMULANTS

Mention should be made here about the medication treatment of ADHD because of the diagnostic confusion between ADHD and bipolar disorder, the high rate of the co-occurrence between bipolar disorder and mood dysregulation, and the risks of using psychostimulants when a person has a bipolar (spectrum) diathesis. We do find a group of people with ADHD that can safely and successfully use psychostimulant medication (e.g., Ritalin, Concerta, Adderall), with no mood destabilization. These are likely people with ADHD and no co-occurring mood dysregulation or clinical diagnosis on the bipolar spectrum. Often, however, we see people with ADHD and an unrecognized mood dysregulation (or bipolar disorder misdiagnosed as ADHD) that only becomes apparent if unmasked when treatment with psychostimulants destabilizes or flips them from a normal, euthymic state to an energized hypomanic/manic state. They become activated, develop mood swings, irritability, and agitation, or euphoria. Akiskal describes this as a soft bipolar spectrum (bipolar III or III ½) disorder. Suicidality can occur and hospitalization may be necessary.

People who come to see us have often been poorly responsive to past treatment efforts. Adding insult to injury, they have not chosen their lot. They did not choose their genes and they did not choose their early environment.

Somehow, their misfortune must be overcome if our efforts to help are to succeed. Treatment resistant depression all too often perpetuates itself when it is embedded in relational trauma that interferes with an individual's capacity to trust us as caregivers. This may manifest in nocebo effects such as stalled psychotherapy and poor response to somatic treatments, including medication. However, it is ours, no matter the modality of treatment, to help such individuals accept the cards life has dealt them and to assist them in taking agency over their lives, so that, with all the tools at hand, they can win the game.

KEYWORDS

- **environmental treatment**
- **neurofeedback**
- **psychoanalytic psychotherapy**
- **mood stabilizers**
- **lithium**

CHAPTER 9

Clinical Illustrations

Unless someone like you cares a whole awful lot,
nothing is going to get better. It's not.
—Dr. Seuss, The Lorax

ABSTRACT

People with mood dysregulation have symptoms with core similarities. Their presentations are variations on a theme. The individuals presented in this chapter came to me with "depression," discouraged by a poor response to multiple antidepressant medications. Some had other diagnoses such as eating disorder, anxiety, alcohol abuse and ADHD. Treatment failures were affecting their lives and relationships and, at times, their willingness to engage in further treatment. But all had dysphoria, which was never previously recognized. It was typically described as "depression" and/or "anxiety" whilst manifest in a state of activation characterized by irritability, restlessness and agitation. These activated/increased energy symptoms resembled the activated hypomania of bipolar disorders more than symptoms of any depressive disorder, but none of these individuals could be diagnosed with bipolar disorder. Perhaps the bipolar connection was in the genetic commonality between people with this symptom constellation and those with bipolar disorders, "an affective temperament" or genetic predisposition to a mood disorder. Each of these individuals, in fact, experienced significant relational stress that could account for development of mood dysregulation if genetically predisposed. Psychotherapy was an important part of the treatment response for those who embraced it. And for each one, mood stabilizers proved uniquely effective in mitigating symptoms of what they had previously described as "depression." In each case, proper identification of their symptoms made a crucial difference in treatment outcome and was key to effective treatment.

Mood Dysregulation: Beyond the Bipolar Spectrum. Deborah A. Deliyannides, MD (Author)
© 2024 Apple Academic Press, Inc. Co-published with CRC Press (Taylor & Francis)

9.1 MADELINE

Madeline came to see me because she felt "depressed." She was 28 at the time, had her MBA and a prestigious job on Wall Street, but was in a troubled relationship with a man she had been dating since college. The dynamic between them was historically tumultuous, which was a large part of her discontent. She had been off and on antidepressants for over 10 years but felt they had never really helped. She had never been in psychotherapy. Madeline had a number of girlfriends but had recently withdrawn from them because she felt so awful. She derived a sense of satisfaction and worth only from her job, but conflict in her relationship with her boyfriend was increasingly distracting her from her work. What did not emerge until later in the treatment was her struggle with binge/purge behavior and preoccupation with her weight, some of which affected her compliance (and possibly lack of success) with prior medication treatment.

Madeline had difficulty trusting her boyfriend. Though he had never given her overt reason for mistrust, neither did he give her reassurances to ease her anxieties. She could never quite put her finger on it but she was often left with a sense that he simply wasn't faithful. She was ever vigilant, and whenever he was delayed coming home from work, she was convinced he was with another woman. She never believed him when he told her he was out with friends from work (she could not mentalize: if it didn't ring true in her mind, she couldn't imagine it could be true for him). When he returned home to her accusative tone, he withdrew from her, creating more anxiety in her. In her anxiety and anger, she would follow him and provoke a fight, which invariably escalated until they both said vicious things they each later regretted. In her state of upset, she turned to her binge/purge behavior, which dysregulated her further. Over time, her boyfriend arrived home from work later and later, giving her more realistic anxieties. In her anxiety, she lashed out. When he finally threatened to leave, she backed down, withdrew, and became more despairing and upset. Upset exacerbated her binging, and even a two-pound weight gain added to her anxiety that her boyfriend would cheat on her because she was unattractive.

When she first came to me, Madeline was overwhelmed, anxious, and irritable. She was aware she was reliving a feeling of separation anxiety she knew from the time she was a little girl and her mother was unavailable to her. She was an only child in what she described as a sad and empty home. Her father was a film director; he was rarely with the family but when he was, he was always in a foul mood. He had a bad temper. Her mother was an actress who was supremely self-involved and consumed with her career. From the time Madeline was an infant, her mother left her to the care of various nannies. Madeline recalls the perpetual image of reaching her arms out for her mother, longing for a hug, only to watch her mother blow her a kiss as she ran out the door. When her mother came back, she continued to leave Madeline in the care of the nanny.

Madeline knew we needed to talk about how her anxiety related to her experience with her mother but we both understood her reactivity with her boyfriend was an urgent concern we ought to try to address with medication. Her reactivity and mood outbursts were provoked in the context of relationships. She had no history of major depression though she had periods of sadness in the past. For the sake of expediency, she chose to start low dose lithium, which she trusted would not cause weight gain and which quickly helped lengthen her fuse with her boyfriend, giving her more reflective space to plan a response rather than a reaction to him. Within a few days, she found she was able to keep herself from exploding when he came home late and to measure her response when she felt angry with him so that their conversations didn't devolve into ugly arguments. Regulating these upsets helped her avoid her binge/purge behavior as well.

Madeline quickly opened up to me but frequently called or e-mailed between sessions. Though I was very responsive, I knew maximal responsivity would never fill the void left by her mother's absence. We talked openly about how her reaching out to me related to her thwarted attempts to reach out to her mother, and we discussed what might happen if I didn't have a perfect response rate. We discussed the importance of her feeling free to tell me when she was disappointed or upset with me (this took her some time to do).

After one rather routine call on her part, I failed to respond until late the following day because I was occupied with an illness in the family. When we met, she erupted with me before there was any chance for discussion. I asked her if she had thought about why I might not have called her. She insisted it was because I must not care sufficiently about her.

This catalyzed our work on helping her mentalize or understand what her boyfriend might be thinking. While she was able (with effort) to imagine other reasons why I might not have called her back immediately, no matter how hard she tried, she seemed unable to imagine that her boyfriend could maintain interest in her while doing something else important to him. After all, her mother was always far more interested in acting than in spending time with her. Though lithium greatly reduced her reactivity, mentalization remained a challenge for her and limited her ability to rein in her anxiety sufficiently to establish a new interactional pattern with her boyfriend, so his behavior continued to provoke her.

While experiencing validation in her relationship with me, she began to regulate herself better in her relationship with her boyfriend, and her binge/purge behavior all but ceased. Over time, she came to understand how anxiety provoking her boyfriend's unreliability had been for her over the years and how dysregulating the relationship had been for her moods. Two years after she began treatment with me, Madeline broke up with her boyfriend. It was a destabilizing time for her because it meant being alone for the first time in her life, which was very frightening. Being alone made her more anxious and upset but she coped with it by throwing herself into her work and increasing her exercise routine. She also spent more time with her girlfriends, whom she had been seeing more while in the process of breaking up. We added an additional weekly session to help support her through this period. All these interventions helped regulate her during a time of situational destabilization. She remained on the same low dose of lithium throughout. Interestingly, her eating disorder symptoms did not recur during this time.

Within the year, she began dating a man who she felt adored her. She wanted to throw herself into the relationship, but we discussed slowing down her engagement with him and giving herself a chance to get to know him. She struggled to pace herself, always eager for attachment, but became fairly seriously involved within five months. Fortunately, it seemed her new boyfriend was a caring person who was reliable and validating, and the relationship regulated her quite well. We talked carefully through situations that came up between them to make sure she was aware of how she was feeling and that she was comfortable with the way she was responding to her boyfriend. She recognized that her current boyfriend, in contrast to her last one, understood, accepted, and was able to reassure her anxieties, qualities she had come to realize were important in a partner for her. Two years later, Madeline and her new boyfriend married.

Madeline remained reasonably well regulated for several years while we continued therapy, during which time she was on medication. But when she became interested in starting a family, she needed to stop taking lithium. Many psychotropic medications may be used with relative safety during pregnancy if necessary, but lithium is not one of them. But because it is a salt and is flushed out of the system immediately, it does not require a washout period. Nevertheless, Madeline decided to stop taking the medication as soon as she decided to try getting pregnant.

Without lithium, her reactivity (and her binge/purge behavior) returned fairly quickly. Her husband, who knew her anxiety, was unprepared for her irritability. After a period of turbulence, they decided to enter couples' therapy, which was very helpful. Interestingly, Madeline was unable to get pregnant until the period of conflict with her husband settled down. Madeline, her husband, and I continued to meet throughout her pregnancy until she could resume medication. She did experience post-partum depression, which we treated with Lamictal, both for its mood regulating and its antidepressant effects, and because it was safe for her to use while breastfeeding.

Madeline delivered a healthy baby boy. She was very concerned about making sure her son didn't feel anxious about his connection with her, given her experience with her mother. She was determined to be an attentive mother and not to create in her son the anxiety her mother had created in her. We talked about the importance of focusing on what her son needed and on her need to imagine his world, not the world she lived in when she was young. Mentalization was always a struggle for her so we spent a great deal of time with this. She was determined to "see" her son. Her son's responses seemed to reflect his sense that she was understanding and anticipating his needs. Her apparent success at creating a strong connection with her son reinforced her attentive behavior and enhanced her relationship with him. It appeared within the first 24 months of her son's life that she had helped him create a secure attachment to her.

While Madeline still struggles with anxiety and remains on low dose medication to regulate her mood, she is in a loving marriage with a man who helps her regulate her mood and is thoughtfully raising a son with guidance and attention to help him maintain a healthy attachment to her. Her binge/purge behavior has been absent since her early pregnancy.

Though Madeline worried she had a treatment resistant depression, she did not. She has a mood dysregulation that responds quite well to mood stabilizing medication. In close relationships, she becomes easily triggered into agitated, dysphoric states by fear of separation. For years she experienced these mood states as "depression," which is what she described to the people who treated her, so she was given medication (for depression rather than dysphoria) that never adequately treated her mood symptoms. Once we identified her mood dysregulation, we were able to intervene with the appropriate medication for these symptoms.

Mood dysregulation may be based on affective temperaments but affective temperaments are not categorical: they are better construed as being on a continuum, often showing mixed or overlapping features. Madeline has an anxious temperament and is never far from irritability. Hers is a good illustration of the overlap we see between features of different affective temperaments, which are seldom as clearly defined in reality as they are in a text. Madeline's temperament has both anxious and irritable features.

Her symptoms also reflect the transdiagnostic nature of mood dysregulation, which can co-occur with diagnoses such as eating disorders. Often, treatment aimed at affective symptoms is effective in treating symptoms of co-occurring diagnoses.

Madeline was eager to engage in psychotherapy because she needed a model for a consistent, caring and engaged partner. She struggled to feel valued within that context, though to her credit, she persisted, and in the psychotherapeutic dyad, made important progress in learning to regulate her moods, enabling her to develop healthier relationships with others, who then helped her regulate her moods in turn.

9.2 DARRYL

Darryl came to me in his early twenties for a medication consultation when he was in college. With no clear explanation, he said he wanted to switch psychiatrists. He brought with him a list of "failed" medications that would intimidate any prospective pharmacologist. His current cocktail was only partially adequate as it left him still feeling quite anxious.

While his anxiety involved overthinking, worrying, ruminating, and obsessing about many of life's ordinary activities, I also learned he could be irritable and agitated. One of the medications he was taking was a neuroleptic, usually prescribed for mood stabilization and agitated anxiety: had his prior psychiatrist wondered whether or not there was a mood disorder on the bipolar spectrum or was he simply being treated for severe anxiety? There were no mood stabilizing medications on the list he brought. He denied that there had ever been mention of anything remotely bipolar and seemed disturbed when I raised the subject. No, he would not consider a mood stabilizer. The issue, after all, was not depression or any other mood. It was anxiety. I spoke with Darryl's former psychiatrist, who indicated she did not think Darryl had any symptoms of a bipolar (or spectrum) disorder.

Darryl was raised by his mother, who had never been married to his father. He had met his father but his father had never been part of his life. He was resentful of this and felt unloved for it. If his father had loved him, he felt, he would have made it a point to be part of his life. Meanwhile, his mother knew no boundaries and treated him as though he was the husband she never had. On the one hand, he loved his mother: she had raised him and he was beholden to her. On the other hand, he resented her: she smothered him and didn't seem to let him grow up.

Clearly, there were dynamic issues to be addressed, but whenever I suggested to Darryl that we talk, he insisted he did not want psychotherapy. He had done it before (for 6 months) and didn't get anything out of it. There was no use. Every once in a while, I engaged him in a reflective moment but on the whole, he was shut down. Darryl had developed a fierce sense of self-sufficiency to make up for the hole left by his father's absence and his mother's failure to mother. It was inconceivable to him at the moment that I could function for him in any helpful or trustworthy way. He assumed I did not have anything useful to say, nor did I have any medication to offer. He alone had to decide which medication he needed.

He came to each meeting prepared with a new medication he had researched on the internet that he wanted me to prescribe with a lengthy apologetic regarding why this should be the next medication trial. If I made an alternative suggestion it was dismissed out of hand (especially "no mood stabilizers"). On one occasion, he decided he wanted to discontinue the neuroleptic started by his prior psychiatrist because of the incidence of tardive dyskinesia, so he found an alternative, safer neuroleptic he wanted to try. I concurred.

We met on a monthly basis to monitor his medication. For a time, the new neuroleptic was effective for his "anxiety" and had no side effects. I always arranged his appointment to be the last in my day in case he wanted to linger. He began staying longer, clearly interested in engaging with me, telling me how sad and angry he felt about his family. But he didn't want psychotherapy.

Eight months later, quite abruptly and for unclear reasons (other than "I don't want to be dependent on pills"), he decided to stop all medication. So he weaned himself off of everything over the course of one week. I did not hear from him for over three months.

Then he called in a panic, explaining that he was so anxious he needed to resume medication right away. But he did not want to take any of the medications he had used before. He wanted something new. And he did not want to talk. He wanted medication.

Knowing that anxious people are generally anxious about taking medication, I prescribed a low dose of propranolol, which generally has no side effects (on ten times the starting dose I gave him it can cause dry mouth and dizziness). On a minimal dose, he experienced ringing in his ears, so he discontinued it after two doses. Ringing in the ears has actually been successfully treated with propranolol (Albertino et al., 2005).

His "anxiety" continued to escalate and was now complicated by depression. I felt in a corner with regard to his treatment: he was unwilling to take nearly every medication which had helped or which I recommended, and I was unable to engage him in psychotherapy.

In the midst of one of his crises, he opened up about the nature of his "anxiety" and "depression" when I asked him to set aside these terms and simply describe what he was experiencing. He said, "I'm just losing my temper, getting out of control, crying, and yelling." As the conversation went on, it became clear that he was describing reactive states of dysphoric hypomania. His "anxiety" and "depression" were characterized by irritability, agitation, racing thoughts, restlessness, inability to get comfortable in his own skin, and dysphoric mood.

Perhaps because he was in a state of desperation (different from when we first met), he was open to my explanation that this type of "anxiety" was symptomatic of a mood dysregulation. I explained to him that we could turn this around quickly, within days, but that he would have to be willing to take lithium. I described the benefits of lithium and told him there were virtually no risks of low dose lithium. I explained that we would know rather quickly if it was going to work and that there would be no withdrawal if he decided to stop using it, so there was little investment in giving it a try. He should look at it as a probe into whether or not a mood stabilizer was in the right ballpark compared with other medications he had used. After some negotiation and much reassurance, he agreed to give it a try. We would talk in a week and he could stop it immediately if he felt it wasn't helping.

Within four days he felt better. He seemed not to want to admit it to me but he was no longer "anxious" or "depressed." And there were no side effects. This was on a very low dose of lithium, 300 mg/d. He liked the medication and wanted to increase the dose, so he began taking 450 mg/d. On this dose, his agitation, irritability, dysphoria and anxiety virtually resolved. He remained stable on this dose of lithium for over three months.

Then somewhat suddenly, I got an e-mail from him: he was reading about Stevens-Johnson Syndrome (SJS) on the internet (a potentially fatal rash associated with some medications but unheard of with lithium) and had developed a rash. He was worried that the rash was associated with lithium and asked me whether or not this could be possible. I responded that SJS has never been known to be associated with lithium, so he should feel confident that he did not have SJS. Nevertheless, he wrote back saying that he was going to discontinue lithium and stop treatment. Darryl has not contacted me since.

Darryl's situation demonstrates, again, the transdiagnostic nature of mood dysregulation. He had symptoms of generalized anxiety as well as depression co-occurring with mood dysregulation. Like Madeline, he appeared to have an irritable affective temperament. Though he presented complaining of "anxiety," he was not articulate in describing his symptoms, but the overall presentation of his "anxiety" was that of agitated dysphoria partially treated with the neuroleptic prescribed by his prior psychiatrist. Given his unwillingness to take a mood stabilizer, it was hard to achieve full symptom control on this medication. Naturally, his symptoms worsened when he stopped his neuroleptic, leaving him vulnerable to an extreme state of agitated dysphoric hypomania. It was only when he reached this state that he was open to my medication recommendation of a mood stabilizer.

While his affective symptoms were quite responsive to lithium, Darryl's traumatic background predisposed him to the nocebo effect to medication and to psychotherapy, interfering with his ability to benefit from both forms of treatment. He (almost longingly) attempted to engage with me but I had the sense he felt he would have to violate himself to do so. He was even more ambivalent about taking medication. Unfortunately, people with Darryl's apprehension about treatment often fail to get the help available to them.

9.3   RENA

Rena was a 63-year-old physician sent to me by her psychoanalyst for a medication consultation, more or less on the orders of her controlling husband, who felt her current psychiatrist was failing her as she remained depressed after years of treatment. She had been on various antidepressants over the course of decades and was currently on Prozac. Curiously, though depression was the chief concern and presenting problem, our initial discussion focused almost entirely on her temper, its effect on her marriage, and its origins in her family. Her husband was apparently losing patience with her snide comments and outbursts at him and was concerned about her drinking.

She rarely smiled and she never felt happy. She had few friends but those she had shared her negative outlook. She had raised two children with whom she was strict and often angry. Her marriage was strained by her chronic irritability, which was typical of her both at home and at work. In her career as a physician, she was unable to keep nurses on staff over the years because she was so short-tempered with them. Her husband felt her drinking exacerbated her irritability.

Rena came from a family in which she witnessed chronic fighting between her parents. Her mother had a history of depression. Rena was the youngest of four children who were overwhelming to her mother and siphoned her attention away from Rena. Her father, it turns out, was verbally abusive and her mother was taken to hitting the children. Rena grew up hungry for attachments but not knowing how to connect with people in any way other than through verbal aggression. Her overtures to others were caustic and she habitually interpreted others' responses to her in a negative light.

In addition to irritability, Rena was also prone to anxiety. When questioned closely, it became clear that what she called "anxiety" was the inability to settle the pressured, ruminative, racing thoughts of an activated state. Alcohol helped slow these thoughts in the evening after a hard day of work. Her sleep was erratic, and she smoked marijuana at night to relax and help her sleep\in what she knew was a self-medicating way (Gruber et al., 2010). Above all, her relationship with her husband was imperiled by her reactivity. Though she found him to be bossy, she had virtually no tolerance for his treatment of her: if she wasn't sniping at him, she was blowing up at him. She was aware of her behavior but didn't think it had anything to do with her alcohol use.

Rena carried the depression diagnosis of dysthymia over the years but when I questioned her carefully, the nature of her "depression" was rarely that of "sadness" or "hopelessness" but primarily one of "anger" and "irritability," which was nevertheless upsetting. She did not distinguish between the two so she always described her mood as "depressed." I explained "dysphoria" to Rena and suggested that the patterns in her mood, sleep, and behavior represented a mood dysregulation, a term she realized better captured her problem than the notion of depression alone.

To assure that Rena's dysregulation was not medication-induced by the antidepressant Prozac, we stopped the medication for three months to see whether or not her irritability would remit. The medication washout made no difference in her level of irritability; in fact, she became more irritable during this time.

In addition, what emerged were periods of tearfulness that had not been present when we first met, and which Rena said she had not experienced for years. These were generally triggered by losses she felt were out of her control. When I questioned her in detail about the nature of her mood, she was clear that during these periods of tearfulness, she experienced a sad, low, blue quality to her mood. Her baseline state was irritable, anxious, and more or less angry, but her new "sad" mood state had emerged several months after discontinuation of Prozac.

Rena appeared to be experiencing intermittent mixed states. Superimposed on a baseline irritable temperament were dysregulated episodes of depressed mood. While antidepressant medication had been helping her more than we knew (because it was treating this underlying depression), she was still not feeling well because her dysphoric hypomania had never been treated. Resumption of an antidepressant and treatment with mood-stabilizers was indicated.

Rena opted to begin a shorter-acting SSRI and to try a tentative course of lithium: after 2 weeks of treatment with lithium, she reported a remarkable reduction in her irritability, which was sustained without side effects. She no longer experienced racing, ruminative thoughts and her sense of feeling anxious had resolved. She had no physical restlessness and was no longer agitated. Within a month, her evening use of alcohol was down to one glass of wine/night and her use of marijuana diminished to once or twice weekly. Her sadness resolved within six weeks. Because of her positive response to lithium, she chose not to switch to a different mood stabilizer.

Her husband noticed a reduction in her reactivity but was concerned about her ongoing critical behavior, and insisted she begin DBT. Primarily out of the fear of losing him, she agreed to a treatment about which she knew nothing. Skeptical and reluctant, she began group DBT while maintaining treatment with her psychoanalyst. To her surprise, she found it was helpful to learn how to look at the way her behavior affected those around her and to learn to make decisions about what she said based on the consequences she anticipated it would have on others' lives. After 12 months of DBT, Rena began reflecting on the way she reacted to her husband and began responding in a constructive way to him rather than reacting to him. He, in turn, became less controlling and their relationship improved.

During this time, her medication management continued. She asked for a dose increase in lithium, but the adjustment left her feeling too cognitively slowed to function properly at work. We opted for addition of Lamictal as an adjunctive medication. After achieving a therapeutic dose of Lamictal, we were able to discontinue lithium altogether. We were also able to discontinue her antidepressant as Lamictal is often itself an effective antidepressant. On Lamictal 250 mg/d, her depression was resolved and her mood was stabilized. On many days she used no alcohol and her marijuana use was very intermittent. After completing 18 months of DBT Rena was only on mood stabilizing medication and in continued treatment with her psychoanalyst. She was fundamentally depression and anxiety-free and felt she was in charge of her moods rather than her moods being in charge of her.

When two treatments are initiated concurrently and a successful outcome results, it is often difficult to know how to construe the success. Rena's reactivity virtually resolved within two years while she was on medication and in DBT. After completing DBT, would she need to remain on medication? At times, an individual has a sense of whether or not the medication is playing a big part in their capacity to regulate their moods. In this case, Rena was not sure of the role of medication as she felt she had gained so many important tools from DBT.

Since Rena remained in psychotherapy and had the support and feedback of her analyst in addition to my supervision, we decided to try to lower her dose of Lamictal to see whether or not she still needed to be on the medication. We gradually reduced her dose of Lamictal and she remained well regulated until we reached a dose of 75 mg/d. Rena then found herself becoming frustrated that she could not employ the tools she knew she should be using to manage her irritability. Once she reached the point where she felt she was struggling to maintain control, we decided she should remain on a maintenance dose of the medication. She has remained comfortable and feeling in charge of her reactivity on Lamictal 100 mg/d for the past 2 years.

Rena's mood dysregulation, again a transdiagnostic phenomenon co-occurring with alcohol abuse, is somewhat more complicated than the others we have discussed but, like the others, is not a treatment resistant depression. Part of her symptomatology involved chronic intermittent periods of depressed mood, though these alternated and were sometimes superimposed on moods of dysphoric hypomania, creating mixed states. These moods are particularly hard to recognize and cannot generally be understood when observed in a single consultation: they require longitudinal observation. Rena has an irritable affective temperament with depressive dysregulation. Hers is an example of the way in which affective temperaments can take any form and combine with any mood at different times. The presence, in Rena's situation, of a mixed state, illustrates the need, at times, for use of both antidepressant medication and mood stabilizers.

Rena's irritability was so extreme when she first came to see me that she was not open to DBT. It was only after she was on medication that she became agreeable enough to engage in a treatment she ended up valuing greatly and which helped her marriage significantly. The productive use of DBT as an adjunctive therapy by Rena and her psychoanalyst was another important feature of Rena's treatment.

9.4 STEPHEN

Stephen was a driven 50-year-old portfolio manager who had been "depressed for as long as I can remember." He was referred to me by his psychotherapist, who felt his irritability and reactivity were interfering with their work together. Though his therapist felt quite engaged with him, no matter how closely she tracked him, he reacted as though she was somehow not on his side. Any attempt she made to contextualize this in terms of his neglectful background was met with anger on his part that she was deflecting his "valid" criticism of her. While he seemed unaware of the extent to which he was reactive, he was quite aware that he felt "depressed," so he accepted a psychiatric referral to discuss medication.

When I first saw him, he told me he felt chronically depressed. He had re-entered psychotherapy in lieu of trying medication again because medication had never really helped. A confluence of things had led to his worsening depression including losing a major account and escalating conflict in his marriage. He felt his therapist had been helpful because she was gentle and steady, but he did not feel she paid sufficient attention to him. Yet he was feeling discouraged lately and was thinking of terminating treatment because his mood was not improving.

He noted that he felt irritable with his wife and that their relationship was very contentious. He complained she was inattentive and dismissive of him. She, in turn, complained he was irresponsible and disorganized, and in fact, blamed him for losing his recent account for those reasons. He acknowledged he often overreacted to her, and that this fed into his own problematic behavior, but he also felt out of control of his temper with her. This led to his depression. His irritability and his temper, it turned out, had been problems at least as long as his depression.

Stephen came from a family in which his parents practiced benign neglect. His mother was generally absent. He described her as a rather vain woman, preoccupied with running various non-profit organizations rather than tending to the family. His father was a workaholic attorney whom he rarely saw and with whom he had a distant relationship. He had an older brother who was out of the house by the time he entered high school. There was no overt history of trauma. When he was in elementary school, he was diagnosed with ADHD and treated with Adderall, but it made him so angry and irritable it had to be stopped. The same thing happened with Focalin, which made him irritable and aggressive. Without medication for ADHD, he struggled through school, maintaining a B average throughout college. He married in his mid-30s: he has always felt his wife has been unsympathetic to his depression. He had three children but felt they were closer to his wife than to him. He had several acquaintances but only one close friend. He described himself as "too busy to be lonely." His mother, brother, and paternal grandmother had histories of depression.

When I asked him to tell me about his depression, he described behaviors, cognitions, and energy states, not his mood. When I asked him to describe his mood, he seemed stumped. When I prompted him with a list of choices, he chose words such as "irritable" and "angry" rather than "sad" or "down." When we pieced things together, it appeared his "depression" was never quite independent of temper flares and irritability, and it was never associated with loss of interest, lack of energy, loss of appetite, suicidality, or other typical signs of a slowed energy state or major depression. In fact, he never seemed to have met criteria for a depressive disorder. His "depression" frequently had the qualities of restlessness, high energy, agitation, and anxiety. He had chronic trouble sleeping at night. During college he was a risk taker, but he never got into serious trouble. There were isolated episodes of sexual indiscretion during activated mood states from time to time but these were unassociated with other symptoms of hypomania. His "depression" followed the pattern of dysphoric activation. In short, he described symptoms of a mood dysregulation which had never previously been recognized, quite different from the diagnosis he had always been given of unipolar depression.

He was not optimistic that I could help him, given his treatment history. Curiously, he was never treated with medication for his "depression" until he was 30. He was subsequently given various antidepressants, none of which seemed to work for any period of time. He was in psychotherapy off and on for several decades but never felt it helped because his therapist never "got" him. He had not been on medication for over five years.

I told him I thought his "depression" had not responded to standard antidepressant treatments because it was not a simple "depression" but a mood dysregulation, and I suggested we take a different approach. Understandably, he was skeptical, but he heard me out when I recommended he start a trial of a mood stabilizer. He was unwilling to take lithium because of the associated stigma, so we opted for Lamictal. For the first time in 20 years, he was treated not for depression but for a mood dysregulation.

Because treatment with Lamictal requires a slow up-titration, we did not begin to see a response until about 6 weeks later, but by the time he was on a dose of 150 mg/d, Stephen began to feel as though he was out in front of his irritability, and his "depression" was much reduced. He began to understand what it meant to feel a sense of enjoyment and to be free of "anxiety" and agita. It proved to be an adjustment for him. Within the next two months, we fine-tuned the medication to achieve what he felt was a vast improvement, suitable for maintenance treatment.

More important was the change the medication introduced into his psychotherapy. With the reduction in his irritability, he became less reactive and less rejection sensitive. He began to recognize the ways in which his therapist was actually his ally. He not only stopped complaining to her that she was not "with him" but he apologized to her for having doubted and confronted her before he was on medication. The change brought about by the medication was somewhat disorienting to him and was a topic in itself to be processed. His psychotherapy, in many ways, was transformed by his response to the medication. As he came to see his behavior with his therapist, she was able to help him understand how the dynamic he had been playing out with her was being played out in his marriage as well.

As the spotlight turned to his relationship with his wife, he began to see how his reactivity toward his wife led to her withdrawal from him, resulting in his feelings of neglect. This realization helped him not to be so reactive, which had an impact on her behavior in turn. This allowed him to experience her in a different way: rather than withdrawing from him, she engaged with him. Over time, he came to feel he had a partner.

Stephen was reassessed for symptoms of ADHD, which had persisted into adulthood. Now with a mood stabilizer, it seemed safe to restart a psychostimulant such as Adderall or Ritalin without the risk of precipitating activated dysphoria. So we initiated a trial of Adderall while he was on Lamictal. On the combination of medications, Stephen experienced improved focus and organizational skills that significantly improved his performance at work with no evidence of irritability/ agitation or activation.

Three years later, Stephen remains more or less "depression"-free and stable on the same medication, engaged productively with a therapist he was about to leave when he and I first met. He is still working on feeling seen in his relationships, but his once alienated marriage is a growing intimacy.

Stephen might have appeared to have a treatment resistant depression on the basis of his history, but once again, his situation illustrates a transdiagnostic mood dysregulation (co-occurring with ADHD) in which dysphoric hypomania was masquerading as depression. Stephen had an irritable affective temperament with flares of reactivity when provoked by feelings of neglect. It is difficult to tease out nature from nurture: Stephen has family members with affective disorders who may have passed along his temperamental predisposition, but he was also raised in an environment in which family dynamics made him vulnerable to a mood dysregulation as well.

For want of any better way to define his negative feelings of rejection and neglect and the associated anger, Stephen came to define his mood states as "depression." Neglect was consistently the exogenous trigger leading to his reactivity and dysregulation. This type of mood dysregulation "beyond the bipolar spectrum" does not respond adequately to antidepressants, which is why Stephen never responded to medication until treated with mood stabilizers. When recognized as a mood dysregulation however, his mood symptoms were readily responsive to medication, which enabled him to engage successfully in psychotherapy for the first time. He worked productively with his therapist to improve his marriage, and began to learn, in the context of his relationships, how to better regulate his moods. In addition, recognition of Stephen's mood dysregulation enabled successful use of psychostimulants for his ADHD for the first time in his life since he was "pre-treated" with a mood stabilizer. Those like Stephen who are predisposed to dysphoric activation on psychostimulant medication are described by Akiskal and colleagues as having bipolar III disorder (on the soft bipolar spectrum, i.e., a mood dysregulation). His childhood history of irritability on Adderall and aggression when re-challenged with Focalin (the dysphoric hypomanic destabilization on psychostimulant medication) are the clues to his mood dysregulation and the need for mood stabilizers for successful psychostimulant use.

People wouldn't come to see us if they weren't hurting in some way. It takes vulnerability to reach out to another person for help, whether it be for medicine, talk therapy, or any other sort of treatment. It is especially difficult when one has lost faith in treatment—and treaters.

Understanding something about mood dysregulation can go a long way toward helping people who have not benefitted from past treatment for "depression" or "anxiety." Once these feeling states are more clearly identified, they can be more successfully approached.

People develop mood dysregulation in the context of relational stress experienced in their developmental years. Just as relational stress adversely affects the wiring of the brain, so can relational repair help rewire the brain, enabling improved mood regulation: the brain is plastic, after all. Medicine may provide a platform for this change, but it is in the relationship with the caregiver, psychiatrist, psychotherapist, or psychoanalyst, that people who seek our help will ultimately feel and function better.

KEYWORDS

- treatment-resistant depression
- irritable affective temperament
- nocebo effect
- mixed state
- co-occurring symptoms

Epilogue

The only new voyage of discovery consists not in seeking new landscapes but in having new eyes.

—Marcel Proust

Problems with mood comprise a disproportionate number of all psychological disturbances, yet our current diagnostic systems account for but a portion of them. Still to be described are those in which moods shift from states of activation to states of low energy with no cyclical rhythmicity, in contrast to bipolar disorders. Such "orphan" mood disorders may masquerade as bipolar disorders, bearing some resemblance to any one of them, but they are distinct. They are beyond the bipolar spectrum and might be better described as mood dysregulations.

Mood dysregulation is typically characterized by externally provoked shifts in physiological states from slowed to activated along with their accompanying moods. The activated state in mood dysregulation is typically associated with a negative (dysphoric) rather than a positive (euphoric) mood. Dysphoria is characterized by irritability, anxiety and agitation. Because dysphoria is a negative mood, it is typically confused with the negative mood of the low energy depressive state (depression) and treated with antidepressants when mood stabilizers should be the first-line treatment instead. Understanding the distinction between mood dysregulation and bipolar disorder enables many people with mood dysregulation, who know they do not have bipolar disorder, better understand themselves and have a rationale for more effective pharmacologic intervention.

Our current diagnostic system has proved inadequate in capturing the more nuanced range of mood symptoms we see in the populations we treat. This has paved the way for newer diagnostic approaches in which phenomena are described not in categorical terms but on a continuum, as they exist in nature. This approach is underway at the NIMH with the Research Domain Criteria (RDoC) initiative, which aims to take a fresh look at our psychiatric diagnostic system.

In the RDoC system, a mood disorder diagnosis must be made on the basis of biological and etiologic factors as well as on the basis of symptoms. Thus, genetic underpinnings of behavioral disturbances are being explored and at some point, the temperamental factors we propose to underlie our mood disorders will be understood in meaningful detail. Physiologic factors are being looked at: the glutamate hypothesis of depression is eclipsing the monoamine hypothesis and the immunologic basis for mood dysregulation is being more clearly delineated. Anatomic changes underlying mood dysregulation are being documented as well.

In addition, the RDoC system accounts for the role of the environment as it interacts with the organism in the development of psychological disturbance. In so doing, there is recognition of the importance of safe and secure attachment relationships in which individuals can develop to achieve optimal psychological functioning, and of the deleterious effects of early childhood/relational trauma, which we have seen lead to mood dysregulation.

By whatever name, mood dysregulation is not new. It is simply a matter of looking at clinical information through new eyes. Even as RDoC's more systematic efforts are underway to refashion our diagnostic approach in a comprehensive manner to aid in our ultimate clinical endeavor, it is hoped that the provisional thoughts presented here will provide a helpful lens through which to assess the symptoms of people with mood disturbances so as to engage them in more effective treatment.

Appendix A: Abbreviations

ACC	Anterior Cingulate Cortex
ACID	Antidepressant Chronic Irritable Dysphoria
ACTH	Adrenocorticotropic Hormone
ADHD	Attention-Deficit Hyperactivity Disorder
AED	Anti-Epileptic Drug
ANS	Autonomic Nervous System
APA	American Psychological Association
BA	Behavior Activation (treatment)
BAD	Bipolar Affective Disorder
BD	Bipolar Disorder
BDNF	Brain Derived Neurotrophic Factor
BLT	Bright Light Therapy
BPD	Borderline Personality Disorder
CBD	Cannabidiol
CBT	Cognitive Behavior Therapy
CNS	Central Nervous System
CRH	Corticotropin Releasing Hormone
DBT	Dialectical Behavior Therapy
DMDD	Disruptive Mood Dysregulation Disorder
DNA	Deoxyribonucleic Acid
DSM	Diagnostic and Statistical Manual (1952)
DSM II	DSM second edition (1968)
DSM III	DSM third edition (1980)
DSM-III-R	DSM third edition, revised (1987)
DSM IV	DSM fourth edition (1994)
DSM-IV-TR	DSM fourth edition, text revision (2000)
DSM-5	DSM fifth edition (2013)
ECT	Electroconvulsive Therapy
EDO	Eating Disorder
EEG	Electroencephalogram
ELA	Early Life Adversity
EMDR	Eye Movement Desensitization and Reprocessing
ERDs	Emotion Regulation Deficits

fMRI	Functional Magnetic Resonance Imaging
GABA	Gamma Amino Butyric Acid
GAD	Generalized Anxiety Disorder
GR	Glucocorticoid Receptor
HPA	Hypothalamic Pituitary Axis
HRV	Heart Rate Variability
ICD	International Classification of Diseases
IED	Intermittent Explosive Disorder
IPT	Interpersonal Therapy
ISTDP	Intensive Short-Term Dynamic Psychotherapy
LCSW	Licensed Clinical Social Worker
LFC	Left Frontal Cortex
LGBTQ	Lesbian Gay Bisexual Transgender Queer
LPFC	Lateral Prefrontal Cortex
MAOI	Monoamine Oxidase Inhibitor
MFT	Marriage and Family Therapist
MDD	Major Depressive Disorder
mPFC	Medial Prefrontal Cortex
MPT	Munich Personality Inventory
MRI	Magnetic Resonance Imaging
NDRI	Norepinephrine Dopamine Reuptake Inhibitor
NIMH	National Institute of Mental Health
NMDA	N-Methyl-D Aspartate
NRI	Norepinephrine Reuptake Inhibitor
ODD	Oppositional Defiant Disorder
OFC	Orbito-Frontal Cortex
OSBD	Other Specified Bipolar Disorder
PAG	Periaqueductal Gray
PD	Panic Disorder
PDD	Persistent Depressive Disorder
PDM	Psychodynamic Diagnostic Manual
PFC	Prefrontal Cortex
PNS	Peripheral Nervous System
PPS	Parasympathetic Nervous System
PTSD	Posttraumatic Stress Disorder
PUD	Pseudo-Unipolar Depression
RDoC	Research Domain Criteria
REM	Rapid Eye Movement

RNA	Ribonucleic Acid
rt-fMRI	Real Time-Functional Magnetic Resonance Imaging
SAM	Sympathetic Adreno-Medullary System
SCN	Suprachiasmic Nucleus
SMD	Severe Mood Dysregulation
SNRI	Serotonin Norepinephrine Reuptake Inhibitor
SNS	Sympathetic Nervous System
SSRI	Selective Serotonin Reuptake Inhibitor
SJS	Stevens Johnson Syndrome
STS	Superior Temporal Sulcus
SUD	Substance Use Disorder
TCA	Tricyclic Antidepressant
TCI	Temperament Character Inventory
TEMH	Treatment-Emergent Mania/Hypomania
TEMPS	Temperament Evaluation in Memphis, Pisa, Paris, and San Diego
TEMPS-A	TEMPS Autoquestionnaire
THC	Tetrahydrocannabinol
TRD	Treatment Resistant Depression
UBD	Unspecified Bipolar Disorder
vmPFC	Ventromedial Prefrontal Cortex
WHO	World Health Organization

Appendix B: Clinical Data

I looked at two cohorts of individuals. The first cohort is a series of the last 400 people seen in my private practice. The second cohort is a series of the 100 most recently seen people currently in treatment with me. All diagnoses follow DSM-IV criteria for the sake of uniformity excepting those for people with activated states in non-cyclical patterns for which I believe there is no suitable DSM diagnosis: in such cases, I used the designation "mood dysregulation."

COHORT #1

266/400	All mood disorders	67%
128/400	Unipolar depressions	32%
45/400	BPDI, BPD II, Cyclothymia, OSBD, and UBD	11%
93/400	Mood dysregulation	23%

Of the 266 individuals with mood disorders,

128/266	Unipolar depression	48%
45/266	BPD I, BPDII, Cyclothymia, OSBD, and UBD	17%
93/266	Mood dysregulation	35%
		100%

Seventy-five of the 79 people with mood dysregulation (95%) had family members with a BPD, a suspected BPD, "tempers," depressive disorders or completed suicides, consistent with the notion that people with mood dysregulation share family history of mood disorders as a phenotypic similarity with bipolar spectrum mood disorders.

Below are the 25 most common presenting complaints of the 79 people I assessed to have mood dysregulation.

1.	Anxiety	56	71%
2.	Irritability	50	63%
3.	Depression	46	58%
4.	Anger	28	35%
5.	Short fuse	27	34%
6.	Moodiness	27	34%
7.	"ADHD"	27	34%
8.	Outbursts	27	34%
9.	Agitation	27	34%
10.	Reactivity	26	33%
11.	Panic attacks	26	33%
12.	Crying jags	20	25%
13.	Impulsivity	19	24%
14.	Activation on stimulants	18	23%
15.	Rejection sensitivity	17	22%
16.	Substance abuse	17	22%
17.	Agitated suicidality	16	20%
18.	Self-harm	13	16%
19.	Not depressed	12	15%
20.	Eating disorder	11	14%
21.	Relationship tumult	11	14%
22.	Recklessness	7	9%
23.	Easily overwhelmed	6	8%
24.	Numbness	5	6%
25.	Euphoria	1	>1%

The most common complaint among these people, all of whom who sought consultation for depression, was "anxiety." This likely represents the experience of being in the hyperaroused state characteristic of dysphoric activation. Not once in over 30 years have I heard anyone come into my office saying, "Hey doc, I'm dysphoric." Instead I hear, "I'm anxious." Or, "I'm irritable." Or, I want to cry and yell and I can't settle down." The anxiety of mood dysregulation appears to be distinct from the anxiety of GAD (generalized anxiety disorder): anxiety symptoms in 52/56 (93%) of these individuals were associated with irritability, anger, agitation, outbursts, short fuse, recklessness, self-harming, or medication activation, none of which are typically seen in GAD.

The second most common complaint was "irritability," which lies at the heart of dysphoric activation. The third most common complaint, "depression," is ambiguous: are these people describing a "sad" mood or are they dysphoric without knowing it? I have no statistic on how often "depression" meant "dysphoria" in this cohort because the difference was lost on me for a number of years during which these data were collected. "ADHD" was a common complaint despite the fact that none of these individuals actually met criteria for ADHD. Racing thoughts and the experience of anxiety often leave people with the sensation that they are unable to concentrate. Mood activation on antidepressant medication was seen in 18/79 (23%) of these individuals as described in the literature with "bipolar III disorder" (which is not a cyclical disorder but on the "soft" bipolar spectrum as per Akiskal). In this sample, 16% of people spontaneously said they did not feel depressed: this number is likely higher when including people who are asked this question directly. Finally, only one person reported any experience of elevated mood at any time: there is little euphoria associated with mood dysregulation.

COHORT # 2:

75/100	Mood disorders	75%
24/100	Unipolar depressions	24%
15/100	BPD I, BPD II, Cyclothymia, OSBD, and UBD	15%
36/100	Mood dysregulation	36%

Of the 75 individuals with mood disorders:

24/75	Unipolar depressions	32%
15/75	BPDI, BPD II, Cyclothymia, OSBD, and UBD	20%
36/75	Mood dysregulation	48%

Undoubtedly, the percentage of people I saw with mood dysregulation increased in later years as I became more willing to see beyond the DSM lens of diagnostic categorization as I made clinical assessments.

Among this population of 100 individuals, 36 had mood dysregulation, representing 48% of all mood disorders in the cohort. Among these 36 individuals, the most common symptoms were:

1.	Dysphoria	30	88%
2.	Irritability	24	71%
3.	Anxiety	24	71%
4.	Depression	19	56%
5.	Anger	19	56%
6.	Agitation	18	53%
25.	Euphoria	1	3%

All 36 people with a diagnosis of mood dysregulation presented initially with a primary complaint of depression but dysphoria was the most common symptom among this group. Many individuals experienced both dysphoria and depression, typically not at the same time, but dysphoria remained the more common complaint: 30/36 (83%) reported dysphoria while 19/36 (53%) reported depression. In addition, 16/36 individuals (44%) had dysphoria without depression, consistent with the notion that people can present complaining of "depression" when their negative mood is not depression but dysphoria. Among the top symptoms in this cohort were irritability, anxiety, anger, and agitation, all hallmarks of the activated state seen with symptoms of hypomania/mania in bipolar disorders, tethering mood dysregulation to the bipolar spectrum. Irritability and anxiety again are among the most frequent symptoms in this population consistent with the state of dysphoric activation. Only 1 individual reported euphoria: euphoria is usually seen in bipolar disorders but almost never seen in mood dysregulation, suggesting that mood dysregulation might be a different entity from bipolar disorder.

Euphoria in BPD I, BPD II, cyclothymia, OSBD, and UBD:	13/16	81%
Euphoria in mood dysregulation:	1/36	3%

Because there is no operationalized definition for "mood dysregulation," I treated these individuals "as if" they had a DSM-5 diagnosis of "unspecified bipolar disorder" according to the treatment guidelines for bipolar disorder. Treatment was often symptom-focused as is common in transdiagnostic models (impulsivity, anxiety, moodiness, and ADHD when present).

NO. OF MEDICATIONS

Antidepressants alone	6
Lithium alone	1
2 medications	12
3 medications	9
4 medications	4

TREATMENT TYPE

Antidepressants	29
Lithium	26
Lamictal	12
Neuroleptic	8
ADHD medication	7
Psychotherapy	17

Of the 36 people treated, 22 (61%) had TRD (defined here as minimal or no response to two trials of antidepressant medications at a minimum of 2/3 recommended maximum PDR dose for a minimum of 6 weeks). *Every person with TRD had dysphoria.* Most people were treated with mood stabilizers in combination with antidepressant medication or with other mood stabilizers or both. In terms of outcome, GAF (global assessment of function) scores to date are the following:

OVERALL

Very much improved:	15	42%
Much improved:	14	39%
Moderately improved:	1	3%
Mildly improved:	3	8%
Unimproved:	1	3%

All 36 individuals were on medication: 17 (47%) of them were also in psychotherapy. If we separate out overall treatment response for the 17 people in psychotherapy, the results are as follows:

PSYCHOTHERAPY
(in combination with medication)

Very much improved:	9	53%
Much improved:	3	18%
Moderately improved:	2	12%
Mildly improved:	2	12%
Unimproved:	1	>1%

Twenty-nine (81%) of 34 index cases with mood dysregulation whose primary symptom was "dysphoria" responded to treatment that was predominantly centered on mood-stabilizer medication though response rates were clearly improved with addition of psychotherapy. Successful medication treatment initiated for symptoms of "depression" or "anxiety" was primarily based on agents typically used for people with cyclical bipolar disorders, tethering mood dysregulation to the bipolar spectrum, albeit beyond its outer boundaries.

Appendix C: Diagnostic Criteria

DSM 1952

Disorders of Psychogenic Origin Without Clearly Defined Physical Cause or Structural Changes in the Brain

Psychotic Disorders

302 Involutional psychotic reaction

301.2 Affective reactions

   301.0 Manic depressive reaction, manic type

   301.1 Manic depressive reaction, depressive type

   301.2 Manic depressive reaction, other

   309.0 Psychotic depressive reaction

300.7 Schizophrenic reactions

303 Paranoid reaction

309.1 Paranoid reaction without clearly defined structural change, other than above

Psychoneurotic Disorders

318.5 Psychoneurotic reactions

   310 Anxiety reaction

   311 Conversion reaction

   314 Depressive reaction

   318.5 Psychoneurotic reaction, other

302.7 Personality pattern disturbance

   320.2 Cyclothymic personality

   320.3 Inadequate personality

   320.0 Schizoid personality

   320.1 Paranoid personality

321.5 Personality trait disturbance

    321.0 Emotionally unstable personality

    321.1 Passive-aggressive personality

    321.5 Compulsive personality

DSM-II 1968

III. Psychoses Not Attributed to Physical Conditions Listed Previously

295.x Schizophrenia

296.x Major Affective Disorders (affective psychoses)

    295.0 Involutional melancholia

    295.1 Manic-depressive illness, manic type

    295.2 Manic-depressive illness, depressed type

    295.3 Manic-depressive illness, circular type

        295.33 Manic-depressive illness, circular

        295.34 Manic-depressive illness, depressed

    295.8 Other major affective disorder (affective psychosis, other)

    295.9 Unspecified major affective disorder

        [affective disorder not otherwise specified]

        [manic-depressive illness not otherwise specified]

IV. NEUROSES

300.x Neuroses

    300.0 Anxiety neurosis

    300.1 Hysterical neurosis

    300.2 Phobic neurosis

    300.3 Obsessive-compulsive neurosis.

    300.4 Depressive neurosis

    300.5 Neurasthenic neurosis

300.6 Depersonalization neurosis

300.7 Hypochondriacal neurosis

V. PERSONALITY DISORDERS AND CERTAIN
OTHER NON-PSYCHOTIC MENTAL DISORDERS

301.x Personality Disorders

301.0 Paranoid personality

301.1 Cyclothymic personality (affective personality)

301.2 Schizoid personality

301.3 Explosive personality

301.5 Hysterical personality

DSM-III 1980

Affective Disorders

Major Affective Disorders

296.xx Bipolar disorder

286.6x Mixed

296.4x Manic

296.5x Depressed

296.xx Major depression

296.2x Single episode

296.3x Recurrent

Other Specific Affective Disorders

301.13 Cyclothymic disorder

300.40 Dysthymia (depressive neurosis)

Atypical Affective Disorders

296.70 Atypical bipolar disorder

296.82 Atypical depression

DSM-III-R (1987)

Mood Disorders

Bipolar Disorders
  296.xx Bipolar disorder
    296.6x Mixed
    296.4x Manic
    296.5x Depressed
  301.13 Cyclothymia
  296.70 Bipolar disorder NOS (not otherwise specified)

Depressive Disorders
  296.xx Major depression
    296.2x Single episode
    296.3x Recurrent
  300.40 Dysthymia (depressive neurosis)
  311.00 Depressive disorder NOS

DSM-IV 1994

Mood Disorders

Depressive Disorders
  296.xx Major Depressive disorder
    296.2x Single episode
    296.3x Recurrent
  300.4 Dysthymic disorder
  311 Depressive disorder NOS

Bipolar Disorders
  296.xx Bipolar I disorder
    296.1x Single manic episode (specify if mixed)
    296.2x Most recent episode hypomanic
    296.3x Most recent episode manic
    296.6x Most recent episode mixed

296.5x Most recent episode depressed

296.7x Most recent episode unspecified

296.89 Bipolar II disorder

301.13 Cyclothymic disorder

296.80 Bipolar disorder NOS

293.83 Mood disorder due to (general) medical condition

296.90 Mood disorder NOS

DSM-V 2013

Bipolar and Related Disorders

296.xx Bipolar I Disorder

SPECPIFY: with anxious destress (mild, moderate, moderate-severe, and severe); with mixed features; with rapid cycling; with melancholic features; with atypical features; with mood-congruent psychotic features; with mood-incongruent psychotic-features; with catatonia; with peripartum onset; with seasonal pattern.

296.4x [F31.x] Current episode manic

296.4x [F31.x] Current episode hypomanic

296.5x [F31.x) Current episode depressed

296.7 [F31.9] Current episode unspecified

296.89 [F31.81] Bipolar II Disorder

SPECIFY: current or most recent episode manic or depressed

SPECIFY: if full criteria are not mete; full or partial remission

SPECIFY severity if full criteria for mood episode not currently met: mild, moderate, and severe

301.13 [F34.0] Cyclothymic Disorder

SPECIFY: with anxious distress

( ) [   ] Substance/medication-induced bipolar/related disorder

293.83 [   ] Bipolar disorder related to another medical condition

296.89 [F31.89] Other specified bipolar disorder

296.80 [F31.9] Unspecified bipolar and related disorder

Depressive Disorders

296.99 [F34.8] Disruptive Mood Dysregulation Disorder

296.xx F3x.xx] Major Depressive Disorder

SPECIFY: with anxious distress; (specify mild, moderate, moderate-severe, and severe); with mixed features; with melancholic features; with atypical features; with mood-congruent psychotic features; with mood-incongruent psychotic features; with catatonia; with peripartum onset; with seasonal features.

296.2x [F32.x] Single episode

296.3x F33.x] Recurrent episode

300.4 [F34.1] Persistent Depressive Disorder (dysthymia)

625.4 [N94.3] Premenstrual Dysphoric Disorder

( ) [ ] Substance/Medication-Induced Depressive Disorder

293.83 {] Depressive Disorder due to another Medical Condition

311 [F32.8] Other Specified Depressive Disorder

311 [F32.9] Unspecified Depressive Disorder

IDC-VI 1949

(300-309) Psychosis

300 Schizophrenic Disorders (Dementia Praecox)

 300.0 Simple type

 300.1 Hebephrenic type

 300.2 Catatonic type

 300.3 Paranoid type

 300.4 Acute schizophrenic reaction

 300.5 Latent schizophrenia

 300.6 Schizo-affective psychosis

 300.7 Other and unspecified

301 Manic-Depressive Reaction

 301.0 Manic and circular

301.1 Depressive

301.2 Other

302 Involutional Melancholia

(310-318) Psychoneurotic Disorders

310 Anxiety Reaction Without Mention of Somatic Symptoms

311 Hysterical Reaction Without Mention of Anxiety Reaction

312 Phobic Reaction

313 Obsessive-Compulsive Reaction

314 Neurotic-Depressive Reaction

315 Psychoneurosis With Somatic Symptoms (Somatization Reaction) Affecting Circulatory System

316 Psychoneurosis With Somatic Symptoms (Somatization Reaction) Affecting Digestive System

317 Psychoneurosis With Somatic Symptoms (Somatization Reaction) Affecting Other Systems

318 Psychoneurotic Disorders, Other, Mixed, and Unspecified Types

   318.0 Hypochondriacal reaction

   318.1 Depersonalisation

   318.2 Occupational neurosis

   318.3 Asthenic reaction

   318.4 Mixed

   *318.5 Of other and unspeci*fied types

(320-326) Disorders of Character, Behavior and Intelligence

320 Pathological Personality

   320.0 Schizoid personality

   320.1 Paranoid personality

   320.2 Cyclothymic personality

   320.3 Inadequate personality

   320.4 Antisocial personality

   320.5 Asocial personality

320.6 Sexual deviation

320.7 Other and unspecified

321 Immature Personality

   321.0 Emotional instability

   321.1 Passive dependency

   321.2 Aggressiveness

   321.3 Enuresis characterizing immature personality

   321.4 Other symptomatic habits except speech impediments

   321.5 Other and unspecified

ICD-10 1994

Mood (Affective) Disorders (F30–F39)

a. **F30 Manic Episode**

   a. **F30.0** *Hypomania*

   b. **F30.1** *Mania without psychotic symptoms*

   c. **F30.2** *Mania with psychotic symptoms*

   d. **F30.8** *Other manic episodes*

   e. **F30.9** *Manic episode and unspecified*

b. **F31 Bipolar Affective Disorder**

   a. **F31.0** *Bipolar affective disorder, current episode hypomanic*

   b. **F31.1** *Bipolar affective disorder, current episode manic without psychotic symptoms*

   c. **F31.2** *Bipolar affective disorder, current episode manic with psychotic symptoms*

   d. **F31.3** *Bipolar affective disorder, current episode mild or moderate depression*

   e. **F31.4** *Bipolar affective disorder, current episode severe depression without psychotic symptoms*

   f. **F31.5** *Bipolar affective disorder, current episode severe depression with psychotic symptoms*

g. **F31.6** *Bipolar affective disorder, current episode mixed*

h. **F31.7** *Bipolar affective disorder, currently in remission*

i. **F31.8** *Other bipolar affective disorders*

    i. Bipolar II disorder

    ii. Recurrent manic episodes NOS

j. **F31.9** *Bipolar affective disorder, unspecified*

c. **F32 Depressive Episode**

    a. **F32.0** *Mild depressive episode*

    b. **F32.1** *Moderate depressive episode*

    c. **F32.2** *Severe depressive episode without psychotic symptoms*

    d. **F32.3** *Severe depressive episode with psychotic symptoms*

    e. **F32.8** *Other depressive episodes*

        i. Atypical depression

        ii. Single episodes of "masked" *depression NOS*

    f. **F32.9** *Depressive episode, unspecified*

d. **F33 Recurrent depressive disorder**

    a. **F33.0** *Recurrent depressive disorder, current episode mild*

    b. **F33.1** *Recurrent depressive disorder, current episode moderate*

    c. **F33.2** *Recurrent depressive disorder, current episode severe without psychotic symptoms*

    d. **F33.3** *Recurrent depressive disorder, current episode severe with psychotic symptoms*

    e. **F33.4** *Recurrent depressive disorder, currently in remission*

    f. **F33.8** *Other recurrent depressive disorders*

    g. **F33.9** *Recurrent depressive disorder, unspecified*

e. **F34 Persistent Mood (Affective) Disorders**

    a. **F34.0 Cyclothymia**

    b. **F34.1 Dysthymia**

    c. **F34.8** *Other persistent mood (affective) disorders*

    d. **F34.9** *Persistent mood (affective) disorder, unspecified*

f. F38 *Other* Mood (Affective) Disorders

 a. **F38.0** *Other single mood (affective) disorders*

  i. Mixed affective episode

 b. **F38.1** *Other recurrent mood (affective) disorders*

  i. *Recurrent brief depressive episodes*

 c. **F38.8** *Other specified mood (affective) disorders*

 d. **F39** Unspecified mood (affective) disorder

DSM-5 (2013) DIAGNOSTIC CRITERIA FOR AFFECTIVE DISORDERS

Bipolar and Related Disorders

Manic Episode

A. A distinct period of abnormally and persistently elevated, expansive or irritable mood and abnormally persistent and goal-directed behavior or energy lasting for least 1 week and present most of the day, nearly every day (or any duration if hospitalization is necessary).

B. During the period of mood disturbance and increased energy or activity, three or more of the following are present (four if mood is only irritable):

 1) inflated self-esteem or grandiosity

 2) decreased need for sleep (e.g., feels rested after only 3 hours of sleep)

 3) more talkative than usual or pressure to keep talking

 4) flight of ideas or subjective experience that thoughts are racing

 5) distractibility (i.e., attention too easily drawn to unimportant or irrelevant external stimuli), as reported or observed

 6) increase in goal-directed activity (either socially, at work, school, or sexually) or psychomotor agitation (i.e., purposeless, non-goal-directed activity)

 7) excessive involvement in activities that have a high potential for painful consequences (e.g., engaging in unrestrained buying sprees, sexual indiscretions, or foolish business investments

C. The mood disturbance is sufficiently severe to cause marked impairment in social or occupational functioning or to necessitate hospitalization to prevent harm to self or others, or there are psychotic features.
D. The episode is not directly attributable to the effects of a substance.

Criteria A–D are necessary to diagnose a manic episode. At least one lifetime manic episode is required for a diagnosis of Bipolar I Disorder. Bipolar I disorder should be coded as mild, moderate, or severe, depending on the nature of the most recent episode. Specifiers are then added to describe subtype (with anxious distress (mild, moderate, moderate-severe, and severe); with mixed features; with rapid cycling; with melancholic features; with atypical features; with mood-congruent psychotic features; with mood-incongruent psychotic-features; with catatonia; with peripartum onset; with seasonal pattern).

Hypomanic Episode

A. A distinct period of abnormally and persistently elevated, expansive or irritable mood and abnormally and persistently increased activity or energy lasting at least 4 consecutive days and present most of the day, nearly every day.
B. During the period of mood disturbance and increased energy or activity, three (or more) of the following are present (four if mood is only irritable) have persisted, represent a notable change from usual behavior, and have been present to a significant degree:
   1) inflated self-esteem or grandiosity
   2) decreased need for sleep (e.g., feels rested after only 3 hours of sleep)
   3) more talkative than usual or pressure to keep talking
   4) flight of ideas or subjective experience that thoughts are racing
   5) distractibility (i.e., attention too easily drawn to unimportant or irrelevant external stimuli), as reported or observed
   6) increase in goal-directed activity (either socially, at work, school, or sexually) or psychomotor agitation
   7) excessive involvement in activities that have a high potential for painful consequences (e.g., engaging in unrestrained buying sprees, sexual indiscretions, or foolish business investments)

C. The episode is associated with an unequivocal change in functioning that is uncharacteristic of the individual when not symptomatic.

D. The disturbance in mood and the change in functioning are observable by others.

E. The episode is not severe enough to cause marked impairment in social or occupational functioning or to necessitate hospitalization. If there are psychotic features, the episode is, by definition, manic.

F. The episode is not attributable to the physiologic effects of a substance (e.g., a drug of abuse, a medication, other treatment).

Major Depressive Episode

A. Five or more of the following symptoms have been present for the same 2-week period and represent a change from previous functioning: at least one of the symptoms is either (1) depressed mood or (2) loss of pleasure.

    1) depressed mood most of the day, nearly every day, as indicated either by subjective report (e.g., feel sad, empty, or hopeless) or by observation made by others (e.g., appears tearful). Note: in children and adolescents can be irritable mood.

    2) markedly diminished interest and pleasure in all, or almost all, activities, most of the day, nearly every day (as indicated either by subjective account or observation).

    3) significant weight loss when not dieting or weight gain (e.g., change of more than 5% of body weight in a month), or decrease in appetite nearly every day (Note: in children consider failure to make expected weight gain).

    4) insomnia or insomnia nearly every day

    5) psychomotor agitation or retardation nearly every day (observable by others, not merely subjective feelings of restlessness or being slowed down).

    6) fatigue or loss of energy nearly every day.

    7) feelings of worthlessness or excessive or inappropriate guilt (which may be delusional) nearly every day (not merely self-reproach or guilt about being sick).

8) diminished ability to think or concentrate, or indecisiveness nearly every day (either by subjective account or observed by others).
9) recurrent thoughts of death (not just fear of dying), recurrent suicidal ideation without a specific plan, or a suicide attempt or a specific plan for committing suicide.

B. The symptoms cause clinically significant distress or impairment in social, occupational, or other important areas of functioning.

C. The episode is not attributable to the effects of a substance or another medical condition.

Bipolar I Disorder

A. A diagnosis of bipolar I disorder requires at least one episode of mania (as above, euphoric or irritable mood with functional impairment, with or without psychosis).

B. The occurrence of the manic and major depressive episodes must not be better explained by schizoaffective disorder, schizophrenia, schizophreniform disorder, delusional disorder, or other psychotic disorder.

Bipolar II Disorder

A. A diagnosis of bipolar II disorder requires at least one episode of hypomania, euphoric or irritable, AND at least one episode of major depression.

B. There must not be a history of a manic episode.

C. The occurrence of the hypomanic and major depressive episodes must not be better explained by schizoaffective disorder, schizophrenia, schizophreniform disorder, delusional disorder, or other psychotic disorder.

D. The symptoms of depression or the unpredictability caused by frequent alternation between periods of depression and hypomania causes clinically significant distress or impairment in social, occupational, or other important areas of functioning.

Bipolar II disorder should be coded as mild, moderate, or severe, depending on the nature of the most recent episode. Specifiers are then added to describe subtype, e.g., with anxious distress (mild, moderate, moderate-severe, and severe); with mixed features; with rapid cycling; with melancholic features; with atypical features; with mood-congruent psychotic features; with mood-incongruent psychotic features; with catatonia; with peripartum onset; with seasonal pattern.

Cyclothymic Disorder

A. For at least 2 years (1 year in children and adolescents), there have been numerous periods with hypomanic symptoms that do not meet criteria for a hypomanic episode and numerous periods with depressive symptoms that do not meet criteria for a major depressive episode.

B. During the above 2-year period (1 year for children and adolescents), the hypomanic and depressive periods have been present for at least half the time and the individual has not been without the symptoms for more than 2 months at a time.

C. Criteria for a major depressive, manic, or hypomanic episode have never been met.

D. The symptoms in criterion A are not better explained by schizoaffective disorder, schizophrenia, schizophreniform disorder, delusional disorder, or other psychotic disorder.

E. The symptoms are not attributable to the psychological effects of a substance (e.g., a drug of abuse and a medication) or another medical condition.

F. The symptoms cause clinically significant distress or impairment in social, occupational, or other areas of functioning.

Other Specified Bipolar and Related Disorder: (examples)

A. Short-duration hypomanic episodes (2–3 days) and major depressive episodes.

B. Hypomanic episodes with insufficient symptoms and major depressive episodes.

C. Hypomanic episode without prior major depressive episode.

D. Short-duration cyclothymia (less than 24 months).

Unspecified Bipolar and Related Disorder

This category applies to presentations in which symptoms characteristic of a bipolar or related disorder cause clinically significant distress or impairment in social, occupational, or other important areas of functioning and predominate but do not meet the full criteria for any of the disorders in the bipolar and related disorders diagnostic class. The unspecified bipolar and related disorder category is used in situations in which the clinician chooses not to specify the reason that the criteria are not met for a specific bipolar and related disorder and includes presentations in which there is insufficient information to make a more specific diagnosis (e.g., in emergency room settings) (DSM-5 2014).

Manic or hypomanic episode with mixed features

A. Full criteria are met for a manic episode or a hypomanic episode and at least three of the following symptoms are present during the majority of days of the current or most recent episode of mania or hypomania:

1. Prominent dysphoria or depressed mood as indicated by either subjective report (e.g., feels sad or empty) or by observation made by others (e.g., appears tearful).
2. Diminished interest or pleasure in all or almost all activities (as indicated by either subjective account or observation made by others).
3. Psychomotor retardation nearly every day (observable by others: not merely subjective feelings of being slowed down).
4. Fatigue or loss of energy.
5. Feelings of worthlessness or excessive or inappropriate guilt (not merely self-reproach or guilt about being sick).
6. Recurrent thoughts of death (not just fear of dying), recurrent suicidal ideation without a specific plan, or a suicide attempt or a specific plan for committing suicide.

B. Mixed symptoms are observable by others and represent a change from the person's usual behavior.

C. For individuals whose symptoms meet full episode criteria for both mania and depression simultaneously, the diagnosis should be manic episode with mixed features due to the marked impairment and the clinical severity of full mania.

D. The mixed symptoms are not attributable to the physiological effects of a substance (e.g., a drug of abuse, a medication, or other treatment).

Depressive episode with mixed features

A. Full criteria are met for a major depressive episode and at least three of the following manic/hypomanic symptoms are present during the majority of days of the current or most recent episode of depression:
   1. Elevated, expansive mood
   2. Inflated self-esteem or grandiosity
   3. More talkative than usual or pressure to keep talking
   4. Flight of ideas or subjective experience that thoughts are racing
   5. Increase in energy or goal-directed activity (either socially, at work or school, or sexually)
   6. Increased or excessive involvement in activities that have a high potential for painful consequences (e.g., engaging in unrestrained buying sprees, sexual indiscretions, or foolish business investments)
   7. Decreased need for sleep (feeling rested despite sleeping less than usual; to be contrasted with insomnia).
B. Mixed symptoms are observable by others and represent a change from the person's usual behavior.
C. For individuals whose symptoms meet full episode criteria for both mania and depression simultaneously, the diagnosis should be manic episode with mixed features.
D. The mixed symptoms are not attributable to the physiological effects of a substance (e.g., a drug of abuse, a medication, or other treatment).

Depressive Disorders

Disruptive Mood Dysregulation Disorder (DMDD)

A. Severe recurrent temper outbursts manifested verbally (e.g., verbal rages) and/or behaviorally (e.g., physical aggression toward people or property) that are grossly out of proportion in intensity or duration to the situation or provocation.

B. The temper outbursts are inconsistent with developmental level.

C. The temper outbursts occur on average, three or more times a week.

D. The mood between temper outbursts is persistently irritable or angry most of the day, nearly every day, and is observable by others (e.g., parents, teachers, and peers).

E. Criteria A–D have been present for 12 or more months. Throughout that time, the individual has not had a period lasting 3 or more consecutive months without all the symptoms in criteria A–D.

F. Criteria A and D are present in at least two of three settings (home, school, and peers) and are severe in at least one of them).

G. The diagnosis should not be made for the first time before the age of 6 years or after the age of 18 years.

H. By history or observation, the age of onset of criteria A–E is before 10 years.

I. There has never been a distinct period lasting more than 1 day during which the full symptom criteria, except duration, for a manic or hypomanic episode have been met. NOTE: Developmentally appropriate mood elevation such as occurs in the context of a highly positive event or its anticipation should not be considered as a symptom of mania or hypomania.

J. The behaviors do not occur exclusively during an episode of major depressive disorder and are not better explained by another mental disorder (e.g., autism spectrum disorder, PTSD, and dysthymia). NOTE: This diagnosis cannot coexist with oppositional defiant disorder, intermittent explosive disorder, or bipolar disorder though it can coexist with others including major depressive disorder, ADHD, conduct disorder, and substance use disorders. Individuals who meet criteria for both disruptive mood dysregulation disorder and oppositional defiant disorder should only be given the diagnosis of disruptive mood dysregulation disorder. If an individual has ever experienced a manic or hypomanic episode, the diagnosis of disruptive mood dysregulation disorder should not be assigned.

K. The symptoms are not attributable to the physiological effects of a substance or to another medical or neurological condition.

Major Depressive Disorder (MDD)

A. Five or more of the following symptoms have been present for the same 2-week period and represent a change from previous functioning: at least one of the symptoms is either (1) depressed mood or (2) loss of pleasure.

   1) depressed mood most of the day, nearly every day, as indicated either by subjective report (e.g., feel sad, empty, or hopeless) or by observation made by others (e.g., appears tearful). Note: children and adolescents can be in irritable mood.

   2) markedly diminished interest and pleasure in all, or almost all, activities, most of the day, nearly every day (as indicated either by subjective account or observation).

   3) significant weight loss when not dieting or weight gain (e.g., change of more than 5% of body weight in a month), or decrease in appetite nearly every day (Note: in children consider failure to make expected weight gain).

   4) insomnia or insomnia nearly every day.

   5) psychomotor agitation or retardation nearly every day (observable by others, not merely subjective feelings of restlessness or being slowed down).

   6) fatigue or loss of energy nearly every day.

   7) feelings of worthlessness or excessive or inappropriate guilt (which may be delusional) nearly every day (not merely self-reproach or guilt about being sick).

   8) diminished ability to think or concentrate, or indecisiveness nearly every day (either by subjective account or observed by others).

   9) recurrent thoughts of death (not just fear of dying), recurrent suicidal ideation without a specific plan, or a suicide attempt or a specific plan for committing suicide.

B. The symptoms cause clinically significant distress or impairment in social, occupational, or other important areas of functioning.

C. The episode is not attributable to the effects of a substance or another medical condition.

   (Note: Criteria A–C represent a major depressive episode).

D. The occurrence of the major depressive episode is not better explained by schizoaffective disorder, schizophrenia, schizophreniform disorder, delusional disorder, or other psychotic disorder.

E. There has never been a manic or hypomanic episode.
    SPECIFIERS: Mild, moderate, severe, with psychotic features, in full remission, in partial remission, with anxious distress, with mixed features, with melancholic features, with mood-congruent psychotic features, with mood-incongruent psychotic features, with catatonia, with peripartum onset, and with seasonal pattern.

Persistent Depressive Disorder (PDD)

A. Depressed mood for most of the days for more days than not as indicated either by subjective account or observation by others for at least 2 years. (NOTE: In children and adolescents, mood can be irritable, and duration can be at least 1 year).
B. Presence, while depressed, of two (or more) of the following:
    1) poor appetite or overeating
    2) insomnia or hypersomnia
    3) low energy or fatigue
    4) low self-esteem
    5) poor concentration or difficulty making decisions
    6) feelings of hopelessness
C. During the 2-year period (1 year for children and adolescents) of the disturbance, the individual has never been without the symptoms in Criteria A and B for more than 2 months at a time.
D. Criteria for a major depressive disorder may be present for 2 years.
E. There has never been a manic episode or a hypomanic episode, and criteria have never been met for cyclothymic disorder.
F. The occurrence of the major depressive episode is not better explained by schizoaffective disorder, schizophrenia, schizophreniform disorder, delusional disorder, or other psychotic disorder.
G. The symptoms are not attributable to the physiological effects of a substance (e.g., a drug of abuse and a medication) or another medical condition.
H. The symptoms cause clinically significant distress or impairment in social, occupational, or other important areas of functioning.
    SPECIFIERS: Mild, moderate, severe, in full remission, in partial remission, with anxious distress, with mixed features, with melancholic features, with mood-congruent psychotic features, with mood-incongruent psychotic features, with peripartum onset, with

pure dysthymic syndrome, and with persistent major depressive episode.

DSM-5 CRITERIA FOR CO-OCCURING and RELATED DISORDERS

Borderline Personality Disorder 301.83 or F60.3

A pervasive pattern of instability of interpersonal relationships, self-image and affects and marked impulsivity, beginning by early adulthood and present in a variety of contexts, as indicated by five or more of the following:

1. Frantic efforts to avoid real or imagined abandonment (NOTE: Do not include suicidal or self-mutilating behavior covered in criterion 5).
2. A pattern of unstable or intense personal relationships characterized by alternating between intense idealization and devaluation.
3. Identity disturbance: markedly and persistently unstable self-image or sense of self.
4. Impulsivity in at least two areas that are potentially self-damaging (e.g., spending sex, substance abuse, reckless driving, and binge eating). (NOTE: Do not include suicidal or self-mutilating behavior covered in criterion 5).
5. Recurrent suicidal behavior, gestures or threats, or self-mutilating behavior.
6. Affective instability due to a marked reactivity of mood (e.g., intense episodic dysphoria, irritability or anxiety, usually lasting a few hours and only rarely more than few days).
7. Chronic feelings of emptiness.
8. Inappropriate and intense anger or difficulty in controlling anger (e.g., frequent display of temper, constant anger, and recurrent physical fights).
9. Transient, stress-related paranoid ideation, or severe dissociative symptoms.

Attention-Deficit/Hyperactivity Disorder 314.xx or F90.x

A. A persistent pattern of inattention and/or hyperactivity-impulsivity that interferes with functioning or development as characterized by (1) and/or (2):

1. Inattention: six (or more) of the following symptoms have persisted for at least 6 months to a degree that is inconsistent with developmental level and that negatively impacts directly on social/academic/occupational activities.

   a. Often fails to give close attention to details or makes careless mistakes in schoolwork, at work, or during other activities.
   b. Often has difficulties sustaining attention in tasks or in play activities.
   c. Often does not seem to listen when spoken to directly.
   d. Often does not follow through on instructions and fails to finish schoolwork, chores, or duties in the workplace.
   e. Often has difficulties in organizing tasks and activities.
   f. Often avoids, dislikes, or is reluctant to engage in tasks that require sustained mental effort.
   g. Often loses things necessary for tasks or activities.
   h. Is often easily distracted by extraneous stimuli.
   i. Is often forgetful in daily activities.

2. Hyperactivity and impulsivity: Six (or more) of the following symptoms have persisted for at least 6 months to a degree that is consistent with developmental level and that negatively impacts directly on social and academic/occupational activities:

   a. Often fidgets with or taps hands or feet or squirms in seat.
   b. Often leaves seat in situations when remaining seated is expected.
   c. Often runs about or climbs in situations in which it is inappropriate.
   d. Often unable to play or engage in leisure activities quietly.
   e. Is often "on the go," acting as if "driven by a motor."
   f. Often talks excessively.
   g. Often blurts out an answer before a question have been completed.
   h. Often has difficulty waiting for his or her turn.
   i. Often interrupts or intrudes on others.

B. Several inattentive or hyperactive/impulsive symptoms were present before the age of 12.

C. Several inattentive or hyperactive/impulsive symptoms are present in two or more settings (home, work, school, with friends, relatives, and other activities).
D. There is clear evidence that the symptoms interfere with or reduce the quality of social, academic, or occupational functioning.
E. The symptoms do not occur exclusively during the course of schizophrenia or another psychotic disorder and are not better explained by another mental disorder.
    (Specifiers are for combined type, inattentive type or hyperactive type)

Binge Eating Disorder 307.51 or F50.8

A. Recurrent episodes of binge eating. An episode of binge eating is characterized by both of the following:
    1. Eating in a discrete period of time (e.g., within any 2-hour period), an amount of food that is definitely larger than what most people would eat in a similar period of time under similar circumstances.
    2. A sense of lack of control of eating during the episode (e.g., a feeling that one cannot stop eating or control what or how much one is eating).
B. The binge eating episodes are associated with three (or more) of the following:
    1. Eating much more rapidly than normal.
    2. Eating until feeling uncomfortably full.
    3. Eating large amounts of food when not feeling physically hungry.
    4. Eating alone because of feeling embarrassed by how much one is eating.
    5. Feeling disgusted with oneself, depressed, or very guilty afterward.
C. Marked distress regarding binge eating is present.
D. The binge eating occurs on average, at least once a week for at least 3 months.
E. The binge eating is not associated with the recurrent use of inappropriate compensatory behavior as in bulimia nervosa and does not occur exclusively during the course of bulimia nervosa or anorexia nervosa.

Posttraumatic Stress Disorder 309.81 or F43.10

A. Exposure to actual or threatened death, serious injury, or sexual violence in one (or more) of the following ways:
   1. Directly experiencing the traumatic event.
   2. Witnessing, in person, the event(s) as it occurs to others.
   3. Learning that the event(s) occurred to a close family member or a close friend. In cases of actual threatened death of a family member or friend, the event(s) must have been violent or accidental.
   4. Experiencing extreme or repeated exposure to aversive details of the traumatic event(s) (e.g., first responders collecting human remains, police officers repeatedly exposed to details of child abuse).
   NOTE: Criterion A4 does not apply to exposure through electronic media, television, movies, or pictures unless this exposure is work-related.

B. Presence of one or more of the following intrusion symptoms associated with the traumatic event(s), beginning after the traumatic event(s) occurred:
   1. Recurrent, involuntary, intrusive, and distressing memories of the traumatic event(s).
   2. Recurrent distressing dreams in which the content and/or affect of the dream are related to the traumatic event(s).
   3. Dissociative reactions (e.g., flashbacks) in which the individual feels or acts as if the traumatic event(s) were recurring (such reactions may occur on a continuum with the most severe expression being a complete loss of awareness of present surroundings).
   4. Intense or prolonged psychological distress at exposure to internal or external cues that symbolize or resemble an aspect of the traumatic event(s).
   5. Marked physiological reactions to internal or external cues that symbolize or resemble an aspect of the traumatic event(s).

C. Persistent avoidance of stimuli associated with the traumatic events, beginning after the event(s) occurred, as evidenced by one or both of the following:

    1. Avoidance of or efforts to avoid distressing memories, thoughts, or feelings about or closely associated with the traumatic event(s).

    2. Avoidance of or efforts to avoid external reminders (people, places, conversations, activities, objects, and situations), which arouse distressing memories, thoughts, or feelings about or associated with the traumatic event.

D. Negative alterations in cognition and mood associated with the traumatic event(s) beginning or worsening after the traumatic event(s) occurred, as evidenced by two or more of the following:

    1. Inability to remember an important aspect(s) of the traumatic event(s), typically due to dissociative amnesia and not to other factors.

    2. Persistent and exaggerated negative beliefs about oneself, others, or the world.

    3. Persistent distorted cognitions about the cause or consequences of the traumatic event(s) that lead the individual to blame him/herself and others.

    4. Persistent negative emotional state.

    5. Markedly diminished interest or participation in significant activities.

    6. Feelings of detachment or estrangement from others.

    7. Persistent inability to experience positive emotions.

E. Marked alterations in arousal and reactivity associated with the traumatic event(s), beginning or worsening after the traumatic event(s) occurred, as evidenced by two (or more) of the following:

    1. Irritable behavior and angry outbursts (with little or no provocation), typically expressed as verbal or physical aggression toward people or objects

    2. Reckless or self-destructive behavior

    3. Hypervigilance

    4. Exaggerated startle response

    5. Problems with concentration

    6. Sleep disturbance

F. Duration of the disturbance is more than 1 month

G. The disturbance causes clinically significant distress or impairment in social, occupational, or other important areas of functioning.

H. The disturbance is not attributable to the physiological effects of a substance or another medical condition.

Developmental Trauma Disorder (not a DSM diagnosis)

A. Exposure

1. Multiple or chronic exposure to one or more forms of developmentally adverse interpersonal trauma (e.g., abandonment, betrayal, physical assaults, sexual assaults, threats to bodily integrity, coercive practices, emotional abuse, witnessing violence, or death).
2. Subjective experience (e.g., rage, betrayal, fear, resignation, defeat, and shame).

B. Triggered Pattern of Repeated Dysregulation in Response to Trauma Cues

Dysregulation (high or low) in presence of cues. Changes persist and do not return to baseline; not reduced in intensity by conscious awareness.

1. Affective
2. Somatic (e.g., physiologic, motoric, and medical)
3. Behavioral (e.g., reenactment and cutting)
4. Cognitive (e.g., thinking that it is happening again, confusion, dissociation, and depersonalization)
5. Relational (e.g., clinging, distrustful, oppositional, and compliant)
6. Self-attribution (e.g., self-hatred and blame)

C. Persistently Altered Attributions and Expectancies

1. Negative self-attribution
2. Distrust of protective caretaker
3. Loss of expectancy of protection by others
4. Loss of trust in social agencies to protect
5. Lack or recourse to social justice/retribution
6. Inevitability of future victimization

D. Functional Impairment

1. Educational

2. Familial
3. Peer
4. Legal
5. Vocational

Generalized Anxiety Disorder 300.02 or F41.1

A. Excessive anxiety and worry (apprehensive expectation) occurring more days than not for at least 6 months about a number of events or activities (such as work or school performance).
B. The individual finds it difficult to control the worry.
C. The anxiety and worry are associated with three (or more) of the following six symptoms (with at least some symptoms having been present more days than not for the past 6 months):
  1. Restlessness or feeling keyed up or on edge
  2. Being easily fatigued
  3. Difficulty concentrating or mind going blank
  4. Irritability
  5. Muscle tension
  6. Sleep disturbance
D. The anxiety, worry, or physical symptoms cause clinically significant distress or impairment in social, occupational, or other important areas of functioning.
E. The disturbance is not attributable to the physiologic effects of a substance or another medical condition.
F. The disturbance is not better explained by another mental disorder.

Panic Disorder 300.01 or F41.0

A. Recurrent unexpected panic attacks. A panic attack is an abrupt surge of intense fear or intense discomfort that reaches a peak within minutes and during which time at least four of the following symptoms appear:
  1. Palpitations, pounding heart, or accelerated heart rate
  2. Seating
  3. Trembling or shaking

4. Sensations of shortness of breath or smothering
5. Feelings of choking
6. Chest pain or discomfort
7. Nausea or abdominal distress
8. Feeling dizzy, unsteady, lightheaded, or faint
9. Chills or heat sensations
10. Paresthesias (numbness or tingling sensations)
11. Derealization (feelings of unreality) or depersonalization (being detached from oneself)
12. Fear of losing control or "going crazy"
13. Fear of dying.

B. At least one of the attacks has been followed by 1 month (or more) of 1 of the following:
C. Persistent concern or worry about additional panic attacks or their consequences.
D. A significant maladaptive change in behavior related to the attacks.
E. The disturbance is not attributable to the physiologic effects of a substance or another medical condition.
F. The disturbance is not better explained by the presence of another mental disorder.

Substance Use Disorder 292.9 or F19.99

A problematic pattern of (substance) use leading to clinically significant impairment or distress, as manifested by at least two of the following, occurring within a 12-month period:

1. (Substance) is often taken in larger amounts or over a longer period of time than intended.
2. There is a persistent desire or unsuccessful efforts to cut down or control (substance) use.
3. A great deal of time is spent in activities necessary to obtain, use, or recover from (substance) effects.
4. Craving or a strong desire or urge to use (substance).
5. Recurrent (substance) use leading to failure to fulfill major role obligations at home, work, or school.
6. Continued (substance) use despite having persistent or recurrent social or interpersonal problems caused or exacerbated by the effects of (substance).

7. Important social, occupational, or recreational activities are given up or reduced because of (substance) use.
8. Recurrent (substance) use in situations in which it is physically hazardous.
9. (Substance) use is continued despite knowledge of having a persistent or recurrent physical or psychological problem that is likely to have been caused or exacerbated by substance.
10. Tolerance, as defined by either of the following:
    a. A need for markedly increased amounts of (substance) to achieve intoxication or the desired effect.
    b. A markedly diminished effect with continued use of the same amount of (substance).
11. Withdrawal (substance-specific).

AFFECTIVE TEMPERAMENT TYPES (NON-DSM)

Hyperthymic Temperament (warm, people-seeking, extroverted, eloquent, jocular, overconfident, self-assured, high-energy, uninhibited, stimulus-seeking, habitual short-sleeper)
1. Cheerful, overoptimistic, and exuberant
2. Naïve, overconfident, self-assured, boastful, bombastic, and grandiose
3. Vigorous, full of plans, improvident, and rushing off with restless impulse
4. Overtalkative
5. Warm, people-seeking, and extroverted
6. Overinvolved and meddlesome
7. Uninhibited, stimulus-seeking, and meddlesome

Cyclothymic Temperament (feeling all emotions intensely; rapid shifts in mood & energy; enjoying people, then losing interest; bubbly, then sluggish; overconfident, then unsure of oneself)
1. Biphasic dysregulation with rare euthymia
2. Marked unevenness in quantity and quality of productivity-associated working hours
3. Lethargy alternating with eutonia
4. Pessimistic brooding alternating with optimism
5. Mental confusion alternating with sharpened and creative thinking

6. Shaky self-esteem alternating between low self-confidence and over-confidence
7. Hypersomnia alternating with decreased need for sleep
8. Introverted self-absorption alternating with uninhibited people-seeking
9. Decreased verbal output alternating with talkativeness
10 Unexplained tearfulness alternating with excessive punning and jocularity

Depressive Temperament (self-blaming; self-denying, rejection-sensitive, uncomfortable with change, low energy, deferential)

1. Gloomy, pessimistic, humorless, or incapable of fun
2. Quiet, passive, and indecisive
3. Skeptical, hypercritical, or complaining
4. Brooding and given to worry
5. Conscientious and given to self-discipline
6. Self-critical, self-reproaching, and self-derogatory
7. Preoccupied with inadequacy, failure & negative events to the point of morbid enjoyment of ones' failures

Irritable Temperament

1. Habitually moody, irritable and choleric, with infrequent euthymia
2. Tendency to brood
3. Hypercritical and complaining
4. Ill-humored and joking
5. Obtrusiveness
6. Dysphoric restlessness
7. Impulsive

Anxious Temperament (worried someone will get ill or have an accident or that there will be bad news about a loved one)

1. Apprehensive cognitive set
2. Autonomic arousal: tension, GI distress
3. Behavior tremulous and irritable

References

Abbass, A. The Emergence of Psychodynamic Psychotherapy for Treatment Resistant Patients: Intensive Short-Term Dynamic Psychotherapy. *Psychodynam. Psych.* **2016**, *44* (2), 245–280.

Abe, R.; Okada, S.; Nakayama, R.; Ikegaya, Y.; Sasaki, T. Social Defeat Stress Causes Selective Attenuation of Neuronal Activity in the Ventromedial Prefrontal Cortex. *Sci. Rep.* **2019**, *9*, 9447.

Abou-Saleh, M. T.; Coppen, A. The Efficacy of Low-Dose Lithium: Clinical, Psychological and Biological Correlates. *J. Psych. Res.* **1989**, *23* (2), 157–164.

Abreu, T.; Braganca, M. The Bipolarity of Light and Dark: A Review on Bipolar Disorder and Circadian Cycles. *J. Aff. Dis.* **2015**, *185*, 219–229.

Adamec, R. E.; Stark-Adamec, C. Behavioral Inhibition and Anxiety: Dispositional, Developmental and Neural Aspects of the Anxious Personality of the Domestic Cat. In *Perspectives on Behavioral Inhibition*; Stevens, J., Ed.; University of Chicago Press: Chicago, 1989; pp 93–124.

Adelman, N. E.; Kayser, R.; Dickstein, D.; Blair, R. J. R.; Pine, D.; Leibenluft, E. Neural Correlates of Reversal Learning in Severe Mood Dysregulation and Pediatric Bipolar Disorder. *J. Am. Acad. Ch. Adol. Psych.* **2011**, *50* (11), 1173–1185.

Ader, R.; Grota, L. J. Effects of Early Experience on Adrenocortical Reactivity. *Physiol. Behav.* **1969**, *4*, 303–305.

Admon, R.; Milad, M. R.; Hendler, T. A Causal Model of Post-Traumatic Stress Disorder: Disentangling Predisposed from Acquired Neural Abnormalities. *Trends Cog. Sci.* **2013**, *17*, 337–347.

Agosti, V.; Stewart, J. W. Efficacy and Safety of Antidepressant Monotherapy in the Treatment of Bipolar II Depression. *Int. Clin. Psychopharm.* **2007**, *22* (5), 309–311.

Ainsworth, M. D.; Blehar, M.; Waters, E.; Wall S: *Patterns of Attachment: A Psychological Study of the Strange Situation*; Lawrence Erlbaum: Hillsdale, NJ, 1978.

Akiskal, H. S. Characterologic Manifestations of Affective Disorders: Toward a New Conceptualization. *Integ Psych* **1984**, *2*, 83–88.

Akiskal, H. S. Delineating Irritable-Choleric and Hyperthymic Temperaments as Variants of Cyclothymia. *J Pers Dis.* **1992a**, 6:326–342.

Akiskal, H. S. Demystifying Borderline Personality: Critique of the Concept and Unorthodox Reflections on Its Natural Kinship with the Bipolar Spectrum. *Acta Psych. Scand.* **2004**, *110* (6), 401–407.

Akiskal, H. S. Developmental Pathways to Bipolarity: Are Juvenile-Onset Depressions Pre-Bipolar? *J. Am. Acad. Ch. Adol. Psych.* 1995, *34* (6), 754–763.

Akiskal, H. S. Dysthymic Disorder: Psychopathology of Proposed Chronic Depressive Subtypes. *Am. J Psych* **1983a**, 140, 11–20.

Akiskal, H. S. Familial Genetic Principles for Validating the Bipolar Spectrum: Their Application in Clinical Practice, 2002. http://www.medscape.com/viewarticle/436384

Akiskal, H. S. Sub-Affective Disorders: Dysthymic, Cyclothymic and Bipolar II Disorders in the "Borderline" Realm. *Psych. Clin. NA.* **1981,** *4,* 25–46.

Akiskal, H. S. The Bipolar Spectrum: New Concepts in Classification and Diagnosis. In *Psychiatry Update: The American Psychiatric Association Annual Review*; Grinspoon, L., Ed., Vol. 2; American Psych Press: Washington, DC, 1983b; pp 271–292.

Akiskal, H. S. The Bipolar Spectrum: New Concepts in Classification and Diagnosis. *Psych Update, APA Annual Rev*; Greenspoon, L., Ed., Vol. I. I.; American Psych Press, Inc.: Washington, DC, 1983; pp 271–292.

Akiskal, H. S. The Clinical Significance of the "Soft" Bipolar Spectrum. *Psych Ann* **1986,** *16* (11), 667–671.

Akiskal, H. S. The Prevalent Clinical Spectrum of Bipolar Disorders: Beyond DSM-IV. *J. Clin. Psych.opharm* **1996a,** *16* (supp 1), 1s–14s.

Akiskal, H. S. The Temperamental Foundations of Affective Disorders. In *Interpersonal Factors in the Origin and Course of Affective Disorders*, Mundt, C., Hahlweg, K., Fiedler, P., Eds.; Gaskell: London, 1996b, pp 3–30.

Akiskal, H. S. Toward a Definition of Generalized Anxiety Disorder as an Anxious Temperament Type. *Act. Scand. Psych.* **1998,** *98* (Supp 393), 66–73.

Akiskal, H. S. Validating 'Hard' and 'Soft' Phenotypes Within the Bipolar Spectrum: Continuity or Discontinuity? *J. Aff. Dis.* **2003,** *73* (1–2), 1–5.

Akiskal, H. S.; Akiskal, K. K.; Haykal, R. F.; Manning, J. S.; Connor, P. D. TEMPS-A: Progress Towards Validation of a Self-Rated Clinical Version of the Temperament Evaluation of the Memphis, Pisa, Paris, and San Diego Autoquestionnaire. *J. Aff. Dis.* **2005a,** *85,* 3–16.

Akiskal, H. S.; Akiskal, K.; Alilaire, J. F.; Azorin, J. M.; Bourgeois, M. L.; Sechter, D.; Fraud, J. P.; Chatenet-Duchene, L.; Lancrenon, S.; Perugi, G.; Hantouche EEG: Validating Affective Temperaments in Their Subaffective and Socially Positive Attributes: Psychometric, Clinical and Familial Data from a French Familial Study. *J. Aff. Dis.* **2005b,** *83,* 29–36.

Akiskal, H. S.; Benazzi, F. Toward a Clinical Delineation of Dysphoric Hypomania— Operational and Conceptual Dilemmas. *Bip. Dis.* **2005,** *7* (5), 456–464.

Akiskal, H. S.; Bourgeois, M. K.; Angst, J.; Post, R.; Moller, H.; Hirschfeld, R. Re-Evaluating the Prevalence of and Diagnostic Composition Within the Broad Clinical Spectrum of Bipolar Disorders. *J. Aff. Dis.* **2000,** *59* s5–s30.

Akiskal, H. S.; Bourgeois, M. L.; Angst, J.; Post, R.; Moller, H.; Hirschfeld, R. Re-Evaluating the Prevalence of and Diagnostic Composition Within the Broad Clinical Spectrum of Bipolar Disorders. *J. Aff. Dis.* **2000,** *59,* s5–s30.

Akiskal, H. S.; Chen, S. E.; Davis, G. C.; Puzantian, V. R.; Kashgarian, M.; Bolinger, J. M. Borderline: An Adjective in Search of a Noun. *J. Clin. Psych.* **1985,** *46* (2), 41–48.

Akiskal, H. S.; Djenderedjian, A. H.; Rosenthal, R. H.; Khani, M. K. Cyclothymic Disorder: Validating Criteria for Inclusion in the Bipolar Affective Group. *Am. J. Psych.* **1977,** *134,* 1227–1233.

Akiskal, H. S.; Hantouche, E. G.; Alilaire, J. F. Bipolar II with and Without Cyclothymic Temperament: "Dark" and "Sunny" Expressions of Soft Bipolarity. *J. Aff. Dis.* **2003,** *73,* 49–57.

Akiskal, H. S.; Khani, M. K.; Scott-Strauss, A. Cyclothymic Temperamental Disorders. *Psych. Clin. NA.* **1979,** *2,* 29–439.

Akiskal, H. S.; Mallya, G. Criteria for the "Soft" Bipolar Spectrum: Treatment Implications. *Psychopharm. Bull.* **1987**, *23* (1), 68–73.

Akiskal, H. S.; Pinto, O. The Evolving Bipolar Spectrum: Phenotypes, I.; 11, 111, and IV. *Psych. Clin. NA.* **1999**, *22* (3), 517–534.

Akiskal, H. S.; Rosenthal, R. H.; Rosenthal, T. L.; Kashgarian, M.; Khani, M. K.; Puzantian, V. R. Differentiation of Primary Affective Illness from Situational, Symptomatic, and Secondary Depressions. *Arch. Gen. Psych.* **1979**, *36*, 635–643.

Akiskal, H. S.; Rosenthal, T. L.; Haykal, R. F.; Lemmi, H.; Rosenthal, R. H.; Scott-Strauss, A. Characterological Depressions: Clinical and Sleep EEG Findings Separating "Subaffective Dysthymias" from "Character Spectrum Disorders." *Arch. Gen. Psych.* **1980**, *7*, 777–783.

Akiskal, H. S.; Walker, P.; Puzantian, V. R.; King, D.; Rosenthal, T. L.; Dranon, M. Bipolar Outcome in the Course of Depressive Illness: Phenomenologic, Familial and Pharmacologic Predictors. *J. Aff. Dis.* **1983**, *5* (2), 115–128.

Akiskal, K. K.; Akiskal, H. S. The Theoretical Underpinnings of Affective Temperaments: Implications for Evolutionary Foundations of Bipolar Disorder and Human Nature. *J. Aff. Dis.* **2005**, *85*, 231–239.

Albertino, S.; deAssuncao, A. R. M.; Souza, J. A. Pulsatile Tinnitus: Treatment with Clonazepam and Propranolol. *Braz. J. Otorhinolaryng.* **2005**, *71* (1), 111–113.

Aldao, A.; Nolen-Hoeksema, S. Specificity of Cognitive Emotion Regulation Strategies: A Transdiagnostic Examination. *Behav. Res. Ther.* **2010**, *48* (10), 974–983.

Alfonso, C. Dynamic Psychopharmacology and Treatment Adherence. *J Amer Acad Psychoanal Dynal Psych* **2009**, *37* (2), 269–285.

Allen, J. G. Commentary on "Plausibility and Possible Determinants of Sudden 'Remissions' in Borderline Patients": What Stabilizes Stable Instability? *Psychiatry* **2003**, *66*, 120–123.

Allen, M. T.; Matthews, K. A.; Kenyon, K. L. The Relationship of Resting Baroreflex Sensitivity, Heart Rate Variability and Measure of Impulse Control in Children and Adolescents. *Int. J. Psychophysiol.* **2000**, *37* (2), 185–194.

Allen, N. D.; Rodysill, B. R.; Bostwick, J. M. A Report of Affective Switching Associated with Ketamine: The Case of Ketamine-Induced Mania Is Not Closed. *Bip. Dis.* **2019**, *21* (2), 176–178.

Alloy, L. B.; Ng, T. N.; Titone, M. K.; Boland, E. M. Circadian Rhythm Disruption in Bipolar Spectrum Disorders. *Curr. Psych. Rep.* **2017**, *19* (4), 19–21.

Althoff, R. R. Dysregulated Children Reconsidered. *J. Am. Acad. Ch. Adol. Psych.* **2010**, *49*, 302–305.

Altshuler, L. L.; Post, R. M.; Leverich, G. S.; Mikalauskas, K.; Rosoff, A.; Ackerman, L. Antidepressant-Induced Mania and Cycle Acceleration: A Controversy Revisited. *Am. J. Psych.* **1995**, *152* (8), 1130–1138.

Amaral, G. The Primate Amygdala and the Neurobiology of Social Behavior: Implications for Understanding Social Anxiety. *Biol. Psych.* **2002**, *51* (1), 11–17.

Amidfar, M.; Kim, Y. K.; Wiborg, O. Effectiveness of Memantine on Depression-Like Behavior, Memory Deficits and Brain mRNA Levels of BDNF and TrkB in Rats Subjected to Repeated Unpredictable Stress. *Pharm. Rep.* **2018**, *70* (3), 600–606.

Amsterdam, J.; Hornig-Rohan, M. Treatment Algorithms in Treatment Resistant Depression. *Psych. Clin. NA.* **1996**, *19* (2), 371–386.

Andreasen, N. C.; Black, D. W. *Introductory Textbook of Psychiatry, 3rd ed.*; American Psych Pub: Washington, DC, 2002.

Angst, J. The Emerging Epidemiology of Hypomania and Bipolar II Disorder. *J. Aff. Dis.* **1988,** *50,* 143–151.

Angst, J. The Etiology and Nosology of Endogenous Depressive Psychoses [1966] Foreign Psychiatry 2, 1–108. [1973 English trans.].

Angst, J. The Evolving Epidemiology of Bipolar Disorder. *World Psych.* **2002,** 1 (3), 146–148.

Angst, J.; Gamma, A.; Clarke, D.; Ajdacic-Gross, V.; Rössler, W.; Regier, D. Subjective Distress Predicts Pretreatment Seeking for Depression, Bipolar, Anxiety, Panic, Neurasthenia and Insomnia Severity Spectra. *Acta Psych. Scand.* **2010,** *122* (6), 488–498.

Angst, J.; Gamma, A.; Sellaro, R.; Zhang, H.; Merikangas, K. Toward Validation of Atypical Depression in the Community: Results of the Zurich Cohort Study. *J. Aff. Dis.* **2002,** *72,* 125–138.

Angst, J.; Perris, C. The Nosology of Endogenous Depression. *Int. J. Mental Health* **1972,** 1 (1–2), 145–158.

Angst, J.; Sellaro, R.; Stassen, H. H.; Gamma, A. Diagnostic Conversion from Depression to Bipolar Disorders: Results of a Long-Term Prospective Study of Hospital Admissions. *J. Aff. Dis.* **2005,** *84,* 149–157.

Ansell, E. B.; Rando, K.; Tuit, K.; Guarnaccia, J.; Sinha, R. Cumulative Adversity and Smaller Gray Matter Volume in Medial Prefrontal, Anterior Cingulate, and Insula Regions. *Biol. Psych.* **2012,** *72* (1), 57–64.

Applebaum, A. A. Supportive Psychoanalytic Psychotherapy for Borderline Patients: An Empirical Approach. *Am. J. Psychoan.* **2006,** *66* (4), 317–332.

Aretaeus. The Extant Works of Aretaeus, the Cappadocian. (Adams, F., Ed. and trans.); Sydenham Society: London, 1856; pp 301–304.

Arnow, B. A.; Constantino, M. J. Effectiveness of Psychotherapy and Combination Treatment for Chronic Depression. *J. Clin. Psychol.* **2003,** *59* (8), 893–905.

Arnsten, A. Development of the Cerebral Cortex: XIV. Stress Impairs Prefrontal Cortical Function. *J. Ch. Adol. Psych.* **1999,** *308,* 220–222.

Arnsten, A. F.; Goldman-Racik, P. S. Noise Stress Impairs Prefrontal Cortical Cognitive Function in Monkeys: Evidence for a Hyperdopaminergic Mechanism. *Arch. Gen. Psych.* **1998,** *55,* 362–368.

Arsenault, L.; Cannon, M.; Witton, J.; Murray, R. M. Causal Association Between Cannabis and Psychosis, Examination of the Evidence. *Br. J. Psych.* **2004,** *184,* 110–117.

Asherson, P. Clinical Assessment and Treatment of Attention Deficit Hyperactivity Disorder in Adults. *Expert Rev. Neurother.* **2005,** 5, 525–539.

Auer, D. P.; Putz, B.; Kraft, E.; Lipinski, B.; Schill, J.; Holsboer, F. Reduced Glutamate in the Anterior Cingulate Cortex in Depression: An In Vivo Proton Resonance Spectroscopy Study. *Biol. Psych.* **2000,** *47,* 305–313.

August, G. J.; Winters, K. C.; Realmuto, G. M.; Fahnhorst, T.; Botzet, A.; Lee, S. Prospective Study of Adolescent Drug Use Among Community Samples of ADHD and Non-ADHD Participants. *J. Am. Acad. Ch. Adol. Psych.* **2006,** *45* (7), 824–832.

Aupperle, R. L.; Veer, J. M.; van Hoof, J. M.; Rombouts, S. A.; van der Wee, N. J.; Vermeiren, R. R. Executive Function and PTSD: Disengaging from Trauma. *Neuropharmacology* **2012,** *62,* 686–694.

Axelson, D. A.; Birmaher, B. J.; Findling, R. L.; Fristad, M. A.; Kowatch, R. A.; Youngstrom, E. A.; Arnold, E. L.; Goldstein, B. I.; Goldstein, T. R.; Chang, K. D.; Delbello, M. P.; Ryan, N. D.; Diler, N. S. Concerns Regarding the Inclusion of Temper Dysregulation Disorder with Dysphoria in the Diagnostic and Statistical Manual of Mental Disorders. *J. Clin. Psych.* **2011**, *72*, 1257–1262.

Aydemir, O.; Deveci, C.; Taneli, F. The Effect of Chronic Antidepressant Treatment on Serum Brain-Derived Neurotrophic Factor Levels in Depressed Patients: A Preliminary Study. *Prog. Neuropsychopharm. Biol. Psych.* **2005**, *29*, 261–265.

Baldessarini, R. J. A Plea for Integrity of the Bipolar Disorder Concept. *Bip. Dis.* **2000**, *2*, 3–7.

Baldessarini, R. J.; VazquezG, Tondo, L. Treatment of Cyclothymic Disorder: Commentary. *Psychother. Psychosom.* **2011**, *180* (3), 131–135.

Baldwin, D. S.; Polkinghorn, C. Evidence-Based Pharmacotherapy of Generalized Anxiety Disorder. *Int. J. Neuropsychopharmacol.* **2005**, 8 (2), 293–302.

Bally, N.; Zullino, D.; Aubry, J. Cannabis Use and First Manic Episode. *J. Aff. Dis.* **2014**, *165*, 103–108.

Bandelow, B.; Michaelis, S.; Wedekind, D. Treatment of Anxiety Disorders. *Dial. Clin. Neurosci.* **2017**, *19* (2), 93–107.

Banwari, G.; Desai, P.; Patidar, P. Ketamine-Induced Affective Switch in a Patient with Treatment-Resistant Depression. *Ind. J. Pharmacol.* **2015**, *47* (4), 454–455.

Barbee, J. G.; Jamhour, N. J. Lamotrigine as an Augmentation Agent in Treatment-Resistant Depression. *J. Clin. Psych.* **2002**, *63* (8), 737–741.

Barcai, A. Lithium in Adult Anorexia Nervosa: A Pilot Report on Two Patients. *Acta Psych. Scand.* **1977**, *55*, 97–101.

Bardeen, J. R.; Tull, M. T.; Dixon-Gordon, K. L.; Stevens, E. N.; Gratz, K. L. Attentional Control as a Moderator of the Relationship Between Difficulties Accessing Effective Emotion Regulation Strategies and Distress Tolerance. *J. Psychopath. Behav. Assess* **2015**, *37* (1), 79–84.

Barkley, R. A. Deficient Emotional Self-Regulation Is a Core Component of ADHD. *J. ADHD Relat. Disord.* **2010**, 5–37.

Barkley, R. A. Emotional Dysregulation is a Core Component of ADHD. In *Attention-Deficit Hyperactivity Disorder: A Handbook for Diagnosis and Treatment*; Barkley, R. A., Ed.; Guilford Press: New York, 2015; pp 81–115.

Barlow, D. H.; Allen, L. B.; Choate, M. L. Toward a Unified Treatment for Emotional Disorders. *Behav. Ther.* **2004**, *35*, 205–230.

Barnard, K. E.; Brazelton, T. B., Eds. *Touch: The Foundations of Experience*; International Universities Press: Madison, CT, 1990.

Barrett, L. F.; Barr, M. See It with Feeling: Affective Predictions During Object Perception. *Phil. Trans. R.Soc. B (Biol. Sci.)* **2009**, *364*, 1325–1334.

Bartholomew, K.; Horowitz, L. M. Attachment Styles Among Young Adults: A Test of a Four-Category Model. *J. Personal Soc. Psychol.* **1991**, *61* (2), 226–244.

Bateman, A.; Fonagy, P. Effectiveness of Partial Hospitalization in the Treatment of Borderline Personality Disorder: A Randomized Controlled Trial. *Am. J. Psych.* **1999**, *156* (10), 15.

Bateman, A.; Fonagy, P. Mentalization Based Treatment for Borderline Personality Disorder. *World Psych.* **2010**, *9* (1), 11–15.

Bauer, M. S.; Mitchener, L. What Is a "Mood Stabilizer"? An Evidence-Based Response. *Am. J. Psych.* **2004,** *161* (1), 3–18.

Bauer, M.; Aida, M.; Priller, J.; Young, L. T. Implications of the Neuroprotective Effects of Lithium for the Treatment of Bipolar and Neurodegenerative Disorders. *Pharmacopsych* **2003,** *36,* 250–254.

Bauer, M.; Bechor, T.; Kunz, D.; Berghofer, A.; Strohle, A.; Mueller-Oerlinghausen, B. Double-Blind, Placebo-Controlled Trial of the Use of Lithium to Augment Antidepressant Medication in Continuation Treatment of Unipolar Major Depression. *Am. J. Psych.* **2000,** *157* (9), 429–1435.

Bauer, M.; Grof, P.; Rasgon, N.; Bschor, T.; Glen, T.; Whybrow, P. C. Temporal Relation Between Sleep and Mood in Patients with Bipolar Disorder. *Bipolar Dis* **2006,** *8* (2), 60–167.

Beaulieu, I.; Godbout, R. Spatial Learning on the Morris Water Maze Test After a Short-Term Paradoxical Sleep Deprivation in the Rat. *Brain Cog.* **2000,** *43* (1–3), 27–31.

Bechtel, W. Circadian Rhythms and Mood Disorders: Are the Phenomena and Mechanisms Causally Related? *Front Psych.* **2015,** *6,* 118.

Beebe, B.; Lachmann, F. M. Co-Constructing Inner and Relational Processes: Self-and Mutual Regulation in Infant Research and Adult Treatment. *Psychoan. Psychol.* **1998,** *15,* 480–516.

Beebe, B.; Lachmann, F. M. *Infancy Research and Adult Treatment*; The Analytic Press: Hillsdale, NJ, 2002.

Beebe, B.; Lachmann, F. M. Representation and internalization in infancy: Three principles of salience. *Psychoanal. Psychol* **1994,** *11,* 127–165.

Benazzi, F. Borderline Personality-Bipolar Spectrum Relationship. *Prog. Neuropsychopharm. Biol. Psych.* **2006,** *30* (1), 68–74.

Benazzi, F. Delineation of the Clinical Picture of Dysphoric/Mixed Hypomania. *Prog Neuro-Psychopharmacol. Biol. Psych.* **2007,** *31* (4), 944–951.

Benazzi, F. Prevalence of Bipolar II Disorder in Outpatient Depression: A 203-Case Study in Private Practice. *J. Aff. Dis.* **1997,** *43,* 163–166.

Benazzi, F. Unipolar Depression with Bipolar Family History: Links with the Bipolar Spectrum. *Psych. Clin. Neurosci.* **2003,** *57,* 497–503.

Benvenuti, A.; Rucci, P.; Miniati, M.; Papasogli, A.; Fagiolini, A.; Cassano, G. B.; Swartz, H.; Frank, E. Treatment-Emergent Mania/Hypomania in Unipolar Patients. *Bip. Dis.* **2008,** *10,* 726–732.

Ben-Zeev, D.; Young, M. A.; Corrigan, P. W. DSM-V and the Stigma of Mental Illness. *J. Mental Health* **2010,** *19* (4), 18–327.

Berking, M.; Lukas, C. A. The Affect Regulation Training (ART): A Transdiagnostic Approach to the Prevention and Treatment of Mental Disorders. *Curr. Opin. Psychol.* **2015,** *3,* 64–69.

Berking, M.; Whitley, B. *Affect Regulation Training: A Practitioner's Manual* 2014 Springer: Berlin, Heidelberg.

Berner, B.; Musalek, M.; Walter, H. Psychopathological Concepts of Dysphoria. *Psychopathology* **1987,** *20,* 93–100.

Berner, P.; Gabriel, E.; Katschnig, H.; Kieffer, W.; Koehler, K.; Lenz, G.; Simhandel, C. *Diagnostic Criteria for Schizophrenic and Affective Psychoses*; American Psychiatric Association: Washington, DC, 1983.

Bertschy, G.; Gervasoni, N.; Favre, S.; Liberek, C.; Ragama-Pardos, E.; Aubry J-M, Gex-Fabry, M.; Dayer, A. Frequency of Dysphoria and Mixed States. *Psychopathology* **2008**, *41*, 187–193.

Biederman, J.; Faraone, S. V.; Wozniak, J.; Mick, E.; Kwon, A.; Aleardi, M. Further Evidence of Unique Developmental Phenotypic Correlates of Pediatric Bipolar Disorder: Findings from a Large Sample of Clinically Referred Preadolescent Children Assessed Over the Last 7 Years. *J. Aff. Dis.* **2004**, *82* (Supp 1), s45–s58.

Biederman, J.; Klein, R. G.; Pine, D. S.; Klein, D. F. Resolved: Mania Is Mistaken for ADHD in Prepubertal Children. *J. Am. Acad. Ch. Adol. Psych.* **1998**, *37*, 1091–1096.

Biederman, J.; Spencer, T.; Lomedico, A.; Day, H.; Petty, C. R.; Faraone, S. V. Deficient Emotional Self-Regulation and Pediatric Attention Deficit Hyperactivity Disorder: A Family Risk Analysis. *Psychol. Med.* 2012, *42*, 639–646.

Birnbaum, R. J. Borderline, Bipolar or Both? *Harv. Rev. Psych.* **2004**, *12* (3), 146–149.

Bishop, S. J. Trait Anxiety and Impoverished Prefrontal Control of Attention. *Nat. Neurosci.* **2009**, *12* (1), 92–98.

Blader, J. C.; Carlson, G. A. Increased Rates of Bipolar Disorder Diagnoses Among US Child, Adolescent and Adult Inpatients, 1996–2004. *Biol. Psych.* **2007**, *37*, 1091–1096. Discussion 1096–1099.

Block, W.; Traber, F.; von Widdern, O.; Metten, M.; Schild, H.; Maier, W.; Zobel, A.; Jessen, F. Proton MR Spectroscopy of the Hippocampus at 3 T in Patients with Unipolar Major Depressive Disorder Correlates and Predictors of Treatment Response. *Int. J. Psychopharmacol.* **2009**, *12*, 415–422.

Bogdan, R.; Hariri, A. R. Neural Embedding of Stress Reactivity. *Nat. Neurosci.* **2012,** *15* (12), 1605–1607.

Borkovec, T. D.; Abel, J. L.; Newman, H. Effects of Psychotherapy on Comorbid Conditions in Generalized Anxiety Disorder. *J. Consult Clin. Psychol.* **1995**, *63* (3), 479–483.

Borkovec, T. D.; Costello, E. Efficacy of Applied Relaxation and Cognitive-Behavioral Therapy in the Treatment of Generalized Anxiety Disorder. *J. Consult Clin. Psychol.* **1993**, *61* (4), 611–619.

Botting, N.; Conti-Ramsden, G. Social and Behavioral Difficulties in Children with Language Impairment. *Child Lang. Teach. Ther.* **2000**, *16* (2), 105–120.

Botvinik, M.; Nystrom, L. E.; Fissell, K.; Carter, C. S.; Cohen, J. D. Conflict Monitoring Versus Selection-For-Action in Anterior Cingulate Cortex. *Nature* **1999**, *402*, 179–181.

Bowden, C. L. A Different Depression: Clinical Distinctions Between Bipolar and Unipolar Depression. *J. Aff. Dis.* **2005**, *84* (2–3), 117–125.

Bowden, C. L. Lithium-Responsive Depression. *Comp. Psych.* **1978**, *19* (3), 224–231.

Bowden, C.; Mitchell, P.; Suppes, T. Lamotrigine in the Treatment of Bipolar Depression. *Eur. Neuropsychopharmacol.* **1999**, *9*, 5113–5117.

Bowlby, J. *Attachment and Loss: Vol 2: Separation*; Basic Books: New York, 1973.

Bowlby, J. *Attachment and Loss: Vol. I: Attachment*; Basic Books: New York, 1969.

Bowlby, J. *Loss: Sadness and Depression*; Basic Books: New York, 1980.

Boyce, W. T.; Ellis, B. J. Biological Sensitivity to Context: I. An Evolutionary-Developmental Theory of the Origins and Functions of Stress Reactivity. *Dev. Psychopath.* **2005**, *17*, 271–301.

Brachman, R. A.; Lehmann, M. L. Maric, D.; Herkenham, M. Lymphocytes from Chronically Stressed Mice Confer Antidepressant-Like Effects to Native Mice. *J. Neurosci.* **2015,** *35,* 530–1538.

Bradley, S. *Affect Regulation and the Development of Psychopathology*; Guilford Press: New York, 2000.

Bradley, S. J. The Relationship of Early Maternal Separation to Borderline Personality in Children and Adolescents: A Pilot Study. *Am. J. Psych.* **1979,** *136* (4A), 424–426.

Brambilla, F.; Bellodi, L.; Arancio, C.; Limonta, D.; Ferrari, E.; Solerte, B. Neurotransmitter and Hormonal Background of Hostility in Anorexia Nervosa. *Neuropsychobiology* **2001,** *43,* 225–232.

Brantigan, C. O.; Brantigan, T. A.; Joseph, N. Effect of Beta Blockade and Beta Stimulation on Stage Fright. *Am. J. Med.* **1982,** *72* (1), 88–94.

Braunstein-Berkovitz, H. Does Stress Enhance or Impair Selective Attention? The Effects of Stress and Perceptual Load on Negative Priming. *Anx. Stress Coping* **2003,** *16* (4), 345–357.

Breggin, P. R. Psychostimulants in the Treatment of Children with ADHD: Risks and Mechanism of Action. *Int. J. Risk Saf. Med.* **1999,** *12* (1), 3–35.

Brietzke, E.; Mansur, R. B.; Subramaniapillai, M.; Balanza-Martinez, V.; Vinberg, M.; Gonzalez-Pinto, A.; Rosenblat, J. D.; Ho, R.; McIntyre, R. S. Ketogenic Diet as a Metabolic Therapy for Mood Disorders: Evidence and Developments. *Neurosci. Biobehav. Rev.* **2018,** *94,* 11–16.

Brockmeyer, T.; Grosse Holtforth, M.; Bents, H.; Herzog, W.; Friederich, H. C. Lower Body Weight Is Associated with Less Negative Emotions in Sad Autobiographical Memories of Patients with Anorexia Nervosa. *Psych. Res.* **2013,** *210* (2), 548–552.

Brockmeyer, T.; Skunde, M.; Wu, M.; Bresslein, E.; Rudofsky, G.; Herzog, W.; Friederich, F-C. Difficulties in Emotion Regulation Across the Spectrum of Eating Disorders. *Comp. Psych.* **2014,** *55* (3), 565–571.

Brotman, M. A.; Rich, B. A.; Guyer, A. E.; Lunsford, J. R.; Horsey, S. E.; Reising, M. M.; Thomas, L. A.; Fromm, S. J.; Towbin, K.; Pine, D. S.; Leibenluft, E. Amygdala Activation During Emotion Processing of Neural Faces in Children with Severe Mood Regulation Versus ADHD or Bipolar Disorder. *Am. J. Psych.* **2010,** *167* (1), 61–69.

Brotman, M. A.; Schmajuk, M.; Rich, B. A.; Dickstein, D. P.; Guyer, A. P.; Costello, E. J.; Egger, H. L.; Angold, A.; Pine, D. S.; Leibenluft, E. Prevalence, Clinical Correlates, and Longitudinal Course of Severe Mood Dysregulation in Children. *Biol. Psych.* **2006,** *60,* 991–997.

Brown, T. A.; Antony, M. M.; Barlow, D. H. Diagnostic Comorbidity in Panic Disorder: Effect on Treatment Outcome and Course of Comorbid Diagnoses Following Treatment. *J. Consult Clin. Psychol.* **1995,** *63* (3), 408–418.

Brugue, E.; Vieta, E. Atypical Antipsychotics in Bipolar Disorder: Neurobiological Basis and Clinical Implications. *Prog. Neuropsychopharm. Biol. Psych.* **2007,** *31* (1), 275–282.

Bruhle, A. B.; Scherpiet, S.; Sulzer, J.; Stampfli, P.; Seifritz, E.; Herwig, U. Real-Time Neurofeedback Using Functional MRI Could Improve Down-Regulation of Amygdala Activity During Emotional Stimulation: A Proof-of-Concept Study. *Brain Topog.* **2014,** *27,* 138–148. DOI: 10.1007/s10548-013-0331-9

Brydon, L.; Harrison, N. A.; Walker, C.; Steptoe, A.; Critchley, H. D. Peripheral Inflammation Is Associated with Altered Substantia Nigra Activity and Psychomotor Slowing in Humans. *Biol. Psych.* **2008,** *63,* 1022–1029.

Buchheim, A.; Viviani, R.; Kessler, H.; Kachele, H.; Cierpka, M.; Roth, G.; George, C.; Kernberg, O.; Bruns, G.; Taubner, S. Changes in Pre-Frontal Limbic Function in Major Depression After 15 Months of Long-Term Psychotherapy. *PLOS One* **2012**, *7* (3), e33745.

Buck, R. Mood and Emotion: A Comparison of Five Contemporary Views. *Psychol. Inq.* **1990**, *1*, 330–336.

Buckholtz, J. W.; Meyer-Lindenberg, A. Psychopathology and the Human Connectome: Toward a Transdiagnostic Model of the Risk for Mental Illness. *Neuron* **2012**, *74* (6), 990–1004.

Bunford, N.; Evans, S. W.; Langberg, J. M. Emotion Dysregulation Is Associated with Social Impairment Among Young Adolescents with ADHD. *J. Atten. Dis.* **2014**, *22* (1), 66–82.

Bunford, N.; Evans, S. W.; Wymbs, F. ADHD and Emotion Dysregulation Among Children and Adolescents. Clin Child Fam Psychol Rev **2015**, *18*, 185–217.

Burghy, C. A.; Stodola, D. E.; Ruttle, P. L.; Molloy, E. K.; Armstrong, J. M.; Oler, J. A.; Fox, M. E.; Hayes, A. S.; Kalin, N. H.; Essex, M. J.; Davidson, R. J.; Birn, R. M. Developmental Pathways to Amygdala-Prefrontal Function and Internalizing Symptoms in Adolescence. *Nat. Neurosci.* **2012**, *15*, 736–1741.

Burkhart, K.; Phelps, J. R. Amber Lenses to Block Blue Light and Improve Sleep: A Randomized Trial. *Chronobiol. Int.* **2009**, *26* (8), 1602–1612.

Butler, G.; Fennell, M.; Robson, P.; Gelder, M. Comparison of Behavior Therapy and Cognitive Behavior Therapy in the Treatment of Generalized Anxiety Disorder. *J. Consult Clin. Psychol.* **1991**, *59* (1), 167–175.

Cai, W.; Khaoustov, V. I.; Xie, Q.; Pan, T.; Le, W.; Yoffe, B. Interferon-Alpha-Induced Modulation of Glucocorticoid and Serotonin Receptors as a Mechanism of Depression. *J. Hepatol.* **2005**, *42*, 880–887.

Caffaro, J. V. *Sibling Abuse Trauma: Assessment and Intervention Strategies for Children, Families and Adults*; Routledge: New York, 2014.

Calabrese, J. R.; Bowden, C. L.; Sachs, G. S.; Ascher, J. A. A Double-Blind Placebo-Controlled Study of Lamotrigine Monotherapy in Outpatients with Bipolar I Depression. *J. Clin. Psych.* **1999**, *60*, 1478–1488.

Caliyurt, O. Role of Chronobiology as a Transdisciplinary Field of Research: Its Applications in Treating Mood Disorders. *Balkan J. Med.* **2017**, *34*, 514–521.

Calkins, S. D. Cardiac Vagal Tone Indices of Temperamental Reactivity and Behavioral Regulation in Young Children. *Dev. Psychobiol.* **1997**, *31*, 125–135.

Calkins, S. D.; Keane, S. P. Cardiac Vagal Regulation Across the Pre-School Period: Stability, Continuity, and Implications for Childhood Adjustment. *Dev. Psychobiol.* **2004**, *45*, 101–112.

Camelo, E. V. M.; Velasques, B.; Ribeiro, P.; Netto, T.; Cheniaux, E. Attention Impairment in Bipolar Disorder: A Systematic Review. *Psychol. Neurosci.* **2013**, 6 (3), 299–309.

Campbell-Sills, L.; Barlow, D. H.; Brown, T. A.; Hofmann, S. G. Acceptability and Suppression of Negative Emotion in Anxiety and Mood Disorders. *Emotion* **2006**, *6* (4), 587–595.

Campos, J. J.; Frankel, C. B.; Camras, L. On the Nature of Emotion Regulation. *Child Dev.* **2004**, *75*, 377–394.

Capuron, L.; Miller, A. H. Immune System to Brain Signaling: Neuropsychopharmacological Implications. *Pharmacol. Ther.* **2011**, *30*, 26–238.

Capuron, L.; Pagnoni, G.; Demetrashvili, M. F.; Lawson, D. H.; Fornwalt, F. B.; Woolwine, B.; Berns, G. S.; Nemeroff, C. B.; Miller, A. H. Basal Ganglia Hypermetabolism and Symptoms of Fatigue during Interferon-α Therapy. *Neuropsychopharmacology* **2007**, *32*, 2384–2392.

Cardinal, R. N.; Parkinson, J. A.; Hall, J.; Everitt, B. J. Emotion and Motivation: The Role of the Amygdala, Ventral Striatum and Pre-Frontal Cortex. *Neurosci. Biobehav. Rev.* **2002**, *26*, 321–352

Carlson, G. A.; Glovinsky, I. The Concept of Bipolar Disorder in Children: A History of the Bipolar Controversy. *Ch. Adol. Psych. Clin. NA.* **2009**, *18*, 257–271.

Carlson, P. J.; Diazgranados, N.; Nugent, A. C.; Ibrahim, L.; Luckenbaugh, D. A.; Brutsche, N.; Herscovitch, P.; Manji, H. K.; Zarate Jr. CA, Drevets, W. C. Neural Correlates of Rapid Antidepressant Response to Ketamine in Treatment-Resistant Unipolar Depression: A Preliminary Positron Emission Tomography Study. *Biol. Psych.* **2013**, *73*, 1213–1221.

Carlson, V.; Ciccetti, D.; Barnett, D.; Braunwald, K. Disorganized/Disoriented Attachment Relationships in Maltreated Infants. *Dev. Psychol.* **1989**, *25* (4), 525–531.

Carr, C. P.; Martins, C. M. S.; Stingel, A. M.; Lemgruber, V. B.; Juruena, M. F. The Role of Early Life Stress in Adult Psychiatric Disorders: A Systematic Review According to Childhood Trauma Subtypes. *J. Nerv. Mental Dis.* **2013**, *201* (12), 1007–1020.

Cartwright, R. Dreaming as a Mood-Regulation System. In *Principles & Practice of Sleep Medicine*; Kryger, M. H., Roth, T., Dement, W. C., Eds., Chapter 54; Saunders: Philadelphia, WB, 1989.

Casper, R. C. Behavioral Activation and Lack of Concern, Core Symptoms of Anorexia Nervosa? *Int. J. Eat Dis.* **1998**, *24*, 381–393.

Cassidy, J. Emotion Regulation: Influences of Attachment Relationships. *Mon. Soc. Res. Child Dev.* **1994**, *59* (2–3), 228–249.

Castanon, N.; Bluthe, R. M.; Dantzer, R. Chronic Treatment with the Atypical Anti-depressant Tianeptine Attenuates Sickness Behavior Produces by Peripheral But Not Central Lipopolysaccharide and Interleukin -1 Beta in the Rat. *Psychopharmacology* **2001**, *154*, 50–60.

Cato, M. A.; Crosson, B.; Gokcay, D.; Soltysik, D, Wierenga, C.; Gopinath, K.; Himes, N.; Belanger, H.; Bauer, R. M.; Fischler, I. S.; Gonzales-Rothi, L.; Briggs, R. W. Processing Words with Emotional Connotation: An fMRI Study of Time Course and Laterality in Rostral Frontal and Retrosplenial Cortices. *J. Cog. Neurosci.* **2004**, *16* (2), 167–177.

Caviccioli, M.; Movalli, M.; Vassena, G.; Ramella, P.; Prudenziati, P.; Maffei, C. The Therapeutic Role of Emotion Regulation and Coping Strategies During a Stand-Alone DBT Skills Training Program for Alcohol Use Disorder and Concurrent Substance Use Disorders. *Add. Behav.* **2019**, *98*, 106035.

Ceraudo, G.; Vannucchi, G.; Perugi, G.; Dell'osso, L. Adult ADHD: Clinical Aspects and Clinical Implications. *Riv. Psych.* **2012**, *47*, 451–464.

Champagne, F. A. Epigenetic Mechanisms and the Transgenerational Effects of Maternal Care. *Front. Neuroendocrinol.* **2008**, *29*, 386–397.

Chattarji, S. Stress-Induced Formation of New Synapses in the Amygdala. *Neuropsycho-pharmacology* **2008**, *33*, 199–200.

Cheetham, A.; Allen, N. B.; Yucel, M.; Lubman, D. I. The Role of Affective Dysregulation in Drug Addiction. *Clin. Psych. Rev.* **2010,** *30* (6), 621–634.

Chen, G.; Rajkowska, G.; Du, F.; Seraji-Bozorgzad, N.; Manji, H. K. Enhancement of Hippocampal Neurogenesis by Lithium. *J. Neurochem.* **2000,** *75* (4), 1729–1734.

Chen, J.; Wang, Z.; Fang, Y. The History, Diagnosis and Treatment of Disruptive Mood Dysregulation Disorder. *Shanghai Arch. Psych.* **2016,** *28* (5), 289–292.

Chengappa, K. R.; Rathore, D.; Levine, J.; Atzert, R.; Solai, L.; Parepally, H.; Levin, H.; Moffa, N.; Delaney, J.; Brar, J. S. Topiramate as Add-on Treatment for Patients with Bipolar Mania. *Bip. Dis.* **2002,** *1* (1), 42–53.

Chiu, C. T.; Chuang, D. M. Neuroprotective Action of Lithium in Disorders of the Central Nervous System. *NIH Mol. Biol.* **2011,** *36* (6), 461–476.

Chiu, C. T.; Scheuing, L.; Liu, G.; Liao, H. M.; Linares, G. R.; Lin, D.; Chuang, D. M. The Mood Stabilizer Lithium Potentiates the Antidepressant-Like Effects and Ameliorates Oxidative Stress Induced by Acute Ketamine in a Mouse Model of Stress. *Int. J. Neuropsychopharm.* **2015,** *18* (6), 1–13.

Christianson, S. A. *Handbook of Emotion and Memory: Current Research and Theory*; Erlbaum: Hillsdale, NJ, 1992.

Chugani, H. T.; Behen, M. E.; Muzik, O.; Juhasz, C.; Nagy, F.; Chugani, D. C. Local Brain Functional Activity Following Early Deprivation: A Study of Post-Institutionalized Romanian Orphans. *NeuroImage* **2001,** *14,* 1290–301.

Ciccetti, D.; White, J. Emotion and Developmental Psychopathology. In *Psychological and Biological Approaches to Emotion*; Stein, N., Leventhal, B., Trebasso, T., Eds.; Lawrence Erlbaum Associates: Hillsdale, NJ, 1990; pp 359–382.

Chiu C-T, Scheuing, L.; Liu, G.; Liao H-M, Linares G-R, Lin, D.; Chuang D-M: The Mood Stabilizer Lithium Potentiates the Antidepressant-Like Effects and Ameliorates Oxidative Stress Induced by Acute Ketamine in a Mouse Model of Stress. *Int. J. Neuropsychopharm.* **2015,** *18* (6), 1–13.

Chun, B.; Dunner, D. L. A Review of Antidepressant Induced Hypomania in Depression: Suggestions for DSM-IV. *Bip. Dis.* **2004,** *6* (1), 32–42.

Cirulli, F.; Alleva, B. E. Early Disruption of the Mother-Infant Relationship: Effects on Brain Plasticity and Implications for Psychopathology. *Neurosci. Biobehav. Rev.* **2003,** *27,* 73–82.

Clark, R. E.; Xie, H.; Brunette, M. F. Benzodiazepine Prescription Practices and Substance Abuse in Persons with Severe Mental Illness. *J. Clin. Psych.* **2004,** *65* (2), 151–155.

Clarkin, J. F.; Foelsch, P. A.; Levy, K. N.; Hull, J. W.; Delaney, J. C.; Kernberg, O. F. The Development of a Psychodynamic Treatment for Patients with Borderline Personality Disorder: A Preliminary Study of Behavioral Change. *J. Pers. Dis.* **2001,** *15* (6), 487–495.

Cloitre, M.; Koenen, K. C.; Cohen, L. R.; Han, H. Skills Training in Affective and Interpersonal Regulation Followed by Exposure: A Phase-Based Treatment for PTSD Related to Childhood Abuse. *J. Consult Clin. Psychol.* **2002,** *70* (5), 1067–1074.

Cludius, B.; Mennin, D.; Ehring, T. Emotion Regulation as a Transdiagnostic Process. *Emotion* **2020,** *20* (1), 37–42.

Cohen, H.; Kaplan, Z.; Kotler, M.; Mittelman, I.; Osher, Y.; Berdusky, Y. Impaired Heart Rate Variability in Euthymic Bipolar Patients. *Bip. Dis.* **2003,** 5 (2), 138–143.

Cohen, R. A.; Grieve, S.; Hoth, K. F.; Paul, R. H.; Sweet, L.; Tate, D.; Gunstad, J.; Stroud, L.; McCaffery, J.; Hitsman, B.; Niaura, R.; Clark, C. R.; McFarlane, A.; Bryant, R.;

Gordon, E.; Williams, L. M. Early Life Stress and Morphometry of the Adult Anterior Cingulate Cortex and Caudate Nuclei. *Biol. Psych.* **2006,** *59* (10), 975–982.

Coie, J. D.; Dodge, K. A.; Kupersmidt, J. B. Peer Group Behavior and Social Status. In *Peer Rejection in Childhood*; Asher, S. R., Coie, J. D., Eds.; Cambridge University Press: New York, 1990; pp 17–59.

Coid, J.; Yang, M.; Tyrer, P.; Roberts, A.; Ulrich, S. Prevalence and Correlates of Personality Disorder in Great Britain. *Br. J. Psych.* **2006,** *188*, 423–431.

Cole, P. M.; Martin, S. E.; Dennis, T. A. Emotion Regulation as a Scientific Construct: Methodological Challenges and Directions for Child Development Research. *Child Dev.* **2004,** *75*, 317–333.

Conklin, C.; Bradly, R.; Western, D. Affect Regulation in Borderline Personality Disorder. *J. Nerv. Mental Dis.* **2006,** *194* (2), 69–77.

Conrad, D.; LeDoux, J. E.; Margarites, A. M.; McEwen, B. S. Repeated Restraint Stress Facilitates Fear Conditioning Independently of Causing Hippocampal CA3 Dendritic Atrophy. *Behav. Neurosci.* **1999,** *113*, 902–913.

Cooper, N. A.; Clum, G. A. Imaginal Flooding as a Supplementary Treatment for PTSD in Combat Veterans: A Controlled Study. *Behav. Ther.* **1989,** *20* (3), 381–391.

Cooper, M. A.; Shaver, P. R.; Collins, N. L. Attachment Styles, Emotion Regulation, and Adjustment in Adolescence. *J. Pers. Soc. Psychol.* **1998,** *74* (5), 1380–1397.

Coppen, A. The Biochemistry of Affective Disorders. *Br. J. Psych.* **1967,** *113* (504), 1237–1264.

Coppen, A.; Shaw, D. M.; Farrell, J. P. Potentiation of the Antidepressive Effect of a Monoamine-Oxidase Inhibitor by Tryptophan. *Lancet* **1963,** 1 (7272), 79–81.

Corbisiero, S.; Stieglitz R-D, Retz, W.; Rösler, M. Is Emotional Dysregulation Part of the Psychopathology of ADHD in Adults? *ADHD Atten. Def. Hyperact Dis.* **2013,** *5*, 83–92.

Correa, R.; Akiskal, H.; Gilmer, W.; Nierenberg, A. A.; Trivedi, A.; Zisook, S. Is Unrecognized Bipolar Disorder a Frequent Contributor to Treatment Resistant Depression? *J. Aff. Dis.* **2010,** *127* (1–3), 10–18.

Correll, C. M.; Rosenkranz, J. A.; Grace, A. A. Chronic Cold Stress Alters Prefrontal Cortical Modulation of Amygdala Neuronal Activity in Rats. *Biol. Psych.* **2005,** *58* (5), 382–391.

Corstorphine, E.; Mountford, V.; Tomlinson, S.; Waller, G.; Meyer, C. Distress Tolerance in the Eating Disorders. *Eating Behav.* **2007,** *8* (1), 91–97.

Coryell, W. Maintenance Treatment in Bipolar Disorder: A Reassessment of Lithium as the First Choice. *Bip. Dis.* **2009,** *11* (s2), 77–83.

Costa, E.; Guidotti, A.; Mao, C. C.; Suria, A. New Concepts on the Mechanism of Action of Benzodiazepines. *Life Sci.* **1975,** *17* (2), 167–185.

Cozolino, L. *The Neuroscience of Psychotherapy: Healing the Social Brain*; W.W. Norton & Co., Inc.: New York, 2010.

Crain, D. P. Kindling in Sensory Systems: Neocortex. *Exp. Neurol.* 1982, *76, 276–283.*

Cuijpers, P.; Dekker, J.; Hollon, S. D.; Andersson, G. Adding Psychotherapy to Pharmacotherapy in the Treatment of Depressive Disorders in Adults: A Meta-Analysis. *Database of Abstracts of Reviews of Effects (DARE): Quality-assessed Reviews [Internet].* Centre for Reviews and Dissemination, UK, 2009.

Cuijpers, P.; Geraedts, A. S.; van Oppen, P.; Andersson, G.; Markowitz, J. C.; van Straten, A. Interpersonal Psychotherapy for Depression: A Meta-Analysis. *J. Clin. Psych.* **2011,** *168* (6), 581–592.

Curtis, C. E.; D'Esposito, M. Success and Failure Suppressing Reflexive Behavior. *J. Cog. Neurosci.* **2003,** *15,* 409–418.

Cushman, T. P.; Johnson, T. B. Understanding "Inattention" in Children and Adolescents. *Ethical Human Sci. Serv.* **2001,** *3* (2), 107–125.

Dager, S. R.; Friedman, S. D.; Parow, A.; Demopulos, C.; Stoll, A. L.; Lyoo, I. K.; Dunner, D. L.; Renshaw, P. F. Brain Metabolic Alterations in Medication-Free Patients with Bipolar Disorder. *Arch. Gen. Psych.* **2004,** *61,* 450–458.

Dales, S.; Jerry, P. Attachment, Affect Regulation, and Mutual Synchrony in Adult Psychotherapy. *Am. J. Psych.other* **2008,** *62* (3), 283–312.

Dalgeish, T.; Black, M.; Johnston, D.; Bevan, A. Transdiagnostic Approaches to Mental Health Problems: Current Status and Future Directions. *J. Consult Clin. Psychol.* **2020,** *88* (3), 179–195.

Dalton, W. S.; Martz, R.; Lemberger, L.; Rodda, B. E.; Forney, R. B. Influence of Cannabidiol on Delta-9-Tetrahydrocannabinol Effects. *Clin. Pharmacol. Ther.* **1975,** *19,* 300–309.

Damasio, A. R. *The Feeling of What Happens: Body and Emotion in the Making of Consciousness*; Harcourt Brace & Co.: New York, 1999.

Damasio, A. R. *Self Comes to Mind: Constructing the Conscious Brain*; Random House: New York, 2010.

Danlowski, U.; Stuhrmann, A.; Beutelmann, V.; Zwangzer, P.; Lenzen, T.; Grotegerd, D.; Domscke, K.; Hohoff, C.; Ohrmann, P.; Bauer, J.; Lindner, C.; Postert, C.; Konrad, C.; Arolt, V.; Heindel, W.; Suslow, T.; Kugel, H. Limbic Scars: Long-Term Consequences of Childhood Maltreatment Revealed by Functional and Structural Magnetic Resonance Imaging. *Biol. Psych.* **2012,** *71* (4), 286–293.

Danner, U. N.; Sternheimm, L.; Evers, C. The Importance of Distinguishing Between the Different Eating Disorders (Sub) Types When Assessing Emotion Regulation Strategies. *Psych. Res.* **2014,** *215* (3), 727–732.

Dantzer, R.; O'Connor, J. C.; Freund, G. G.; Johnson, R. W.; Kelley, K. W. From Inflammation to Sickness and Depression: When the Immune System Subjugates the Brain. *Nat. Rev. Neurosci.* **2008,** *9,* 46–56.

Darwin, C. *The Expression of Emotions in Man and Animals*; D Appleton: New York, 1872.

Davidson, J. R. T. Pharmacotherapy of Generalized Anxiety Disorder. *J. Clin. Psych.* **2001,** *62* (Suppl. 11), 46–50.

Davies, J. A. Mechanism of Action of Antiepileptic Drugs. *Seizure* **1995,** *4* (4), 267–271.

Davis, J. M.; Janicak, P. G.; Hogan, D. M. Mood-Stabilizers in the Prevention of Affective Disorders: A MetaAnalysis. *Act Psych. Scand.* **1999,** *100* (6), 406–417.

DeBellis, M. D.; Hooper, S. R.; Sapia, J. L. Early Trauma Exposure and the Brain; In *Neuropsychology of PTSD: Biological, Cognitive, and Clinical Perspectives*; Vasterling, J. J., Brewin, C. R., Eds.; The Guilford Press, 2005; pp 153–177.

de Jonghe, F.; Kool, S.; van Aalst, G.; Dekker, J.; Peen, J. Combining Psychotherapy and Antidepressants in the Treatment of Depression. *J. Aff. Dis.* **2001,** *64* (2–3), 217–229.

DeLeon, O. A. Antiepileptic Drugs for the Acute and Maintenance Treatment of Bipolar Disorder. *Harv. Rev. Psych.* **2001,** *9* (5), 209–222.

Deltito, J.; Martin, L.; Riefkohl, J.; Austria, B.; Kissilenko, A.; Corless, P.; Morse, C. Do Patients with Borderline Personality Disorder Belong to the Bipolar Spectrum? *J. Aff. Dis.* **2007,** *67,* 221–228.

Denson, T. F.; Creswell, J. D.; Terides, M. D.; Blundell, K. Cognitive Reappraisal Increases Neuroendocrine Reactivity to Acute Social Stress and Physical Pain. *Psychoneuroendocrin* **2014,** *49,* 69–78.

Deveney, C. M.; Connolly, M. E.; Jenkins, S. E.; Kim, P.; Fromm, S. J.; Pine, D. S.; Leibenluft, E. Neural Recruitment During Failed Motor Inhibition Differentiates Youth with Bipolar Disorders and Severe Mood Dysregulation. *Biol. Psych.* **2012,** *89,* 148–155.

Diamond, D. M.; Rose, G. M. Stress Impairs LTP and Hippocampal-Dependent Memory. *Ann. NY Acad. Sci.* **1994,** *746* (1), 411–414.

Diamond, L. M.; Hicks, A. M. Attachment Style, Current Relationship Security, and Negative Emotions: The Mediating Role of Physiological Regulation. *J Soc Pers Rel* **2005,** *22,* 499–518.

Diamond, P. R.; Famery, A. D.; Atkinson, S.; Haldar, J.; Williams, N.; Cowen, P. J.; Geddes, J. R.; McShane, R. Ketamine Infusions for Treatment Resistant Depression: A Series of 28 Patients Treated Weekly or Twice Weekly in an ECT Clinic. *J. Psychopharm.* **2014,** *28* (6), 536–544.

DiazGranados, N.; Ibrahim, L.; Brutsche, N. A.; Newberg, A.; Kronstein, P.; Khalife, S.; Kammerer, W. A.; Quezado, Z.; Luckenbaugh, D. A.; Slavadore, G.; Machado-Vieira, S.; Manji, H. K.; Zarate Jr, C. A. A Randomized Add-on Trial of an N-Methyl-D Aspartate Antagonist in Treatment-Resistant Bipolar Depression. *JAMA Psych.* **2010a,** *67* (8), 793–802.

DiazGranados, N.; Ibrahim, L.; Brutsche, N. A.; Ameli, R.; Henter, I. D.; Luckenbaugh, D. A.; Machado-Vieira, R.; Zarate Jr. CA: Rapid Resolution of Suicidal Ideation after a Single Infusion of an NMDA Antagonist in Patients with Treatment-Resistant Major Depressive Disorder. *J. Clin. Psych.* **2010b,** *71* (12), 1605–1611.

Diler, R. S.; Birmaher, B.; Axelson, D.; Goldstein, B.; Gill, M.; Strober, M.; Kolko, D. J.; Goldstein, T. R.; Hunt, J.; Yang, M.; Ryan, N. D.; Iyengar, S.; Dahl, R. E.; Dorn, L. D.; Keller, M. B. The Child Behavior Checklist (CBCL) and the CBCL-Bipolar Phenotype Are Not Useful in Diagnosing Pediatric Bipolar Disorder. *J. Ch. Adol. Psychopharm.* **2009,** *19,* 23–20.

Dingemans, A.; Danner, U.; Parks, M. Emotion Regulation in Binge Eating Disorder: A Review. *Nutrients* **2017,** *9* (11), 1274.

Dobbing, J. The Later Development of the Brain and Its Vulnerability. In *Scientific Foundations of Pediatrics*; Davis, J. A., Dobbing, J., Eds.; Heinemann Med. Books: London, 1981; pp 744–759.

Dodd, S.; Schacht, A.; Kelein, K.; Duenas, H.; Reed, V. A.; Williams, L. J.; Quirk, F. H.; Malhi, G. S.; Franz, C. P.; Berk, M. Nocebo Effects in the Treatment of Major Depression: results from an individual study participant-level meta-analysis of the placebo arm of duloxetine clinical trials. *J Clin Psych* **2015** Jun 24;76(6):12818.

Dodson, W. Newe Insights into Rejection Sensitive Dysphoria, 2021. www.ADDitude.com

Dolcos, F.; LaBar, K. S.; Cabeza, R. Interaction Between the Amygdala and the Medial Temporal Lobe Memory System Predicts Better Memory for Emotional Events. *Neuron* **2004,** *42,* 855–863.

Depression: Results From an Individual Study Participant–Level Meta-Analysis of the Placebo Arm of Duloxetine Clinical Trials. *J. Clin. Psych.* **2015,** *76* (6), 702–711.

Dodell-Feder, D.; Koster-Hale, J.; Bedny, M.; Saxe, R. fMRI Item Analysis in a Theory of Mind Task. *Neuroimage* **2011,** *55* (2), 705–712.

Drake, C. L.; Roehrs, T.; Richardson, G.; Walsh, J. K.; Roth, T. Shift Work Sleep Disorder: Prevalence and Consequences Beyond That of Symptomatic Day Workers. *Sleep* **2004,** *27,* 1453–1462.

DSM. *American Psychiatric Association,* 1952.

DSM-II. *American Psychiatric Association,* 1968.

DSM-III. *American Psychiatric Association,* 1980.

DSM-IV. *American Psychiatric Association,* 1994.

DSM-IV-TR. *American Psychiatric Association,* 2000.

DSM-5. *American Psychiatric Association,* 2014.

Duman, R. S.; Monteggia, L. M. A Neurotrophic Model for Stress-Related Mood Disorders. *Biol. Psych.* **2006,** *59,* 1116–1127.

Dunner, D. L.; Gershon, E. S.; Goodwin, F. K. Heritable Factors in the Severity of Affective Illness. *Sci. Proc. Am. Psych Assoc* **1970,** *123,* 187–188.

Dutta, A.; McKie, S.; Deakin, J. F. Ketamine and Other Potential Glutamate Antidepressants. *Psych. Res.* **2012,** *225,* 1–13.

Easton, J. D.; Sherman, D. G. Somatic Anxiety Attacks and Propranolol. *Arch. Neurol.* **1976,** *33* (10), 689–691.

Eckersdorf, B.; Gol Biewski, H.; Konopacki, J. Kainic Acid Versus Carbachol Induced Emotional-Defensive Response in the Cat. *Behav. Br. Res.* **1996,** *77,* 201–210.

Edelman, G. M. *Neural Darwinism: The Theory of Neuronal Group Selection;* Basic Books, 1987.

Edwards, C. R.; Benediktsson, R.; Lindsay, R. S.; Seckl, J. R. Dysfunction of Placental Glucocorticoid Barrier: Link Between Fetal Environment and Adult Hypertension? *Lancet* **1993,** *341,* 355–357.

Ehring, T.; Watkins, E. R. Repetitive Negative Thinking as a Transdiagnostic Process. *Int. J. Cog. Ther.* **2008,** *1,* 192–205.

Einat, H.; Yuan, P.; Gould, T. D.; Li, J.; Du, J.; Zhang, L.; Manji, H. K.; Chen, G. The Role of the Extracellular Signal-Regulated Kinase Signaling Pathway in Mood Modulation. *J. Neurosci.* **2003,** *23,* 7311–7316.

Eisenberger, N. I.; Lieberman, M. D. Why Rejection Hurts: A Common Neural Alarm System for Physical and Social Pain. *Trends Cog. Sci.* **2004,** *8,* 294–300.

Eisenberger, N. I.; Lieberman, M. D.; Williams, K. D. Does Rejection Hurt? An fMRI Study of Social Exclusion. *Science* **2003,** *302,* 290–292.

El-Mallakh, R. S.; Karippot, A. Antidepressant Associated Chronic Irritable Dysphoria (Acid) in Bipolar Disorders: A Case Series. *J. Aff. Dis.* **2005,** *84,* 267–272.

El-Mallakh, R. S.; Paskitti, M. E. The Ketogenic Diet May Have Mood-Stabilizing Properties. *Med. Hypoth.* **2001,** *57* (6), 724–726.

Emrich, H. M.; Vogt, P.; Herz, A. Possible Antidepressive Effects of Opioids: Actions of Buprenorphine. *Ann. NY Acad. Sci.* **1982,** *398* (1), 108–112.

Erfurth A, Sachs, G.; Lesch, O. Addiction: The Burden of Cyclothymic Temperament. *Eur. Psych.* **2015,** *30* (Supp. 1), 516.

Engel S G, Wonderlich Sa, Crosby R D, Mitchell, J. E.; Crow, S.; Peterson, C. B.; Le Grange, D.; Simonich, H. K.; Cao, L.; Lavender, J. M.; Gordon, K. H. The Role of Affect in the Maintenance of Anorexia Nervosa: Evidence from a Naturalistic Assessment of Momentary Behaviors and Emotion. *J. Abnorm. Psychol.* **2013**, *122* (3), 709–719.

Esaki, Y.; Kitajima, T.; Ito, Y.; Koike, S.; Nakao, Y.; Tsuchiya, A.; Hirose, M.; Iwata, N. Wearing Blue-Light Blocking Glasses in the Evening Advances Circadian Rhythms in Patients with Delayed Sleep Phase Disorder: An Open-Label Trial. *Chronobiol. Int.* **2016**, *33* (8), 1037–1044.

Even, C.; Schroder, C. M.; Friedman, S.; Rouillon, F. Efficacy of Light Therapy in Non-Seasonal Depression: A Systematic Review. *J. Aff. Dis.* **2008**, *108* (1–2), 11–23.

Evers, C.; Marijn Stok, F.; de Ridder DTD: Feeding Your Feelings: Emotion Regulation Strategies and Emotional Eating. *Personal. Soc. Psychol. Bull.* **2010**, *36* (6), 792–804.

Faedda, G. L.; Baldessarini, R. T.; Glovinsky, I. P.; Austin, N. P. Treatment Emergent Mania in Pediatric Bipolar Disorder: A Retrospective Case Review. *J. Aff. Dis.* **2004**, *82* (1), 149–158.

Fagiolini, M.; Jensen, C.; Champagne, F. Epigenetic Influences on Brain Development and Plasticity. *Curr Op Neurobiol* **2009**, *19* (2), 207–212.

Faraone, S. V.; Biederman, J.; Mennin, D.; Wozniak, D.; Spencer, T. Attention-Deficit Hyperactivity Disorder with Bipolar Disorder: A Familial Sub-Type? *J. Am. Acad. Ch. Adol. Psych.* **1997**, *36*, 1387–1390.

Farb, N. A. S.; Anderson, A. K.; Segal, Z. V. The Mindful Brain and Emotion Regulation in Mood Disorders. *Can. J. Psych.* **2010**, *57* (2), 70–77.

Falret, J. Principes à suivre dans la classification des maladies mentales. *Ann. Méd.* 1861 Psychol. XIX, 145.

Falret, J. P. *Des Maladies Mentales et des Asiles D'aliénés*; J. B. Bailliere: Paris, 1864.

Faraone, S. V.; Biederman, J.; Mennin, D.; Wozniak, J.; Spencer, T. Attention Deficit Hyperactivity Disorder with Bipolar Disorder: A Familial Subtype? *J. Am. Acad. Child Adol. Psych.* **1997**, *36* (10), 1390.

Faravelli, C.; Lo Sauro, C.; Godini, L.; Lelli, L.; Benni, L.; Pietrini, F.; Lazzeretti, L.; Talamba, G. A.; Fioravanti, G.; Ricca, V. Childhood Stressful Events, HPA Axis and Anxiety Disorders. *World J. Psych.* **2012**, *2* (1), 13–25.

Fargason, R. E.; Preston, T.; Hammond, E.; May, R.; Gamble, K. L. Treatment of Attention Deficit Hyperactivity Disorder Insomnia with Blue Wavelength Light Blocking Glasses. *Chronophysiol. Ther.* **2013**, *3*, 1–8.

Fava, M. Diagnosis and Definition of Treatment-Resistant Depression. *Biol. Psych.* **2003**, *53* (8), 649–659.

Fava, M.; Anderson, K.; Rosenbaum, J. F. Anger Attacks: Possible Variants of Panic and Major Depressive Disorders. *Am. J. Psych.* **1990**, *147* (7), 867–870.

Fava, M.; Davidson, K. G. Definition and Epidemiology of Treatment-Resistant Depression. *Psych. Clin. NA.* **1996**, *19* (2), 179–200.

Feldman, R.; Singer, M.; Zagoory, O. Touch attenuates infants' physiological reactivity to stress. *Dev Sci* **2010**, 3 (2), 271–278.

Ferguson, J. M.; Shingleton, R. N. An Open-Label, Flexible-Dose Study of Memantine in Major Depressive Disorder. *Clin. Neuropharmacol.* **2007**, *30*, 136–144.

Ficino, M: Sulla Vita. Libro I. Milano (trans) Rusconi, 1995.

Fiedorowicz, J. G.; Swarz, K. L. The Role of Monoamine Oxidase Inhibitors in Current Psychiatric Practice. *J. Psych. Prac.* **2004,** *10* (4), 239–248.

Field, T.; Hernandez-Reif, M.; Diego, M.; Feeijo, L.; Vera, Y.; Gil, K.; Sanders, C. Still-Face and Depressed Separation Effects on Depressed Mother-Infant Interactions. *Infant Mental Health J.* **2007,** *28* (3), 314–323.

Field, T. M. The Therapeutic Effect of Touch. In *The Undaunted Psychologists: Adventures in Research*; Branningan, G., Merrens, M., Eds.; McGraw Hill, Inc.: New York, 1993.

Fiorentini, A.; Rosi, M. C.; Grossi, C.; Luccarini, I.; Casamenti, F. Lithium Improves Hippocampal Neurogenesis, Neuropathology and Cognitive Functions in APP Mutant Mice. *Plos ONE* **2010,** *5* (12), e14382.

Fischer, W. Anticonvulsant Profile and Mechanism of Action of Propranolol and Its Two Enantiomers. *Seizure* **2002,** *11* (5), 285–302.

Fischer, W.; Lasek, R.; Muller, M. Anticonvulsant Effects of Propranolol and Their Pharmacological Modulation. *Pol. J. Pharmacol. Pharm.* **1985,** *37* (6), 883–896.

Fleming, A. S.; Ruble, D. N.; Flett, G. L.; Shaul, D. L. Post-Partum Adjustment in First-Time Mothers: Relations Between Mood, Maternal Attitudes, and Mother-Infant Interactions. *Dev. Psychol.* **1988,** *24*, 71–81.

Flinn, M. V.; England, B. G. Childhood Stress and Family Environment. *Curr. Anthropol.* **1995,** *36*, 854–866.

Fonagy, P. The Interpersonal Interpretive Mechanism: The Confluence of Genetics and Attachment Theory in Development. In *Emotional Development in Psychoanalysis, Attachment Theory and Neuroscience*; Green, V., Ed.; Routledge: Brunner, 2003; pp 107–128.

Fonagy, P.; Gergely, G.; Jurist, E. L.; Target, M. *Affect Regulation, Mentalization and the Development of the Self*; Routledge: New York, 2018.

Fonagy, P.; Rost, F.; Carlyle, J.; McPherson, S.; Thomas, R.; Pasco Fearon, R. M.; Goldberg, D.; Taylor, D. Pragmatic Randomized Controlled Trial of Long-Term Psychoanalytic Psychotherapy for Treatment-Resistant Depression: The Tavistock Adult Depression Study (TADS). *World Psych.* **2015,** *14* (3), 312–321.

Fonagy, P.; Target, M. Attachment and Reflective Function: Their Role in Self-Organization. *Dev. Psychopathol.* **1997,** *9* (4), 679–700.

Fonagy, P.; Target, M. Bridging the Transmission Gap: An End to an Important Mystery of Attachment Research? *Attach Hum. Dev.* **2005,** *7* (3), 333–343.

Forbes, E. E.; Fox, N. A.; Cohn, J. F.; Galles, S. F.; Kovacs, M. Children's Affect Regulation During a Disappointment: Psychophysiologic Responses and Relation to Parent History of Depression. *Biol. Psych.ol* **2006,** *71*, 264–277.

Forlenza, O. V.; DePaula, V. J. R.; Diniz BSO: Neuroprotective Effects of Lithium: Implications for the Treatment of Alzheimer's Disease and Related Neurodegenerative Disorders. *ACS Chem. Neurosci.* **2014,** *5*, 443–450.

Fosshage, J. L. An Expansion of Motivational Theory: Lichtenberg's Motivational Systems Model. *Psychoan. Inq.* **1995,** *15* (4), 421–436.

Fosshage, J. L. The Meanings of Touch in Psychoanalysis: A Time for Reassessment. *Psychoan. Inq.* **2000,** *20* (1), 21–43.

Fox, N. A., Ed. The Development of Emotion Regulation: Biological and Behavioral Considerations. *Monogr. Soc. Res. Child Dev.* **1994,** *59* (2–3),264–277.

Fox, N. A.; Kimmerly, N. L.; Schafere, W. D. Attachment to Mother/Attachment to Father: A Meta-Analysis. *Ch. Dev.* **1991,** *62* (1), 210–25.

Francesca, M. M.; Efisia, L. M.; Alessandra, G. M.; Mariana, A.; Giovanni, A. M. Misdiagnosed Hypomanic Symptoms in Patients with Treatment-Resistant Major Depressive Disorder in Italy: Results from the Improve Study. *Clin. Pract. Epidemic Mental Health* **2014,** *10,* 42047.

Francis, D.; Diorio, J.; Liu, D.; Meany, M. Nongenomic Transmission Across Generations of Maternal Behavioral and Stress Response in the Rat. *Science* **1999,** *286* (5442), 1155–1158.

Frank, D. W.; Dewitt, M.; Hudgens-Haney, M.; Schaeffer, D. J.; Ball, B. H.; Schwarz, N. F.; Hussein, A. A.; Smart, L. M.; Sabatinelli, D. Emotion Regulation: Quantitative Meta-Analysis of Functional Activation and Deactivation. *Neurosci. Biobehav. Rev.* **2014,** *45,* 202–211.

Frank, E.; Swartz, H. A.; Kupfer, D. J. Interpersonal and Social Rhythm Therapy: Managing the Chaos of Bipolar Disorder. *Biol. Psych.* **2000,** *48,* 593–604.

Freese JL. Amaral, D. G. Neuroanatomy of the Primate Amygdala. In *The Human Amygdala*; Whalen, J. P., Phelps, E. A., Eds.; Guilford Press: New York, 2009.

Friedman, M. J.; Donnelly, C. L.; Mellman, T. A. Pharmacotherapy for PTSD. *Psych. Ann.* **2003,** *33* (1), 57–62.

Friedman, M. J.; Davidson, J. R. T. Pharmacotherapy for PTSD. In *Handbook of PTSD: Science and Practice*; Friedman, M. J., Keane, T. M., Resick, P. A., Eds.; The Guilford Press, 2007; pp 376–405.

Frith, C. D.; Frith, U. The Neural Basis of Mentalizing. *Neuron* **2006,** *50,* 531–534.

Fuerino, L.; Silk, K. R. State of the Art in the Pharmacologic Treatment of Borderline Personality Disorder. *Curr. Psych. Rep.* **2011,** *13,* 69–75.

Gabbard, G. O. Psychodynamic Psychotherapy of Borderline Personality Disorder: A Contemporary Approach: Treatment Approaches in the New Millennium. *Bull. Menn. Clin.* **2001,** *65,* 41–57, 41–57.

Gabriel, A. Lamotrigine Adjunctive Treatment in Resistant Unipolar Depression: An Open, Descriptive Study. *Depression* **2006,** *23* (8), 485–488.

Ganguly, P.; Brenhouse, H. C. Broken or Maladaptive? Altered Trajectories in Neuroinflam-mation and Behavior After Early Life Adversity. *Dev. Cog. Neurosci.* **2015,** *11,* 18–30.

Gao, Y.; Payne, R. S.; Schurr, A.; Hougland, T.; Lord, J.; Herman, L.; Lei, Z.; Banerjee, P.; El-Mallakh, R. S. Memantine Reduces Mania-Like Symptoms in Animal Models. *Psych. Res.* **2011,** *188* (3), 366–371.

Gardner, D. L.; Cowdry, R. W. Positive Effects of Carbamazepine on Behavioral Dyscontrol in Borderline Personality Disorder. *Am. J. Psych.* **1986,** *143,* 519–522.

Gardner, E. L. Cannabinoid Interaction with the Reward System. In *Marihuana and Medicine*; Nahas, G. G., Sutin, K. M., Harvey, D. J., Agurell S., Eds.; Humana Press: Totowa, NJ, 1999; pp 187–205.

Garrison, F. H. *History of Medicine*; W.B. Saunders Co.: Philadelphia, 1996.

Gawin, L. H.; Ellinwood, E. H. Cocaine and Other Stimulants. *NE J. Med.* **1988,** *318,* 1173–1182.

Gazzaniga, M. S.; LeDoux, J. E.; Wilson, D. H. Language, Praxis, and the Right Hemisphere. *Neurology* **1977,** *27* (12), 1144–1147.

Geisler, F. C. M.; Vennewald, N.; Kubiak, T.; Weber, H. The Impact of Heart Rate Variability on Emotional Well-Being Is Mediated by Emotion Regulation. *Person. Individual Diff.* **2010**, *49* (7), 723–728.

Geller, B.; Williams, M.; Zimerman, B.; Frazier, J.; Beringer, L.; Warner, K. I. Prepubertal and Early Adolescent Bipolarity Differentiate from ADHD by Manic Symptoms, Grandiose Delusions, Ultra-Rapid or Ultradian Cycling. *J. Aff. Dis.* **1998**, *51*, 81–91.

George, E. L.; Miklowitz, D. L.; Richards, J. A.; Simoneau, T. L.; Taylor, D. O. The Comorbidity of Bipolar Disorder and Axis II Disorders: Prevalence and Clinical Correlates. *Bip. Dis.* **2003**, *5* (2), 115–122.

Gerhard, T.; Devanand, D. P.; Huang, C.; Crystal, S. Lithium Treatment and Risk for Dementia in Adults with Bipolar Disorder: Population-Based Cohort Study. *Br. J. Psych.* **2015**, *207* (1), 46–51.

Gershon, E. S. Bipolar Illness and Schizophrenia as Oligogenic Diseases: Implications for the Future. *Biol. Psych.* **2000**, *47*, 240–244.

Ghaemi, S. N.; Boiman, E. E.; Goodwin, F. K.; Diagnosing Bipolar Disorder and the Effect of Antidepressants: A Naturalistic Study. *J. Clin. Psych.* **2000**, *61*, 804–808.

Ghaemi, S. N.; Hsu, D. J.; Soldani, F.; Goodwin, F. K. Antidepressants in Bipolar Disorder: The Case for Caution. *Bip. Dis.* **2003**, 5 (6), 321–433.

Gianino, A.; Tronick, E. The Mutual Regulation Model: The Infant's Self and Interactive Regulation and Coping and Defensive Capacities. In *Stress and Coping*; Field, T., McCabe, P., Schneiderman, N., Eds.; Lawrence Erlbaum Associates: Hillsdale, NJ, 1988; pp 47–58.

Gigante, A. D.; Bond, D. J.; Lafer, B.; Lam, R. W.; Young, T.; Yatham, L. N. Brain Glutamate Levels Measured by Magnetic Resonance Spectroscopy in Patients with Bipolar Disorder: A Meta-Analysis. *Bip. Dis.* **2012**, *4* (5), 478–487.

Gill, N.; Bayes, A.; Parker, G. A Review of Antidepressant-Associated Hypomania in Those Diagnosed with Unipolar Depression—Risk Factors, Conceptual Models, and Management. *Bip. Dis.* **2020**, *22* (20), 1–8.

Glayser, D. Child Abuse and Neglect and the Brain—A Review. *J. Ch. Psychol. Psych.* **2000**, *41* (1), 97–116.

Glenn, C.; Klonsky, D. Emotional Dysregulation as a Core Feature of Borderline Personality Disorder. *J. Pers. Dis.* **2009**, *23* (1), 20–28.

Gold, M. S.; Pottash, C.; Sweeney, D.; Martin, D.; Extein, I. Antimanic, Antidepressant and Antipanic Effects of Opiates: Clinical, Neuroanatomical and Biochemical Evidence. *Ann. NY Acad. Sci.* **1982**, *398*:1, 140–150.

Goldberg, E. *The Executive Brain: The Frontal Lobes and the Civilized Mind.* Oxford University Press: New York, 2001.

Goldberg, J. F.; Truman, C. J. Antidepressant-Induced Mania: An Overview of Current Controversies. *Bip. Dis.* **2003**, *5* (6), 407–420.

Goldberg, J. S.; Bell, C. E.; Pollard, D. A. Revisiting the Monoamine Hypothesis of Depression: A New Perspective. *Persp. Med. Chem.* **2014**, 61–68.

Goldin, P. R.; McRae, K.; Ramel, W.; Gross, J. J. The Neural Basis of Emotion Regulation: Reappraisal and the Suppression of Negative Emotion. *Biol. Psych.* **2008**, *63* (6), 577–586.

Gonda, X.; Eszlari, N.; Sutori, S.; Aspan, N.; Rihmer, Z.; Juhasz, G.; Bagdy, G. Nature and Nurture: Effects of Affective Temperaments on Depressive Symptoms Are Markedly Modified by Stress Exposure. *Front. Psych.* **2020**, *11*, Art 599.

Gonul, A. S.; Akdeniz, F.; Taneli, F.; Donat, O.; Eker, C.; Vahip, S. Effect of Treatment on Serum Brain-Derived Neurotrophic Factor Levels in Depressed Patients. *Eur. Arch. Psych. Clin. Neurosci.* **2005,** *255*, 381–386.

Goodwin, F. K.; Zis, A. P. Lithium in the Treatment of Mania. *Arch. Gen. Psych.* **1979,** *36* (8), 840–844.

Goodwin, F. K.; Jamison, K. R. *Manic Depressive Illness*; Oxford University Press: New York, 1990.

Gorman, J. M.; Docherty, J. P. A hypothesized Role for Dendritic Remodeling in the Etiology of Mood and Anxiety Disorders. *J. Neuropsych. Clin. Neurosci.* **2010,** *22*, 256–264.

Gotos, Terao, T.; Hoaki, N.; Wang, Y. Cyclothymic and Hyperthymic Temperaments May Predict Bipolarity in Major Depressive Disorder: A Supportive Evidence for Bipolar II1/2 and IV. *J. Aff. Dis.* 2011, *129* (1–3), 34–38.

Gould, R. A.; Otto, M. W.; Pollack, M. H.; Yap, L. Cognitive Behavioral and Pharmacological Treatment of Generalized Anxiety Disorder: A Preliminary Meta-Analysis. *Behav. Ther.* **1997,** *28* (2), 285–305.

Graham, Y. P.; Heim, C.; Goodman, S. H.; Miller, A. H.; Nemeroff, C. B. The Effects of Neonatal Stress on Brain Development: Implications for Psychopathology. *Dev. Psychopath.* **1999,** *11*, 545–565.

Grant, B. F.; Hasin, D. S.; Stinson, F. S.; Dawson, D. A.; Chou, S. P.; Ruan, W. J.; Pickering, R. P. Prevalence, Correlates and Disability of Personality Disorders in the United States: Results from the National Epidemiologic Survey on Alcohol and Related Conditions. *J. Clin. Psych.* **2004,** *65*, 948–958.

Grant, B. F.; Stinson, F. S.; Dawson, D. A.; Chou, S. P.; Dufour, M. C.; Compton, W.; Pickering R. P.; Kaplan K. Prevalence and Co-Occurrence of Substance Use Disorders and Independent Mood and Anxiety Disorders: Results from the National Epidemiologic Survey on Alcohol and Related Conditions. *Arch. Gen. Psych* **2004,** *61*, 807–816.

Graziano, P. A.; Garcia, A. Attention-Deficit Hyperactivity Disorder and Children's Emotion Dysregulation: A Meta-Analysis. *Clin. Psychol. Rev.* **2016,** *46*, 106–123.

Greenberg, B. R.; Harvey, P. D. Affective Lability Versus Depression as Determinants of Binge Eating. *Addict. Behav.* **1987,** *12*, 357–361.

Greydanus, D. E. Psychopharmacology for ADHD in Adolescents: Quo Vadis? *Psych. Times* **2003,** *20* (5), 44.

Grinspoon, L.; Bakalar, J. The Use of Cannabis as a Mood Stabilizer in Bipolar Disorder: Anecdotal Evidence and the Need for Clinical Research. *J. Psychoact. Dr.* **1998,** *30* (2), 171–177.

Gross, J. J. Antecedentand Response-Focused Emotion Regulation: Divergent Consequences for Experience, Expression and Physiology. *J. Person. Soc. Psychol.* **1998,** *74* (1), 224–237.

Gross, J. J.; John, O. P. Individual Differences in Two Emotion Regulation Processes: Implications for Affect, Relationships and Well-Being. *J. Person. Soc. Psychol.* **2003,** *85* (2), 348–362.

Gross, J. J.; Thompson, R. A. Emotion Regulation: Conceptual Foundations for the Field. In *Handbook of Emotion Regulation*; Gross, J. J., Ed.; Guilford Press: New York, 2007.

Gruber, S. A.; Sagar, K. A.; Dahlgren, M. K.; Olson, D. P.; Centorrion, F.; Lukas, S. Marijuana Impacts Mood in Bipolar Disorder: A Pilot Study. *Mental Health Sub Use* **2012,** *5* (3), 228–239.

Giuseppe, T. The Temperaments and Their Role in Early Diagnosis of Bipolar Spectrum Disorders. *Psychiatria Danubina* **2010**, *22* (Suppl), 15–17.

Guille, C.; Sachs, G. Clinical Outcome of Adjunctive Topiramate Treatment in a Sample of Refractory Bipolar Patients with Comorbid Conditions. *Prog. Neuropsychopharm. Biol. Psych.* **2002**, *26* (6), 1035–1039.

Gunderson, J. G.; Singer, T. Defining Borderline Patients: An Overview. *Am. J. Psych.* **1975**, *132*, 1–10.

Gunderson, J. G.; Shea, M. T.; Skodol, A. E.; McGlashan, T. H.; Morey, L. C.; Stout, R. L.; Zanarini, M. C.; Grilo, C. M.; Oldham, J. M.; Keller, M. B. The Collaborative Longitudinal Personality Disorders Study: Development, Aims, Design and Sample Characteristics. *J. Pers. Dis.* **2000**, *14* (4), 300–315.

Gunderson, J. G.; Weinberg, M. T.; Daversa, K. T.; Kuppenbender, K. D.; Zanarini, M. C.; Shea, T.; Skodol, A. E.; Stanislow, E. A.; Yen, S.; Morey, L. C.; Grilo, C. M.; McGlashan, T. H.; Stout, R. L.; Dyck, I. Descriptive and Longitudinal Observations on the Relationship Between Borderline Personality and Bipolar Disorder. *Am. J. Psych.* **2006**, *163* (7), 1173–1178.

Gunnar, M. Integrating Neuroscience and Psychosocial Approaches in the Study of Early Experiences. In *Roots of Mental Illness in Children*; King, J. A.; Ferris, C. F.; II Lederhendler; New York Academy of Sciences, 2003; pp 238–247.

Gunnar, M.; Brodersen, L.; Nachmias, M.; Buss, K.; Rigatuso, J. Stress Reactivity and Attachment Security. *Dev. Psychobiol.* **1996**, *29* (3), 191–204.

Gunnar, M.; Quevedo, K. The Neurobiology of Stress and Development. *Ann. Rev. Psychol.* **2007**, *58*, 145–173.

Gunnar, M.; Vasquez, D. Stress Neurobiology and Developmental Psychopathology. In *Developmental Psychopathology: Vol III. Risk, Disorder and Adaptation*; Cicchetti, D., Cohen, D., Eds., 2nd ed.; Wiley: New York, 2006; pp 533–537.

Gunnar, M. R.; Donzella, B. Social Regulation of the Cortisol Levels in Early Human Development. *Psychoneuroendocrinology* **2002**, *27*, 199–220.

Hafeman, D. M.; Rooks, B.; Merranko, J.; Liao, F.; Gill, M. K.; Goldstein, T. R.; Diler, R.; Ryan, N.; Goldstein, B. I.; Axelson, D. A.; Strober, M.; Keller, M.; Hunt, J.; Hower, H.; Weinstock, L. M.; Yen, S.; Birmaher, B. Lithium Versus Other Mood-Stabilizing Medications in a Longitudinal Study of Youth Diagnosed With Bipolar. *J. Am. Acad. Child Adol. Psych.* Online July 20, **2019**.

Haefely, E. GABA and the Anticonvulsant Action of Benzodiazepines and Barbiturates. *Br. Res. Bull.* **1980**, *5* (2), 873–878.

Hall, W. Is Cannabis Use Psychotogenic? *Lancet* **2006**, *367*, 193–195.

Hansen, L. Olanzepine in the Treatment of Anorexia Nervosa. *Br. J. Psych.* **1999**, *175*, 592.

Hantouche, E. G.; Akiskal, H. S.; Lancrenon, S.; Aliliare, J. F.; Sechter, D.; Azorin, J. M.; Bougeois, M.; Fraud, J. P.; Chatenet-Duchene, L. Systematic Clinical Methodology for Validating Bipolar-II Disorder: Data in Midstream from a French National Multisite Study (EPIDEP). *J. Aff. Dis.* **1998**, *50*, 163–173.

Hardingham, G. E.; Fukunaga, Y.; Bading, H. Extrasynaptic NMDAR's Oppose Synaptic NMDAR's by Triggering CREB Shut-Off and Cell Death Pathways. *Nat. Neurosci.* **2002**, *5*, 405–414.

Harrison, Y.; Horne, J. A. Sleep-Loss and Temporal Memory. *Quart. J. Exp. Psychol.* **2000**, *53* (1), 271–279.

Hart, B. L. Biological Basis of the Behavior of Sick Animals. *Neurosci. Biobehav. Rev.* **1988,** *12,* 123–137.

Hashimoto, R.; Fujimaki, K.; Jeong, M. R.; Christ, L.; Chuang, D. M. Lithium-Induced Inhibition of SRC Tyrosine Kinase in Rat Cerebral Cortical Neurons: A Role in Neuroprotection Against *N*-Methyl-D-Aspartate Receptor-Mediated Excitotoxicity. *FEBS Lett.* **2003,** *538* (1–3), 145–148.

Hawke, L. D.; Parikh, S. V.; Michalak, E. E. Stigma and Bipolar Disorder: A Review of the Literature. *J. Aff. Dis.* **2013,** *150* (2), 181–191.

Haydon, P. G.; Carmignoto, G. Astrocyte Control of Synaptic Transmission and Neurovascular Coupling. *Physiol. Rev.* **2006,** *86,* 1009–1031.

Haynes, P. L.; Gengler, D.; Kelly, M. Social Rhythm Therapies for Mood Disorders: An Update. *Curr. Psych. Rep.* **2016,** *18,* 1–8.

Healy, D.; Herxheimer, A.; Menkes, D. B. Antidepressants and Violence: Problems at the Interface of Medicine and Law. *Int. J. Risk Saf. Med.* **2007,** *19* (1–2), 17–33.

Hearing, C. M.; Chang, W. C.; Szuhany, K. L.; Deckersback, T, Nierenberg, A. A.; Sylvia, L. G. Physical Exercise for Treatment of Mood Disorders: A Critical Review. *Curr. Behav. Neuroschi. Rep.* **2016,** *3,* 350–359.

Heim, C.; Nemeroff, C. The Impact of Early Adverse Experiences on Brain Systems Involved in the Pathophysiology of Anxiety and Affective Disorders. *Biol. Psych.* **1999,** *46,* 1509–1522.

Heim, C.; Newport, D. J.; Bonsall, R.; Miller, A. H.; Nemeroff, C. B. Altered Pituitary-Adrenal Axis Responses to Provocative Challenge Tests in Adult Survivors of Childhood Abuse. *Am. J. Psych.* **2001,** *158* (4), 575–581.

Hennevin, E.; Hars, B. Is Increase in Post-Learning Paradoxical Sleep Modified by Cueing? *Behav. Brain. Res.* **1987,** *24* (3), 243–249.

Henry, C.; Bailara, K. M.; Mathieu, F. M.; Poinsot, R.; Falissard, B. Construction and Validation of a Dimensional Scale Exploring Mood Disorders: MAThyS (Multidimensional Assessment of Thymic States). *BMC Psych.* 2008, *8,* 82, 1–8.

Henry, C.; Mitropoulou, V.; New, A. S.; Koenigsberg, H. W.; Silverman, J.; Siever, L. J. Affective Instability and Impulsivity in Borderline Personality and Bipolar II Disorders: Similarities and Differences. *J Psych. Res.* **2001,** *35* (6), 307–312.

Henry, C.; Sorbara, F.; Lacoste, J.; Gindre, C.; Leboyer, M. Antidepressant-Induced Mania in Bipolar Patients: Identification of Risk Factors. *J. Clin. Psych.* **2001,** *62* (4), 249–255.

Herman, J. P.; Cullinam, W. E. Neurocircuitry of Stress: Central Control of the Hypothalamo-Pituitary-Adrenocortical Axis. *Trends Neurosci.* **1997,** *20,* 78–84.

Hermens, D. F.; Scott, E. M.; White, D.; Lynch, M.; Lagopoulos, J.; Whitwell, B. G.; Naismith, S. L.; Hickie, I. B. Frequent Alcohol, Nicotine or Cannabis Use Is Common in Young Persons Presenting for Mental Healthcare: A Cross-Sectional Study. *BMJ* 2013 Open, 3.

Herringa, R. J.; Birn, R. M.; Ruttle, P. L.; Burghy, C. A.; Stodola, D. E.; Davidson, R. J.; Essex, M. J. Childhood Maltreatment Is Associated with Altered Fear Circuitry and Increased Internalizing Symptoms by Late Adolescence. *Proc. Nat. Acad. Sci. USA* **2013,** *110* (47), 19119–19124.

Hertsgaard, L.; Gunnar, M. R.; Erickson, M.; Nachmias, M. Adrenocortical Responses to the Strange Situation in Infants with Disorganized/Disoriented Attachment Relationships. *Child Dev.* **1995,** *66,* 1100–1106.

Het, S.; Schoofs, D.; Rohleder, N.; Wolf, O. T. Stress-Induced Cortisol Level Elevations Are Associated with Reduced Negative Affect After Stress: Indications for a Mood-Buffering Cortisol Effect. *Psychosom. Med.* **2012**, *74* (1), 23–32.

Het, S.; Wolf, O. T. Mood Changes in Response to Psychosocial Stress in Healthy Young Women: Effects of Pre-Treatment with Cortisol. *Behav. Neurosci.* **2007**, *121* (1), 11–20.

Hett, W. S., Trans. *Aristotle: Problems, I. I.; Books XXII–XXXVIII*; Harvard University Press/Loeb Classical Library: Cambridge, MA, 1965; pp 155–169.

Higgit, A.; Fonagy, P.; Toone, B.; Shine, P. The Prolonged Benzodiazepine Withdrawal Syndrome: Anxiety or Hysteria? *Acta Psych. Scand.* **1990**, *82* (2), 165–168.

Hill, C. L. M.; Updegraff, J. A. Mindfulness and Its Relation to Emotional Regulation. *Emotion* **2012**, *12* (1), 81–90.

Hill, D. *Affect Regulation Theory: A Clinical Model*; W. W. Norton, 2015.

Himmelhoch, J. M.; Mulla, D.; Neil, J. F.; Detre, T. P.; Kupfer, D. J. Incidence and Significance of Mixed Affective States in a Bipolar Population. *Arch. Gen. Psych.* **1976**, *33* (9), 1062–1066.

Hirschfeld, R. M. A. History and Evolution of the Monoamine Hypothesis of Depression. *J. Clin. Psych.* **2000**, *61* (supp. 6), 4–6.

Hofer, M. Relationships as Regulators: A Psychobiologic Perspective on Bereavement. *Psychosom. Med.* **1984**, *46* (3), 182–197.

Hofer, M. A. Hidden Regulators: Implications for a New Understanding of Attachment, Separation, and Loss. In *Attachment Theory: Social, Developmental, and Clinical Perspectives*; Golberg, S., Muir, R., Kerr, J., Eds.; Analytic Press: Hillsdale, NJ, 1995; pp 203–230.

Hoffman, C.; Crnic, K. A.; Baker, J. K. Maternal Depression and Parenting: Implications for Children's Emergent Emotion Regulation and Behavioral Functioning. *Parent. Sci. Pract.* **2006**, *6*, 271–295.

Horne, J. A. Sleep Function, with Particular Reference to Sleep Deprivation. *Ann. Clin. Res.* **1985**, *17* (5), 199–208.

Hostinar, C. E.; Gunnar, M. R. Future Directions in the Study of Social Relationships as Regulators of the HPA Axis Across Development. *J. Clin. Ch. Adol. Psych.* **2013**, *42* (4), 564–575.

Howland, R. H. Induction of Mania with Serotonin Reuptake Inhibitors. *J. Clin. Psychopharm* **1996**, *16* (6), 425–427.

Hsu, L. K. G. Treatment of Bulimia with Lithium. *Am. J. Psych.* **1984**, *141*, 1260–1262.

Hsu, L. K. G. Lithium in the Treatment of Eating Disorders. In *The Role of Drug Treatments for Eating Disorders*; Garfinkel, P. E., Gardner, D. M., Eds.; Brunner-Mazel: New York, 1987.

Hudson, J. I.; Pope Jr. H. G.; Jonas, J. M.; Yurgelun-Todd, D. Treatment of Anorexia Nervosa with Antidepressants. *J. Clin. Psychopharm.* **1985**, *5*, 17–23.

Ibrahim, L.; Diazgranados, N.; Luckenbaugh, D. A.; Machado-Vieira, R.; Baumann, J.; Mallinger, A. G.; Zarate Jr. C. A. Rapid Decrease in Depressive Symptoms with an N-Methyl-D-Aspartate Antagonist in ECT-Resistant Major Depression. *Prog. Neuropsychopharm. Biol. Psych.* **2011**, *35*, 1155–1159.

Infortuna, C.; Silvestro, S.; Crenshaw, K.; Muscatello, M. R. A.; Bruno, A.; Zocalli, R. A.; Chusid, E.; Intrator, J.; Han, Z.; Battaglia, F. Affective Temperament Traits and

Age-Predicted Recreational Cannabis Use in Medical Students: A Cross-Sectional Study. *Int. J. Env. Res. Pub. Health* **2020,** *17* (13), 4836.

Inoue, T.; Nakagawa, S.; Kitaichi, Y.; Izumi, T.; Tanaka, T.; Masui, T.; Kusumi, I.; Denda, K.; Koyama, T. Long-Term Outcome of Antidepressant-Refractory Depression: The Relevance of Unrecognized Bipolarity. *J. Aff. Dis.* **2006,** *95* (1–3), 61–67.

Insel, T.; Cuthbert, B.; Garvey, M. et al. Research Domain Criteria (RDoC): Toward a New Classification Framework for Research on Mental Disorders. *Am. J. Psych.* **2010,** *167* (&), 748–751.

Insel, T. Director's Blog: Transforming Diagnosis. National Institute of Mental Health, April 29, 2013.

Ivanov, I.; Newcorn, J.; Morton, K.; Tricamo, M. Inhibitory Control Deficits in Childhood: Definition, Measurement, and Clinical Risk for Substance Use Disorders. In *Inhibitory Control and Drug Abuse Prevention*; Springer: New York, 2011.

Ivkovic, M.; Damjanovic, A.; Jovanovic, Cvetic, T.; Jasovic-Gasic, M. Lamotrigine Versus Lithium Augmentation of Antidepressant Therapy in Treatment-Resistant Depression: Efficacy and Tolerability. *Psychiatria Danubina* **2009,** *21* (2), 187–193.

Ivy, A. S.; Rex, C. S.; Chen, Y.; Dube, C.; Maras, P. M.; Grigoriadis, D. E.; Gall, C. M.; Lynch, G.; Baram, T. Z. Hippocampal Dysfunction and Cognitive Impairments Provoked by Chronic Early-Life Stress Involve Excessive Activation of CRH Receptors. *J. Neurosci.* **2010,** *30* (39), 13005–13015.

Jacobson, L.; Sapolsky, R. The Role of the Hippocampus in Feedback Regulation of the Hypothalamo-Pituitary-Adrenocortical Axis. *Endocrinol. Rev.* **1991,** *12*, 118–134.

Jaeger, V.; Esplin, B.; Capek, R. The Anticonvulsant Effects of Propranolol and β-Adrenergic Blockade. *Experientia* **1979,** *35* (1), 80–81.

Jamison, K. R. *An Unquiet Mind*; A. Knopf: New York, 1996.

Jauhar, P.; Weller, M. P. Psychiatric Morbidity and Time Zone Changes: A Study of Patients from Heathrow Airport. **1982,** *140* (3), 231–235.

Jensch, V. L.; Merz, C. J.; Wolf, O. T. Restoring Emotional Stability: Cortisol Effects on the Neural Network of Cognitive Emotion Regulation. *Behav. Br. Res.* **2019,** *374*, 111880.

Johnston, S. J.; Boehm, S. G.; Healy, D.; Goebel, R.; Linden, D. E. J. Neurofeedback: A Promising Tool for the Self-Regulation of Emotion Networks. *Neuroimage* **2010,** *49* (1), 1066–1072.

Jones, B. T.; Corbin, W.; Fromm, K. A Review of Expectancy Theory and Alcohol Consumption. *Addiction* **2001,** *96* (1), 57–72.

Jones, S. Psychotherapy of Bipolar Disorder: A Review. *J. Aff. Dis.* **2004,** *80* (2–3), 101–114.

Jope, R. S. Anti-Bipolar Treatment: Mechanism of Action of Lithium. *Mol. Psych.* **1999,** *4* (2), 117–128.

Judd, I. J.; Akiskal, H. S. The Prevalence and Disability of Bipolar Spectrum Disorders in the US Population: Re-Analysis of the ECA Database Taking Into Account Subthreshold Cases. *J. Aff. Dis.* **2003,** *73*:123–131.

Jung, C. G. On Manic Mood Disorder. In *Psychiatric Studies*, Vol. 1, *Collected Works*, 2nd ed., 1970, trans. Hull, R. F. C.; Routledge and Kegan Paul, 1903; pp 109–111.

Kalivas, P. W.; Sorg, B. A.; Hooks, M. S. The Pharmacology and Neural Circuitry of Sensitization to Psychostimulants. *Behav. Pharm.* **1993,** *4*, 315–334.

Kamaradova, D.; Latalova, K.; Prasko, J.; Kubinek, R.; Vrbova, K.; Mainerova, B.; Cinculova, A.; Ocislova, M.; Holubova, M.; Smoldasova, J.; Tichakova, A. Connection

Between Self-Stigma, Adherence to Treatment, and Discontinuation of Medication. *Patient Prefer Adherence* **2016**, *10*, 1289–1298.

Karege, F.; Perret, G.; Bondolfi, G.; Schwald, M.; Bertschy, G.; Aubry, J. M. Decreased Serum Brain-Derived Neurotrophic Factor Levels in Major Depressed Patients. *Psych. Res.* **2002**, *109*, 143–148.

Karniol, I. G.; Carlini, E. A. Pharmacological Interaction Between Cannabidiol and Delta 9Tetrahydrocannabinol. *Psychopharmacology* **1973**, *33*, 53–70.

Karst, H.; Berger, S.; Turiault, M.; Tronche, F.; Schutz, G.; Joels, M. Mineralocorticoid Receptors Are Indispensable for Nongenomic Modulation of Hippocampal Glutamate Transmission by Corticosterone. *Proc. Nat. Acad. Sci. USA* **2005**, *102* (52), 19204–19207.

Katzow, J. J.; Hsu, D. J.; Nassir Ghaemi, S. The Bipolar Spectrum: A Clinical Perspective. *Bip. Dis.* 2003, *5* (6), 436–442.

Kebets, V.; Favre, P.; Houenou, J.; Polosan, M.; Aubry J-M, Van De Ville,D, Piguet, C. Fronto-Limbic Neural Variability as a Transdiagnostic Correlate of Emotion Dysregulation. *mMedRx iv* **2020**, *12*.18.20248457.

Keck, P. E.; McElroy, S. L. Redefining Mood Stabilization. *J. Aff. Dis.* **2003**, *73* (3), 163–169.

Keenan, J.; Rubio, J.; Raciorri, C.; Johnson, A.; Barnacz, A. The Right Hemisphere and the Dark Side of Consciousness. *Cortex* **2005**, *41* (5), 695–704.

Keijsers, G. P. J.; Hoogduin, C. A. L.; Schaap, C. P. D. R. Prognostic Factors in the Behavioral Treatment of Panic Disorder with and Without Agoraphobia. *Behav. Therap.* **1994**, *25*, 689–708.

Kelly, D. Clinical Review of Beta-Blockers in Anxiety. *Pharmakopsych. Neuropsychopharmakol.* **1980**, *13* (5), 259–266.

Kelley, W. M.; Macrae, C. N.; Wyland, C. L.; Caglar, S.; Inati, S.; Heatherton, T. F. Finding the Self? An Event-Related fMRI Study. *J. Cog. Neurosci.* **2002**, *14* (5), 785–794.

Kendell, S. F.; Krystal, J. H.; Sanacora, G. GABA and Glutamate Systems as Therapeutic Targets in Depression and Mood Disorders. *Expert Opin. Therap. Targets* **2005**, *9* (1), 53–168.

Kendler, K. S. Explanatory Models for Psychiatric Illness. *Am. J. Psych.* **2008**, *165* (6), 695–702.

Kent, S.; Blurhe, R. M.; Kelley, K. W.; Dantzere, R. Sickness Behavior as a New Target for Drug Development. *Trends Pharmacol. Sci.* **1992**, *13*, 24–28.

Kernberg, O. Borderline Personality Organization. *J. Am. Psychoan. Assoc.* **1967**, *15* (3), 641–685.

Kernberg, OF: The Management of Affect Storms in the Psychoanalytic Psychotherapy of Borderline Patients. *J. Am. Psychoan. Assoc.* **2003**, *51* (2), 517–545.

Kernberg, O. F.; Selzer, M. A.; Koenigsberg, H. W.; Carr, A. C.; Appelbaum, A. H. Psychodynamic Psychotherapy of Borderline Patients; Basic Books, 1989.

Kessing, L. V.; Forman, J. L.; Andersen, P. K. Does Lithium Protect Against Dementia? *Bip. Dis.* **2010**, *12* (1), 87–94.

Kessler, R. C.; Chui, W. T.; Dernier, O.; Walters, E. Prevalence, Severity and Co-Morbidity of 12-Month DSM-IV Disorders in the National Comorbidity Survey Replication. *Gen. Psych.* **2005**, *62*, 617–627.

Kessler, R. C.; Merikangas, K. R. The National Comorbidity Survey Replication (NCS-R): Background and aims. *Int. J. Meth. Psych. Res.* **2004**, *13* (2), 60–68.

Khantzian, E. J. The Self-Medication Hypothesis of Substance Use Disorders: A Reconsideration and Recent Applications. *Harv. Rev. Psych.* **1997,** *4,* 231–244.

Kilpatrick, L.; Cahill, L. Amygdala Activation of Parahippocampal and Frontal Regions During Emotionally Influenced Memory Storage. *NeuroImage* **2003,** *20* (4), 2091–2099.

Kim, J. J.; Diamond, D. M. The Stressed Hippocampus, Synaptic Plasticity, and Lost Memories. *Nat. Rev. Neurosci.* **2002,** *3,* 453–462.

Kim, S. H.; Hamann, S. Neural Correlates of Positive and Negative Emotion Regulation. *J. Cog. Neurosci.* **2007,** *19,* 776–798.

Kish, T.; Matsunaga, S.; Iwata, N. A Meta-Analysis of Memantine for Depression. *J. Alzheimer Dis.* **2017,** *57* (1), 113–121.

Kitagami, T.; Yamada, K.; Miura, H.; Hashimoto, R.; Nabeshima, T.; Ohta, T. Mechanism of Systematically Injected Interferon-Alpha Impeding Monoamine Biosynthesis in Rats: Role of Nitric Oxide as a Signal Crossing the Blood-Brain Barrier. *Brain Res.* **2003,** *978,* 104–114.

Klerman, G. The Spectrum of Mania. *Comp. Psych.* **1981,** *22* (1), 1–20.

Klufas, A.; Thompson, D. Topiramate-Induced Depression. *Am. J. Psych.* **2001,** *158* (10), 1736.

Koenigsberg, H. W.; Harvey, P. D.; Mitropoulou, V.; Schmeidler, J.; New, A. S.; Goodman, S.; Silverman, J. M.; Serby, M.; Schopick, F.; Siever, L. J. Characterizing Affective Instability in Borderline Personality Disorder. *Am. J. Psych.* **2002,** *159* (5), 784–788.

Koob, G. F. Neuroadaptive Mechanisms of Addiction: Studies on the Extended Amygdala. *Eur. Neuropsychopharm. Special Issue: Neuropsychopharmacol. Addict.* **2003,** *13* (6), 442–452.

Koob, G. F. The Neurobiology of Addiction: A Neuroadaptational View Relevant for Diagnosis. *Addiction* **2006,** *101* (Suppl 1), 23–30.

Korngaonkar, M. S.; Antees, C.; Williams, L. M.; Gatt, J. M.; Bryant, R. A.; Cohen, R.; Paul, R.; O'Hara, R.; Grieve, S. M. Early Exposure to Traumatic Stressors Impairs Emotional Brain Circuitry. *PloS One* **2013,** *8,* e75524.

Kotov, R.; Kreuger, R. F.; Watson, D. A Paradigm Shift in Psychiatric Classification: The Hierarchical Taxonomy of Psychopathology (HiTOP). *World Psych.* **2018,** *17* (1), 4–25.

Koukopoulos, A. Ewald Hecker's Description of Cyclothymia as a Cyclical Mood Disorder: Its Relevance to the Modern Concept of Bipolar II. *J. Aff. Dis.* **2003,** *73,* 199–205.

Koukopoulos, A.; Sani, G.; Koukopoulos, A.; Girardi, P. Cyclicity and Manic-Depressive Illness. In *Bipolar Disorders: 100 Years after Manic Depressive Insanity;* Marneros, A.; Angst, J., Eds.; Dordrecht, The Netherlands: Kluwer Academic Publishers, 2000; pp 315–334.

Kovacs, M.; Sherill, J.; George, C. J.; Pollack, M.; Tumuluru, R. V.; Ho, V. Contextual Emotion-Regulation Therapy for Childhood Depression: Description and Pilot Testing of a New Intervention. *J. Am. Acad. Ch. Adol. Psych.* **2006,** *45,* 892–903.

Kovacs, M.; Joormann, J.; Gotlib, I. H. Emotion (Dys)regulation and Links to Depressive Disorders. *Child Dev. Persp.* **2008,** *2* (3), 149–155.

Kraepelin, E. *Manic-Depressive Insanity and Paranoia;* Trans. Livingstone, E. and S.; Edinburgh, 1899a/1921.

Kraepelin, E. Psychiatrie, 6th ed.; J. A. Barth: Leipzig, 1899b.

Kraepelin, E. Psychiatrie, 7th ed.; J. A. Barth: Leipzig, 1904.

Kraepelin, E. Manic Depressive Insanity and Paranoia; Barclay, R. M., Trans., Robertson, G. M., Ed.; Reprinted 1976, Ayer Company: Salem, NH, 1921.

Kretschmer, E. *Physique and Character*; Macmillan: New York, 1936.

Krieger, F. V.; Pheula, G. F.; Coelho, R.; Zeni, T.; Tramontina, S.; Zeni, T. P.; Rohde, L. A. An Open-Label Trial of Risperidone in Children and Adolescents with Severe Mood-Dysregulation. *J. Ch. Adol. Psychopharm.* **2011,** *21* (3), 237–243.

Kring, A. M.; Sloan, D. M. Emotion Regulation and Psychopathology: A Transdiagnostic Approach to Etiology and Treatment; Guilford Press: New York, 2010.

Kron, L.; Katz, J. L.; Gorsynski, G.; Weiner, H. Hyperactivity in anorexia nervosa: a fundamental clinical feature. *Comp. Psych.* **1978,** *19,* 433–440.

Krystal, J. H.; Perry Jr. EB, Geuorguieva, R.; Belger, A.; Madonick, S. H.; Abi-Dargham, A.; Cooper, T. B.; MacDougall, L.; Abi-Saab, W.; D'Souza, C. Comparative and Interactive Human Psychopharmacologic Effects of Ketamine and Amphetamine Implications for Glutamatergic and Dopaminergic Model Psychoses and Cognitive Function. *Arch. Gen. Psych.* **2005,** *62* (9), 985–995.

Kupfer, D. J.; Carpenter, L. L.; Frank, E. Possible Role of Antidepressants in Precipitating Mania and Hypomania in Recurrent Depression. *Am. J. Psych.* 1988, *145* (7), 804–808.

Kupfer, D. J.; Pickar, D.; Himmelhoch, J. M. Are There Two Types of Unipolar Depression? *Arch. Gen. Psych.* **1975,** *32* (7), 855–871.

Kwan, P.; Sills, G. J.; Brodie, M. J. The Mechanism of Action of Commonly Used Antiepileptic Drugs. *Pharm. Therap.* **2001,** *90* (1), 21034.

Lachmann, F.; Beebe, B. Three Principles of Salience in the Organization of the Patient-Analyst Interaction. *Psychoan. Psychol.* **1996a,** *13,* 1–22.

Lachmann, F.; Beebe, B. The Contribution of Selfand Mutual Regulation to Therapeutic Action: A Case Illustration. *Prog. Self Psych.* **1996b,** *12,* 123–140.

Lam, R. W.; Levitt, A. J. Canadian Consensus Guidelines for the Treatment of Seasonal Affective Disorder. *Clin. Acad. Pub.* 1999.

Lam, S.; Dickerson, S. S.; Zoccola, P. M.; Zaldivar, P. M. Emotion Regulation and Cortisol Reactivity to a SocialEvaluative Speech Task. *Psychoneeuroendocrinology* **2009,** *34* (9), 1355–1362.

Lara, D. R.; Bisol, L. W.; Munari, L. R. Antidepressant, Mood-Stabilizing and Precognitive Effects of Very Low Dose Sublingual Ketamine in Refractory Unipolar & Bipolar Depression. *Int. J. Neuropsychopharm.* **2013,** *16* (9), 2111–2117.

Lashansky, G.; Saenger, P.; Kishman, K.; Gautier, T.; Mayes, D.; Berg, G.; DiMartino-Nardi, J.; Reiter, E. Normative Data for Adrenal Steroidogenesis in a Healthy Pediatric Population: Ageand Sex-Related Changes after Adrenocorticotropin Stimulation. *J. Clin. Endocrinol. Metab.* **1991,** *73* (3), 674–686.

Latalova, K.; Prasko, J.; Diveky, T.; Grambal, A.; Kamaradova, D.; Velartova, H.; Salinger, J.; Opavsky, J. Autonomic Nervous System in Euthymic Patients with Bipolar Affective Disorder. *Neuro Endocrinol. Lett.* **2010,** *31* (6), 829–836.

Laufer, M. W.; Denhoff, E. Hyperkinetic Behavior Syndrome in Children. *J Pediat.* **1957,** *50,* 463–474.

Laverdure, B.; Boulanger, J. P. [Beta-Blocking Drugs and Anxiety: A Proven Therapeutic Value]. *L'Encephale* **1991,** *17* (5), 481–492.

Lecrubier, Y. The Impact of Comorbidity on the Treatment of Panic Disorder. *J. Clin. Psych.* **1998,** *59* (8), 11–14.

LeDoux, J. E. Emotion Circuits in the Brain. *Ann. Rev. Neurosci.* **2000**, *23*, 155–184.

Leibenluft, E.; Charney, D. S.; Towbin, K. E.; Bhangoo, R. K.; Pine, D. S. Defining Clinical Phenotypes of Juvenile Mania. *Am. J. Psych.* **2003**, *160* (3), 430–470.

Leibenluft, E. Severe Mood Dysregulation, Irritability, and the Diagnostic Boundaries of Bipolar Disorder in Youths. *Am. J. Psych.* **2011**, *168* (2), 129–142.

Leibowitz, M. R.; Klein, D. F. Hysteroid Dysphoria. *Psych. Clin. NA.* **1979**, *2*, 555–575.

Leibowitz, M. R.; Quitkin, F. M.; Stewart, J. W.; McGrath, P. J.; Harrison, W. M.; Markowitz, J. S.; Rabkin, J. G.; Tricamo, E.; Goetz, D. M.; Klein, D. F. Antidepressant Specificity in Atypical Depression. *Arch. Gen. Psych.* **1988**, *45* (2), 129–137.

Leichsenring, F.; Salzer, S.; Jaeger, U.; Kächele, H.; Kreische, R.; Leweke, F.; Rüger, U.; Winkelbach, C.; Leibing, E. Short-Term Psychodynamic Psychotherapy and Cognitive-Behavioral Therapy in Generalized Anxiety Disorder: A Randomized, Controlled Trial. *Am. J. Psych.* **2009**, *166* (8), 875–881.

Le Moal, M. Drug Abuse: Vulnerability and Transition to Addiction. *Pharmacopsychiatry* **2009**, *42* S42–S55.

Leonhard, K: Aufteilung der endogenen Psychosen und ihre differenzierte Atiologie (1957); Akademie-Verlag: Berlin; English Trans 1979 by Berman, R.; New York: Irvington, 1957.

Leonhard, K. *Classification of Endogenous Psychoses and Their Differentiated Etiology*, 2nd ed.; Beckmann, H., Ed.; Springer-Verlag; New York/Wien, 1999.

Leuner, B.; Gould, E. Structural Plasticity and Hippocampal Function. *Ann. Rev. Psychol.* **2010**, *6*, 111–140 C 1–3.

Levesque, J.; Eugene, F.; Joanette, Y.; Paquette, V.; Mensour, B.; Beaudoin, G.; Leroux, J-M.; Bourgoin, P.; Beauregard, M. Neural Circuitry Underlying Voluntary Suppression of Sadness. *Biol. Psych.* **2003**, *532* (6), 502–510.

Levine, S.; Haltemeyer, G. C.; Karas, G. G.; Denenberg, V. H. Physiological and Behavioral Effects of Infant Stimulation. *Physiol. Behav.* **1967**, *2*, 55–63.

Levy, D.; Kimhi, R.; Barak, Y.; Viv, A.; Elizur, A. Antidepressant-Associated Mania: A Study of Anxiety Disorders Patients. *Psychopharmacology* **1998**, *136*, 243–246.

Lewis, M. D.; Todd, R. M. The Self-Regulating Brain: Cortical-Subcortical Feedback and the Development of Intelligent Action. *Cog. Dev.* **2007**, *22*, 406–430.

Leyba, C. M.; Gold, D. D. The Relation Between Rapid-Cycling Cyclothymia and Bulimia: Case Reports of Two Women. *SDJ Med.* **1988**, *41*, 21–22.

Li, C. S. R.; Sinha, R. Inhibitory Control and Emotional Stress Regulation: Neuroimaging Evidence for Frontal-Limbic Dysfunction in Psycho-Stimulant Addiction. *Neurosci. Biobehav. Rev.* **2008**, *32* (3), 581⁻597.

Licht, R. W. Drug Treatment of Mania: A Critical Review. *Acta Psych. Scand.* **1998**, *97* (6), 387–397.

Lichtenberg, J. D.; Lachmann, F. M.; Fosshage, J. L. *Psychoanalysis and Motivational Systems: A New Look*; Routledge Taylor & Francis Group: New York, 2011.

Lieberman, M. D. Social Cognitive Neuroscience: A Review of Core Processes. *Ann. Rev. Psychol.* **2007**, *58*, 259–289.

Linden, R. D.; Pope HG Jr, Jonas, J. M. Pathological Gambling and Major Affective Disorder: Preliminary Findings. *J. Clin. Psych.* **1986**, *47* (4), 201–203.

Lindquist, K. A.; Barrett, L. F. Constructing Emotion: The Experience of Fear as a Conceptual Act. *Psychol. Sci.* **2008**, *19* (9), 898–903.

Linehan, M. M. *Cognitive-Behavioral Treatment of Borderline Personality Disorder*; Guilford Press: New York, London, 1993.

Linehan, M. M.; Bohus, M.; Lynch, T. R. Handbook of Emotion Regulation: Ch 29 Dialectical Behavior Therapy for Pervasive Emotional Dysregulation, 2007, 581–609.

Lingiardi, V.; McWilliams, N. *Psychodynamic Diagnostic Manual*, 2nd ed.; Guilford Press, 2017.

Linhartova, P.; Latalova, A.; Kosa, B.; Kasparek, T.; Schmahl, C.; Paret, C. fMRI Neurofeedback in Emotion Regulation: A Literature Review. *Neuroimage* **2019**, *193*, 75–92.

Link, B. G.; Phelan, J. C. Conceptualizing Stigma. *Ann. Rev. Sociol.* **2001**, *27*, 363–385.

Links, P. S.; Steiner, M.; Boiago, I.; Irwin, D. Lithium Therapy for Borderline Patients: Preliminary Findings. *J. Pers. Dis.* **1990**, *4* (2), 173–181.

Liotti, G. Trauma, Dissociation and Disorganized Attachment: Three Strands of a Single Braid. *Psychother. Theor. Res. Pract. Train.* **2004**, *41*, 472–486.

Lisj, J.; Dime-Meenan, S.; Whybrow, P. C.; Price, R. A.; Hirschfeld, R. M. The National Depressive and Manic-Depressive Association (National DMDA) Survey of Bipolar Members. *J. Aff. Dis.* **1994**, *31*, 281–294.

Lochman, J. E.; Evans, S. C.; Burke, J. D.; Roberts, M. C.; Fite, P. J.; Reed, G. M.; de la Pena, F. R.; Matthys, W.; Ezpeleta, L.; Siddiqui, S.; Garralda, M. E. An Empirically Based Alternative to DSM-5's Disruptive Mood Dysregulation Disorder for ICD-11. *World Psych.* **2015**, *14* (1), 30–33.

Lopresti, A. L.; Jacka, F. N. Diet and Bipolar Disorder: A Review of its Relationship and Potential Therapeutic Mechanism of Action. *J. Alt. Comp. Med.* **2015**, *21* (12), 733–739.

Loranger, A. W.; Oldham, J. M.; Tullis, E. H. Familial Transmission of DSM-III Borderline Personality Disorder. *Arch. Gen. Psych.* **1982**, *39*, 795–799.

Lowy, M.; Gault, L.; Yamamoto, B. Adrenalectomy Attenuates Stress Induced Elevation in Extracellular Glutamate Concentration in Hippocampus. *J. Neurochem.* **1993**, *61*, 1957–1960.

Lozano, A. M.; Mayberg, H. S.; Giacobbe, P.; Haman, C.; Craddock, R. C.; Kennedy, S. H. Subcallosal Cingulate Gyrus Deep Brain Stimulation for Treatment-Resistant Depression. *Biol. Psych.* **2008**, *64*, 461–467.

Lu, Y. Y.; Lin, C. H.; Lane, H. Y. Mania Following Ketamine Abuse. *Neuropsychiatr Dis. Tr.* **2016**, *12*, 237–239.

Lunde, A. V.; Fasmer, O. B.; Akiskal, K. K.; Akiskal, H. S.; Oedegaard, K. J. The Relationship of Bulimia and Anorexia Nervosa with Bipolar Disorder and Its Temperamental Foundations. *J. Aff. Dis.* **2009**, *115* (3), 309–914.

Lupien, S. J.; Gillin, C. J.; Hauger, R. L. Working Memory Is More Sensitive Than Declarative Memory to the Acute Effects of Corticosteroids: A Dose-Response Study in Humans. *Behav. Neurosci.* **1999**, *113*, 420–430.

Lupien, S. J.; Maheu, F.; Tu, M.; Fiocco, A.; Schramek, T. E. The Effects of Stress and Stress Hormones on Human Cognition: Implications for the Field of Brain and Cognition. *Brain Cog.* **2007**, *65* (3), 209–237.

Lyons-Ruth, K. Contributions of the Mother-Infant Relationship to Dissociative, Borderline, and Conduct Symptoms in Young Adulthood. *Inf. Mental Health J.* **2008**, *29*, 203–218.

Lyons-Ruth, K.; Connell, D. B.; Zoll, D.; Stahl, J. Infants at Social Risk: Relations Among Infant Maltreatment, Maternal Behavior, and Infant Attachment Behavior. *Dev. Psychol.* **1987**, *23* (2), 223–232.

Maccari, S.; Krugers, H. J.; Morley-Fletcher, S.; Szyf, M.; Brunton, P. J. The Consequences of Early-Life Adversity: Neurobiological, Behavioural and Epigenetic Adaptations. *J. Neuroendocrinol.* **2014**, *26* (10), 707–723.

MacDonald, G.; Leary, M. R. Why Does Social Exclusion Hurt? The Relation Between Social and Physical Pain. *Psychol. Bull.* **2005**, *131* (2), 202–223.

Machado-Vieira, R.; Salvadore, G.; DiazGranados, N.; Zarate, C. A. Ketamine and the Next Generation of Antidepressants with a Rapid Onset of Action. *Pharma. Therapeut.* **2009**, *123*, 143–150.

MacKinnon, D. F.; Pies, R. Affective Instability as Rapid Cycling: Theoretical Clinical Implications for Borderline Personality and Bipolar Spectrum Disorders. *Bip. Dis.* **2006**, *8* (1), 1–14.

Mackinnon, G. F.; Parker, W. A. Benzodiazepine Withdrawal Syndrome: A Literature Review and Evaluation. *Am. J. Drug Alcol. Abuse* **1982**, *9* (1), 19–33.

Maestripieri, D.; Carroll, K. A. Child abuse and Neglect: Usefulness of the Animal Data. *Psychol. Bull.* **1998**, *123*, 211–223.

Magill, C. The Boundary Between Borderline Personality Disorder and Bipolar Disorder: Current Concepts and Challenges. *Can. J. Psych.* **2004**, *49* (8), 551–556.

Mahli, G. S.; Byrow, Y.; Outhred, T.; Fritz, K. Exclusion of Overlapping Symptoms in DSM-5 Mixed Features Specifier: Heuristic Diagnostic and Treatment Implications. *CNS Spect.* **2017**, *22* (2), 126–133.

Main, N.; Solomon, J. Procedures for Identifying Infants as Disorganized/Disoriented During the Strange Infant Situation. In *Attachment in the Preschool Years: Theory, Research, and Intervention*; Greenberg, M. T., Cicchetti, C., Cummings, E. M., Eds.; University of Chicago Press: Chicago, 1990.

Maj, M.; Akiskal, H. S.; Lopez-Ibor, J. J.; Sartorius, N., Eds. Bipolar *Disorder: World Psychiatric Association Series Evidence and Experience in Psychiatry*; John Wiley and Sons: London, 2002.

Malejko, K.; Abler, B.; Plener, P.; Straub, J. Neural Correlates of Psychotherapeutic Treatment of Post-Traumatic Stress Disorder: A Systematic Literature Review. *Front Psych.* **2017**, *8*, Art 85.

Malhi, G.; Tanious, M.; Das, P.; Coulston, C. M.; Berk, M. Potential Mechanisms of Action of Lithium in Bipolar Disorder. *CNS Drugs* **2013**, *27*, 135–153.

Malina, A.; Gaskill, J.; McConaha, C.; Frank, G. K.; LaVia, M.; Scholar, L.; Kaye, W. H. Olanzapine Treatment of Anorexia Nervosa: A Retrospective Study. *Int. J. Eat Dis.* **2003**, *33*, 234–237.

Mallinger, A. G.; Frank, E.; Thase, M. E.; Barwell, M. M.; DiazGranados, N.; Luckenbaugh, D. A.; Kupfer, D. J. Revisiting the Effectiveness of Standard Antidepressants in Bipolar Disorder: Are Monoamine Oxidase Inhibitors Superior? *Psychopharmacol. Bull.* **2009**, *42* (2), 64–74.

Mallo, C. J.; Mintz, D. L. Teaching all the Evidence Bases: Reintegrating Psychodynamic Aspects of Prescribing into Psychopharmacology Training. *Psychodynam. Psych.* **2013**, *41* (1), 13–37.

Manber, R.; Kraemer, H. C.; Arnow, B. A.; Trivedi, M. H.; Rush, A. J.; Thase, M. E.; Rothbaum, B. O.; Klein, D. N.; Kocsis, J. A.; Gelenberg, A. J.; Keller, M. E. Faster Remission of Chronic Depression with Combined Psychotherapy and Medication Than with Each Therapy Alone. *J. Consult Clin. Psychol.* **2008**, *76* (3), 459–467.

Manji, H. S.; Moore, G. J.; Chen, G. Lithium at 50: Have the Neuroprotective Effects of This Unique Cation Been Overlooked? *Biol. Psych.* **1999**, *46* (7), 929–940.

Manning, J. S.; Haykal, R. F.; Connor, P. D.; Akiskal, H. S. On the Nature of Depressive and Anxious States in a Family Practice Setting: The High Prevalence of Bipolar II and Related Disorders in a Cohort Followed Longitudinally. *Comp. Psych.* **1997**, *38*, 102–108.

Mannuzza, S.; Klein, R. G.; Moultonill, J. L. Does Stimulant Treatment Place Children at Risk for Adult Substance Abuse? A Controlled, Prospective Follow-Up Study. *J. Ch. Adol. Psychopharm* **2004**, *13* (3), 273–282.

Marazziti, D.; Consoli, G.; Picchetti, G.; Carlini, M.; Faravelli, L. Cognitive Impairment in Major Depression. *Eur. J. Pharmacol.* **2010**, *626* (1), 83–86.

Marcus, SV, Marquis, P.; Sakai, C. Controlled Study of Treatment of PTSD Using EMDR in an HMO Setting. *Psychother. Theor. Res. Pract. Train.* **1997**, *34* (3), 307–315.

Maremmani, I.; Pacini, M.; Popovic, D.; Romano, A.; Maremmani, A. G. I.; Perugi, G.; Deltito, J.; Akiskal, K.; Akiskal, H. S. Affective Temperaments in Heroin Addiction. *J. Aff. Dis.* **2009**, *117* (3), 186–192.

Maremmani, I.; Perugi, G.; Pacini, M.; Akiskal, H. S. Toward a Unitary Perspective on the Bipolar Spectrum and Substance Abuse: Opiate Addiction as a Paradigm. *J. Aff. Dis.* **2006**, *93* (1–3), 1–12.

Marneros, A. Origin and Development of the Concepts of Mixed Bipolar States. *J. Aff. Dis.* **2001**, *67* (1–3), 229240.

Marneros, A.; Angst, J. Bipolar Disorders: Roots and Evolution. In *Bipolar Disorders: 100 Years After Manic-Depressive Insanity*; Marneros, A., Angst, J., Eds. Kluwer Academic Publishers: New York, Boston, Dordrecht, London, Moscow, 2000.

Marshall, J. R. Comorbidity and Its Effects on Panic Disorder. *Bull. Men. Clin.* **1996**, *60* (2), supp A 39–53.

Mason, W. A. Early Social Deprivation in Non-Human Primates: Implications for Human Behavior. In *Biology and Behavior: Environmental Influences*; Glass, D.C., Ed.; Rockefeller University Press: New York, 1968; pp 70–101.

Mason, W. A. Early Socialization. In *Primates, the Road to Self-Sustaining Populations*; Benirschke, K., Ed.; Springer-Verlag: New York, 1982; pp 321–329.

Mather, M.; Thayer, J. F. How Heart Rate Variability Affects Emotion Regulation Brain Networks. *Curr. Opin. Behav. Sci.* **2018**, *19*, 98–104.

Matthew, S. J.; Murrough, J. W.; aan het Rot, M.; Collins, K. A.; Reich, D. I.; Charney, D. S. Iluzole for Relapse Prevention Following Intravenous Ketamine in Treatment-Resistant Depression: A Pilot, Randomized, Placebo-Controlled Trial. *Int. J. Neuropsychopharm.* **2010**, *13*, 71–82.

Matthews, K.; Schwartz, J, Cohen, S.; Seeman, T. Diurnal Cortisol Decline Is Related to Coronary Calcification: CARDIA Study. *Psychosom. Med.* **2006**, *68*, 657–661.

Maughan, A.; Cicchetti, D.; Toth, S. L.; Rogosch, F. A. Early-Occurring Maternal Depression and Maternal Negativity in Predicting Young Children's Emotion Regulation and Socioeconomic Difficulties. *J. Ab. Ch. Psychol.* **2007**, *35*, 685–703.

Mayberg, H. S. Limbic-Cortical Dysregulation: A Proposed Model of Depression. *J. Neuro. Psych. Clin. Neurosci.* **1997**, *9*, 471–481.

McClung, C. A. Circadian Genes, Rhythms and the Biology of Mood Disorders. *Pharmacol. Therap.* **2007**, *114*, 222–232.

McCombs, J. S.; Ahn, J.; Tencer, T.; Shi, L. The Impact of Unrecognized Bipolar Disorders Among Patients Treated for Depression with Antidepressants in the Fee-for-Services California Medicaid (Medi-Cal) Program: A 6-Year Retrospective Analysis. *J. Affect. Disorder* **2007**, *97*, 171–179.

McCormick, R. A.; Russo, A. M.; Ramirez, L. F.; Taber, J. I. Effective Disorders Among Gamblers Seeking Treatment. *J. Psych.* **1984**, *141* (2), 215–218.

McCraw, S.; Parker, G.; Graham, R.; Synnott, H.; Mitchell, P. B. The Duration of Undiagnosed Bipolar Disorder: Effect on Outcomes and Treatment Response. *J. Aff. Dis.* 2014; 168:422–429.

McElroy, S. L.; Freeman, M. P.; Akiskal, H. S. The Mixed Bipolar Disorders. In *Bipolar Disorders*; Marneros, A., Angst, J., Eds.; Kluwer, 2000.

McElroy, S. L.; Hudson, J. I.; Pope Jr, H. G.; Keck Jr, P. E.; Aizley, H. G. The DSM-III Impulse Control Disorders Not Elsewhere Classified: Clinical Characteristics and Relationship to Other Psychiatric Disorders. *Am. J. Psych.* **1992**, *149* (3), 318–327.

McElroy, S. L.; Keck, P. E.; Pope, H. G.; Hudson, J. I.; Faedda, G. L.; Swann, A. C. Clinical and Research Implications of the Diagnosis of Dysphoric or Mixed Mania or Hypomania. *Am. J. Psych.* **1992**, *149* (12), 1633–1644.

McElroy, S. L.; Suppes, T.; Keck Jr. PE, Frye, M. A.; Denicoff, K. D.; Altshuler, L. L.; Brown, E. S.; Nolen, W. A.; Kupka, R. W.; Rochussen, J.; Leverich, G. S.; Post, R. M. Open-Label Adjunctive Topiramate in the Treatment of Bipolar Disorders. *Biol. Psych.* **2000**, *47* (12), 1025–1033.

McEwan, B. S. Glucocorticoids, Depression and Mood Disorders: Structural Remodeling and the Brain. *Metab* **2005**, *54*, 20–23.

McEwan, B. S. Protective and Damaging Effects of Stress Mediators. *NEJM* **1998**, *338*, 171–179.

McEwen, B. S.; Chattarji, S. In *Handbook of Neurochemistry and Molecular Neurobiology: Behavioral Neurochemistry and Neuroendocrinology*; Lajtha, A.; Blaustein, J. D., Eds., Springer, 2007; pp 571–594.

McGilchrist, I. *The Master and His Emissary*; Yale University Press: New Haven, 2009.

McGrath, P. J.; Stewart, J. W.; Nunes, E. V.; Ocepek-Welikson, K.; Rabkin, J. G.; Quitkin, F. M.; Klein, D. F. A Double-Blind Crossover Trial of Imipramine and Phenelzine for Outpatients with Treatment-Refractory Depression. *Am. J. Psych.* **1993**, *150*, 118–123.

McGreer, P. L.; Eccles, J. C.; McGreer, E. G. *Molecular Neurobiology of the Mammalian Brain*; Plenum Press: New York, 1987.

McGrew, R. E. *Encyclopedia of Medical History*; MacMillan Press: London, 1985; p 280.

McIntyre, R. S.; Lipsitz, O.; Rodrigues, N. B.; Lee, Y.; Cha, D. S.; Vinberg, M.; Lin, K.; Malhi, G. S.; Subramaniapillai, M.; Kratiuk, K.; Fagiolini, A.; Gill, H.; Nasri, H.; Mansur, R. B.; Suppes, T.; Ho, R.; Rosenblat, J. D. The Effectiveness of Ketamine on Anxiety, Irritability, and Agitation: Implications for Treating Mixed Features in Adults with Major Depressive or Bipolar Disorder. *Bip. Dis.* **2020**, *00*, 1–10. https://doi.org/10.1111/bdi.12941

McKee, T. E. Peer Relationships in Undergraduates with ADHD Symptomatology: Selection and Quality of Friendships. *J. Atten. Dis.* **2017**, *21* (12), 1020–1029.

McMain, S.; Korman, L. M.; Dimeff, L. Dialectical Behavior Therapy and the Treatment of Emotion Dysregulation. *Psychother. Pract.* **2001**, *57* (2), 183–196.

McRae, K.; Hughes, B.; Chopra, S.; Gabrieli, J. D. E.; Gross, J. J.; Ochsner, K. N. The Neural Bases of Distraction and Reappraisal. *J. Cog. Neurosci.* **2010**, *22* (2), 248–262.

Meaney, M.; Szyf, M. Environmental Programming of Stress Responses Through DNA Methylation: Life at the Interface Between a Dynamic Environment and a Fixed Genome. *Dialog. Clin. Neurosci.* **2005,** *7,* 101–123.

Meany, M. Maternal Care, Gene Expression and the Transmission of Individual Differences in Stress Reactivity Across Generations. *Ann. Rev. Neurosci.* **2001,** *24* (1), 1161–1192.

Meinz, P.; Main, M. The Evolution of Mary Ainsworth's Understanding of Attachment: Changes in Her Conceptualization of Security and Its Precursors. Poster Presented at the Biennial Meeting of the Society for Research in Child Development, Montreal, April 3, 2011.

Melo, M. C. A.; Daher, E. D. F.; Albuquerque, S. G. C.; deBruin, S.V. M. Exercise in Bipolar Patients: A Systematic Review. *J. Aff. Dis.* **2016,** *198* (1), 32–38.

Meltzer, H. Y.; Massey, B. W. The Role of Serotonin Receptors in the Action of Atypical Antipsychotic Drugs. *Curr. Opin. Psychopharm.* **2011,** *11* (1), 59–67.

Mendels, J. Lithium in the Treatment of Depression. *Am. J. Psych.* **1976,** *133,* 373–378.

Mendoza, S. P.; Smotherman, W. P.; Miner, M. T.; Kaplan, J.; Levine, S. Pituitary-Adrenal Response to Separation in Mother and Infant Squirrel Monkeys. *Dev. Psychobiol.* **1978,** *11,* 169–175.

Menin, D. S.; McLaughlin, K. A.; Flanagin, T. J. Emotion regulation deficits in Generalized Anxiety Disorder, Social Anxiety Disorder, and their co-occurrence. *J. Anx. Dis.* **2009,** 3 (7), 866–871.

Mennin, D. S. Emotion Regulation Therapy for Generalized Anxiety Disorder. *Clin. Psychol. Psychother.* **2004,** *11* (1), 17–29.

Mennin, D. S.; Heimberg, R. G.; Turk, C. L.; Fresco, D. M. Preliminary Evidence for an Emotion Regulation Deficit Model of Generalized Anxiety Disorder. *Behav. Res. Therap.* **2005,** *43,* 1281–1310.

Mercer, D.; Douglass, A. B.; Links, P. S. Meta-Analyses of Mood Stabilizers, Antidepressants and Antipsychotics in the Treatment of Borderline Personality Disorder: Effectiveness for Depression and Anger Symptoms. *J. Pers. Dis.* **2009,** 23 (2), 156–174.

Merikangas, K.; Akiskal, H. S.; Angst, J.; Greenberg, P. E.; Hirschfeld, R. M. A.; Petukhova, M.; Kessler, R. C. Lifetime and 12-Month Prevalence of Bipolar Spectrum Disorder in the National Comorbidity Survey Replication. *Arch. Gen. Psych.* **2007,** *64,* 543–552.

Meyers, E.; DeSerisy, M.; Krain Roy, A. Disruptive Mood Dysregulation Disorder (DMDD): An RDoC Perspective. *J. Aff. Dis.* **2017,** *216,* 117–122.

Mick, E.; Spencer, T.; Wozniak, J.; Biederman, J. Heterogeneity of Irritability in Attention-Deficit/Hyperactivity Disorder Subjects with and Without Mood Disorders. *Biol. Psych.* **2005,** *58* (7), 576–582.

Mikulincer, M.; Shaver, P. R.; Pereg, D. Attachment Theory and Affect Regulation: The Dynamics, Development, and Cognitive Consequences of Attachment-Related Strategies. *Motive Emot.* **2003,** *27,* 77–102.

Miller, E. K.; Cohen, J. D. An Integrative Theory of Prefrontal Cortex Function. *Ann. Rev. Neurosci.* **2001,** *24,* 167–202.

Miller, N. S.; Gold, M. S. Dependence Syndrome: A Critical Analysis of Essential Features. *Psych. Ann.* **1991,** *21* (5), 280–288.

Miller, S.; Suppes, T.; Mintz, J.; Hellerman, G.; Frye, M. A.; McElroy, S. L.; Nolen, W. A.; Kupka, R.; Leverich, G. S.; Grunze, H.; Altshuler, L.; Keck, Jr. P. E.; Post, R. M. Mixed Depression in Bipolar Disorder: Prevalence Rate and Clinical Correlates During

Naturalistic Follow-Up in the Stanley Bipolar Network. *Am. J. Psych.* **2016,** *173* (10), 1015–1023.

Mintz, D.; Belnap, B. A. What Is Psychodynamic Psychopharmacology? An Approach to Pharmacologic Treatment Resistance. In *Treatment Resistance and Patient Authority: The Austin Riggs Reader*; Plakum, E. M., Ed., Norton Professional Books: New York, London, 2011.

Mitchell, C. P.; Chen, Y.; Kundakovic, M.; Costa, E.; Grayson, D. R. Histone Deacetylase Inhibitors Decrease Reeling Promoter Methylation In Vitro. *J. Neurochem.* **2005,** *93,* 483–492.

Mitsikostas, D. D.; Mantonakis, L.; Chalaorakis, N. Nocebo in Clinical Trials for Depression: A Meta-Analysis. *Psych. Res.* **2014,** *215* (1), 82–86.

Moffitt, T. E.; Arseneault, L.; Belsk, D.; Dickson, N.; Hancox, R. J.; Harrington, H.; Houts, R.; Poulton, R.; Roberts, B. W.; Ross, S.; Sears, M. R.; Thompson, W. M.; Caspi, A. A Gradient of Childhood Self-Control Predicts Health, Wealth and Public Safety. *PNAS* **2011,** *108* (7), 2693–2698.

Mohamed, S, Rosenheck, R. A. Pharmacotherapy of PTSD in the U.S. Department of Veterans Affairs: Diagnostic-and Symptom-Guided Drug Selection. *J. Clin. Psych.* **2008,** *69* (6), 959–965.

Mohseni, G.; Ostadhadi, S.; Imran-Khan, M.; Norouzi-Javidan, A.; Zolfaghari, S.; Haddadi N-S, Dehpour A-R. Agmatine Enhances the Antidepressant-Like Effect of lithium in Mouse Forced Swimming Test Through NMDA Pathway. **2017,** *88,* 931–938.

Monkul, E. S.; Matsuo, K.; Nicoletti, M.; Dierschke, N.; Hatch, J. P.; Dalwani, M.; Brambilla, P.; Caetano, S.; Sassi, R. B.; Mallinger, A. G.; Soares, J. C. Prefrontal Gray Matter Increases in Healthy Individuals After Lithium Treatment: A Voxel-Based Morphometry Study. *Neurosci. Lett.* **2007,** *429,* 7–11.

Moon, E.; Lee, S.; Kim, D.; Hwang, B. Comparative Study of Heart Rate Variability in Patients with Schizophrenia, Bipolar Disorder, Post-Traumatic Stress Disorder, or Major Depression. *Clin. Psychopharm. Neurosci.* **2013,** *11* (3), 137–143.

Moore, G. J.; Bebchuk, J. M.; Wilds, I. B.; Chen, G.; Menji, H. K. Lithium-Induced Increase in Human Brain Grey Matter. *Lancet* **2000,** *356* (9237), 1241–1242.

Moreno, C.; Laje, G.; Blanco, C.; Jiang, H.; Schmidt, A. B.; Olfson, M. National Trends in the Outpatient Diagnosis and Treatment of Bipolar Disorder in Youth. *Arch. Gen. Psych.* **2007,** *64,* 1032–1039.

Morris, G. O.; Williams, H. L.; Lubin, L. Misperception and Disorientation During Sleep. *Arch. Gen. Psych.* **1960,** *2,* 247–254.

Morris, S. E.; Cuthbert, B. N. Research Domain Criteria: Cognitive Systems, Neural Circuits, and Dimensions of Behavior. *Dial. Clin. Neurosci.* **2012,** *14* (1), 29–37.

Mosquera, D.; Leeds, A. M.; Gonzalez, A. M. Application of EMDR Therapy for Borderline Personality Disorder. *J EMDR Prac Res* **2014,** 8 (2), 74–89.

Mountcastle, V. B. The Columnar Organization of the Neocortex. *Brain* **1997,** *120,* 701–722.

Mueller, S. C.; Maheu, F. S.; Dozier, M.; Peloso, E.; Mandell, D.; Leibenluft, E.; Pine, D. S.; Ernst, M. Early-Life Stress Is Associated with Impairment in Cognitive Control in Adolescence: An fMRI Study. *Neuropsychologia* **2010,** *48* (10), 3037–3044.

Munakata, Y.; Pfaffly, J. Hebbian Learning and Development. *Dev. Sci.* **2004,** *7* (2), 141–148.

Mundo, E.; Cattaneo, E.; Russo, M.; Altamura, A. C. Clinical Variables Related to Antidepressant-Induced Hypomania in Bipolar Disorder. *J. Aff. Dis.* **2006,** *92* (2–3), 227–230.

Murray, G.; Harvey, A. Circadian Rhythms and Sleep in Bipolar Disorder. *Bip. Dis.* **2010,** *12,* 459–472.

Murray, L.; Arteche, A.; Fearon, P.; Halligan, S.; Goodyer, I.; Cooper, P. Maternal Postnatal Depression and the Development of Depression in Offspring Up to 16 Years of Age. *J. Am. Acad. Child Adol. Psych.* **2011,** *50,* 460–470.

Murray, R. M.; Englund, A.; Abi-Dargham, A.; Lewis, D. A.; DiForti, M.; Davies, C.; Sherif, M.; McGuire, P.; D'Souza, D. C. Cannabis-Associated Psychosis: Neural Substrate and Clinical Impact. *Neuropharmacology* **2017,** *124,* 89–104.

Murrough, J. W.; Perez, A. M.; Pillemer, S.; Stern, J.; Parides, M. K.; Aan Het Rot, M.; Collins, K. A.; Mathew, S. J.; Charney, D. S.; Iosifescu, D. C. Rapid and Longer-Term Antidepressant Effects of Repeated Ketamine Infusions in Treatment-Resistant Major Depression. *Biol. Psych.* **2013,** *74* (4), 50–256.

Musty, R. E. Natural Cannabinoids: Interactions and Effects. In *The Medicinal Uses of Cannabis and Cannabinoids*; Guy, G. W., Whittle, B. A., Robson, P. J., Eds.; Pharmaceutical Press: London, 2004; pp 165–204.

Na, K. S.; Chang, H. S.; Won, E.; Han, K. M.; Choi, S.; Tae, W. S.; Yoon, H. K.; Kim, Y. K.; Joe, S. H.; Jung, I. K.; Lee, M. S.; Ham, B. J. Association Between Glucocorticoid Receptor Methylation and Hippocampal Subfields in Major Depressive Disorder. *PloS One* **2014,** *9* e85425.

Nathanson, D. L. *The Many Faces of Shame*; The Guilford Press: New York, 1987.

Neacsiu, A. D.; Bohus, M.; Linehan, M. M. Dialectical Behavior Therapy: An Intervention for Emotional Dysregulation. *Handbook Emot. Reg.* **2014a,** *2,* 491–507.

Neacsiu, A. D.; Rizvi, S. L.; Linehan, M. M. Dialectical Behavior Therapy Skills Use as a Mediator and Outcome of Treatment for Borderline Personality Disorder. *Behav. Res. Ther.* **2010,** *48* (9), 832–839.

Nebes, R. D. Superiority of the Minor Hemisphere in Commissurotomized Man for the Perception of Part-Whole Relations. *Cortex* **1971,** *7* (4), 333–349.

Nelson, C. A.; Carver, L. J. The Effects of Stress and Trauma on Brain and Memory: A View from Developmental Cognitive Neuroscience. *Dev. Psychopath.* **1998,** *10* (4), 793–809.

Nemeroff, C. B. Early-Life Adversity, CRF Dysregulation, and Vulnerability to Mood and Anxiety Disorders. *Psychopharm. Bull.* **2004,** *38* (s1), 4–20.

Neubauer, H.; Bermingham, P. A Depressive Syndrome Responsive to Lithium: An Analysis of 20 Cases. *J. Nerv. Mental Dis.* **1976,** *163* (4), 276–281.

Newman-Toker, J. Risperidone in Anorexia Nervosa. *J. Am. Acad. Child Adol. Psych.* **2000,** *39,* 941–942.

Nicholson, A. A.; Rabellino, D.; Densmore, M.; Frewen, P. A.; Paret, C.; Kleutsch, R.; Schmahl, C.; Theberge, C.; Neufeld, R. W. J.; McKinnon, M. C.; Reiss, J. P.; Jetly, R.; Lanius, R. A. The Neurobiology of Emotion Regulation in Posttraumatic Stress Disorder: Amygdala Downregulation via Real-Time fMRI Neurofeedback. *Human Brain Map.* **2017,** *38,* 541–560.

Nieacsiu, A. D.; Eberle, J. W.; Kramer, R.; Weismann, T.; Linehan, M. M. Dialectical Behavior Therapy Skills for Transdiagnostic Emotion Dysregulation: A Pilot Randomized Controlled Trial. *Behav. Res. Ther.* **2014b,** *59,* 40–51.

Nierenberg, A. A.; Ostracher, M. J.; Calabrese, J. R.; Ketter, T. A.; Marangell, L. B.; Miklowitz, D. J.; Miyahara, S.; Bauer, M. S.; Thase, M. E.; Wisniewski, S. R.; Sachs,

G. S. Treatment-Resistant Bipolar Depression: A STEP-BD Equipoise Randomized Effectiveness Trial of Antidepressant Augmentation With Lamotrigine, Inositol, or Risperidone. *Am. J. Psych.* **2006**, *163* (2), 210–216.

Nieuwenhuys, R. The neocortex. An Overview of Its Evolutionary Development, Structural Organization, and Synaptology. *Anat. Embryol.* (Berlin), **1994**, *190*, 307–337.

Nolte, T.; Bolling, D. Z.; Hudac, C. M.; Fonagy, P.; Mayes, L.; Pelphrey, K. A. Brain Mechanisms Underlying the Impact of Attachment-Related Stress on Social Cognition. *Front. Human Neurosci.* **2013**, *7* (816), 1–12.

Nooner, K. B.; Mennes, M.; Brown, S.; Castellanos, F. X.; Leventhal, B.; Milham, M. P.; Colcombe, S. J. Relationship of Trauma Symptoms to Amygdala-Based Functional Brain Changes in Adolescents. *J. Trauma Stress* **2013**, *26* (6), 784–787.

Normala, I, Abdul, H. A.; Azlin, B.; Nik, R. N. J.; Hazli, Z.; Shah, S. A. Executive Function and Attention Span in Euthymic Patients with Bipolar I Disorder. *Med. J. Malaysia* **2010**, *65* (3), 199–203.

Nusslock, R.; Frank, E. Subthreshold Bipolarity: Diagnostic Issues and Challenges. *Bip. Dis.* **2011**, *13* (7–8), 587–603.

Ochsner, K. N.; Bunge, S. A.; Gross, J. J.; Gabrielle JDE: Rethinking Feelings: An fMRI Study of the Cognitive Regulation of Emotion. *J. Cog. Neurosci.* **2002**, *14* (8), 1215–1229.

Ochsner, K. N.; Gross, J. J. Cognitive Emotion Regulation: Insights from Social Cognitive and Affective Neuroscience. *Curr. Dir. Psychol. Sci.* **2008**, *17* (2), 153–158.

Ochsner, K. N.; Gross, J. J. The Cognitive Control of Emotion. *Trends Cog. Sci.* **2005**, *9*, 242–249.

Ochsner, K. N.; Gross, J. J. The Neural Architecture of Emotion Regulation. In *Handbook of Emotion Regulation*; Guilford Press: New York, 2007.

Ochsner, K. N.; Knierim, K.; Ludlow, D. H.; Hanelin, J.; Ramachandran, T.; Glover, G.; Mackey, S. C. Reflecting upon Feelings: An fMRI Study of Neural Systems Supporting the Attribution of Emotion to Self and Other. *J. Cog. Neurosci.* **2004**, *16* (10), 1746–1772.

Ochsner, K. N.; Ray, R. D.; Cooper, J. C.; Robertson, E. R.; Chopra, S.; Gabrieli, J. D. E.; Gross, J. J. For Better or for Worse: Neural Systems Supporting the Cognitive Downand Up-Regulation of Negative Emotion. *NeuroImage* **2004**, *23*, 483–499.

Ochsner, K. N.; Ray, R. R.; Hughes, B.; McRae, K.; Cooper, J. C.; Weber, J.; Gabrieli, J. D. E.; Gross, J. J. Bottom-Up and Top-Down Processes in Emotion Generation: Common and Distinct Neural Mechanisms. *Psychol. Sci.* **2009**, *20* (11), 1322–1331.

Oskis, A.; Loveday, C.; Hucklebridge, F.; Thorn, L.; Clow, A. Anxious Attachment Style and Salivary Cortisol Dysregulation in Healthy Female Children and Adolescents. *J. Ch. Psychol. Psych.* **2011**, *52* (2), 111–118.

Ostadhadi, S.; Ahangari, M.; Nikoui, V.; Norouzi-Javidan, A.; Zolfaghari, S.; Jazeri, F.; Chamanara, M.; Akbarian, R.; Dehpour A-R: Pharmacological Evidence for the Involvement of the NMDA Receptor and Nitric Oxide Pathway in the Antidepressant-Like Effect of Lamotrigine in the Mouse Forced Swimming Test. *Biomed. Pharmacother.* **2016**, *82*, 713–721.

Packer, C. D.; Packer, D. M. [Beta]-Blockers, Stage Fright, and Vibrato: A Case Report. **2005**, *20* (3), 126+

Pagliaccio, D.; Luby, J. L.; Bogdan R. Agrawal, A.; Gaffrey, M. S.; Belden, A. C.; Botteron, K. N.; Harms, M. P.; Barch, D. M. Amygdala Functional Connectivity, HPA Axis Genetic

Variation, and Life Stress in Children and Relations to Anxiety and Emotion Regulation. *J. Abnorm. Psychol.* **2015,** *124* (4), 817–833.

Pagliaccio, D.; Luby, J. L.; Bogdan R. Agrawal, A.; Gaffrey, M. S.; Belden, A. C.; Botteron, K. N.; Harms, M. P.; Barch, D. M. Stress-System Genes and Life Stress Predict Cortisol Levels and Amygdala and Hippocampal Volumes in Children. *Neuropsychopharm* **2014,** *39,* 1245–1253.

Pampallona, S.; Bollini, P.; Tibaldi, G.; Kupelnick, B.; Munizza, C. Combined Pharmacotherapy and Psychological Treatment for Depression: A Systematic Review. *Arch. Gen. Psych.* **2004,** *61* (7), 714–719.

Panksepp, J. *Affective Neuroscience: The Foundations of Animal and Human Emotions*; Oxford University Press: New York, 1998.

Panksepp, J. The Psychoneurology of Fear: Evolutionary Perspectives and the Role of Animal Models in Understanding Anxiety. In *Handbook of Anxiety. Vol 3, The Neurobiology of Anxiety*; Burrows, G. D., Roth, M., Noyes, R., Jr. Eds.; Elsevier: Amsterdam, 1990; pp. 3–58.

Panksepp, J. Why Does Separation Distress Hurt? *Biol. Bull.* **2005,** *131* (2), 224–230.

Panksepp, J.; Bean, N. J.; Bisop, P.; Vilberg, T.; Sahley, T. L. Opioid Blockade and Social Comfort in Chicks. *Pharmacol. Biochem. Behav.* **1980,** *13,* 673–683.

Panksepp, J.; Watt, D. Why Does Depression Hurt? Ancestral Primary-Process Separation Distress (PANIC/GRIEF) and Diminished Brain Reward (SEEKEING) Processes in the Genesis of Depressive Affect. *Psychiatry* **2011,** *74* (1), 5–13.

Pare, D. Role of the Basolateral Amygdala in Memory Consolidation. *Prog. Neurobiol.* **2003,** *70* (5), 409–420.

Paredes, R.; Haller, A. E.; Manero, M. C.; Alvarado, R.; Agmo, A. Medial Optic Area Kindling Induces Sexual Behavior in Sexually Inactive Male Rats. *Brain Res.* **1990,** *515,* 20–26.

Parens, E.; Johnston, J. Controversies Concerning the Diagnosis and Treatment of Bipolar Disorder in Children. *Child Adol. Psych. Mental Health* **2010,** *4,* 9.

Paret, C.; Kleutsch, R.; Ruf, M.; Demirakca, T.; Hoesterey, S.; Ende, G.; Schmahl, C. Down-Regulation of Amygdala Activation with Real-Time fMRI Neurofeedback in a Healthy Female Sample. *Front. Behav.. Neurosci.* **2014,** *8,* 299. DOI: 10.3389/fnbeh. 2014.00299.

Paret, C.; Kleutsch, R.; Zaehringer, J.; Ruf, M.; Demirakca, T.; Bohus, M.; Ende, G.; Schmahl, C. Alterations of Amygdala-Prefrontal Connectivity with Real-Time fMRI Neurofeedback in BPD Patients. *Soc. Cog. Aff. Neurosci.* **2016,** *11* (6), 952–960.

Pariante, C. M.; Lightman, S. L. The HPA Axis in Major Depression: Classical Theories and New Developments. **2008,** *31* (9), 464–468.

Paris, J. Borderline or Bipolar? Distinguishing Borderline Personality Disorder from Bipolar Spectrum Disorders. *Harv. Rev. Psych.* **2004,** *12* (3), 140–145.

Paris, J. The Nature of Borderline Personality Disorder: Multiple Dimensions, Multiple Symptoms, But One Category. *J. Pers. Dis.* **2007b,** *21* (5), 457–473.

Paris, J. Why Psychiatrists Are Reluctant to Diagnose Borderline Personality Disorder. *Psychiatry* **2007a,** *4* (1), 35–39.

Park, C. R.; Zoladz, P. R.; Conrad, C. D.; Fleshner, M.; Diamond, D. M. Acute Predator Stress Impairs the Consolidation and Retrieval of Hippocampus-Dependent Memory in Male and Female Rats. *Learn. Mem.* **2008,** *15,* 271–280.

Park, R. D. Progress, Paradigms and Unresolved Problems: A Commentary on Our Recent Understanding of Children's Emotions. *Merrill-Palmer Quart.* **1994,** *40,* 57–169.

Parkin, A. J. Human Memory: The Hippocampus Is the Key. *Curr. Biol.* **1996,** *6* (12), 1583–1585.

Paunovic, N.; Ost, L. Cognitive-Behavior Therapy vs Exposure Therapy in the Treatment of PTSD in Refugees. *Behav. Res. Ther.* **2001,** *39* (10), 1183–1197.

Perlick, D. A.; Rosenheck, R. A.; Clarkin, J. F.; Sirey, J. A.; Salahi, J.; Elmer, B. S.; Streuning, L.; Link, B. G. Stigma as a Barrier to Recovery: Adverse Effects of Perceived Stigma on Social Adaptation of Persons Diagnosed with Bipolar Affective Disorder. *Psych. Serv.* **2001,** *52* (12), 1627–1632.

Perlis, M. L.; Nielsen, T. A. Mood-Regulation, Dreaming and Nightmares: Evaluation of a Desensitization Function for REM Sleep. *Dreaming* **1993,** *3* (4), 243–257.

Perris, C. A Study of Bipolar (Manic-Depressive) and Unipolar Recurrent Depressive Psychoses. *Acta Psych. Scand.* **1966,** *194,* 9–14.

Perugi, G.; Akiskal, H. S. The Soft Bipolar Spectrum Redefined: Focus on the Cyclothymic, Anxious-Sensitive, Impulse-Dyscontrol, and Binge-Eating Connection in Bipolar II and Related Conditions. *Psych. Clin. NA.* **2002,** *25,* 713–737.

Perugi, G.; Akiskal, H. S.; Lattanzi, L.; Cecconi, D.; Mastrocinque, C.; Patronelli, A.; Vignoli, S.; Bemi, E. The High Prevalence of "Soft" Bipolar (II) Features in Atypical Depression. *Comp. Psych.* **1998,** *39,* 63–71.

Perugi, G.; Akiskal, H. S.; Micheli, C.; Musetti, L.; Paiano, A.; Quilici, C.; Rossi, L.; Cassano, G. B. Clinical Subtypes of Bipolar Mixed States: Validating a Broader European Definition in 143 Cases. *J. Aff. Dis.* **1997,** *43,* 169–180.

Perugi, G.; Toni, C.; Passino, M. C. S.; Akiskal, K. K.; Kaprinis, S.; Akiskal, H. S. Bulimia Nervosa in Atypical Depression: The Mediating Role of Cyclothymic Temperament. *J. Aff. Dis.* **2006,** *92,* 91–97.

Perugi, G.; Toni, C.; Traviersa, M. G.; Akiskal, H. S. The Role of Cyclothymia in Atypical Depression: Toward a Data-Based Reconceptualization of the Borderline–Bipolar II Connection. *J. Aff. Dis.* **2003,** *73* (1–2), 87–98.

Pettijohn, T. F. Reaction of Parents to Recorded Infant Guinea Pig Distress Vocalizations. *Behav. Biol.* **1977,** *21,* 438–432.

Petursson, H. The Benzodiazepine Withdrawal Syndrome. *Addiction* **1994,** *89* (11), 1455–1459.

Phabphal, K.; Udomratn, P. Topiramate-Induced Depression in Cases Using Topiramate for Migraine Prophylaxis. *Neurosci. Neurol. Psych.* **2009,** *30* (6), 747–749.

Phan, K. L.; Fitzgerald, D. A.; Nathan, P. J.; Tancer, M. E. Association Between Amygdala Hyperactivity to Harsh Faces and Severity of Social Anxiety in Generalized Social Phobia. *Biol. Psych.* **2006,** *59* (5), 424–429.

Phelps, J. R. Agitated Dysphoria After Late-Onset Loss of Response to Antidepressants: A Case Report. *J. Aff. Dis.* **2005,** *86* (2–3), 277–280.

Phelps, J. R.; Siemers, S. V.; El-Malakh, R. S. The Ketogenic Diet for Type II Bipolar Disorder. *Cephalgia* **2013,** *19* (5), 423–426.

Phelps, L. E.; Brutsche, N.; Moral, J. R.; Luckenbaugh, D. A.; Manji, H. K.; Zarate Jr. CA: Family History of Alcohol Dependence and Initial Antidepressant Response to an N-Methyl-D Aspartate Antagonist. *Biol. Psych.* **2009,** *65,* 181–184.

Phiel, C. J.; Zhang, F.; Huang, E. Y.; Guenther, M. G.; Lazar, M. A.; Klein, P. S. Histone Deacetylase Is a Direct Target of Valproic Acid, a Potent Anticonvulsant, Mood Stabilizer, and Teratogen. *J. Biol. Chem.* **2001**, *276*, 36734–36741.

Phillips, M. L.; Travis, M. J.; Fagiolini, A.; Kupfer, D. Medication Effects in Neuroimaging Studies of Bipolar Disorder. *Am. J. Psych.* **2008**, *165*, 313–320.

Pichot, P. Tracing the Origins of Bipolar Disorder: From Falret to DSM-IV and ICD-10. *J. Aff. Dis.* **2006**, *96* (3), 145–148.

Pies, R. The Historical Roots of the "Bipolar Spectrum": Did Aristotle Anticipate Kraepelin's Broad Concept of Manic-Depression? *J. Aff. Dis.* **2007**, *100* (1–3), 7–11.

Pitt, D.; Nageimeier, I. E.; Wilson, H. C.; Raine, C. S. Glutamate Uptake by Oligodendrocytes: Implications for Excitotoxicity in Multiple Sclerosis. *Neurology* **2003**, *61* (8), 1113–1120.

Pittenger, C.; Duman, R. S. Stress, Depression and Neuroplasticity: A Convergence of Mechanisms. *Neuropsychopharmacology* **2008**, *33*, 88–109.

Plakun, E. Treatment Resistance and Psychodynamic Psychiatry: Concepts Psychiatry Needs from Psychoanalysis. *Psychodynam. Psych.* **2012**, *40* (2), 183–209.

Plotsky, P. M.; Meany, M. J. Early Postnatal Experience Alters Hypothalamic, C. R. F.; mRNA, Median Eminence CRF Content, and Stress-Induced Release in Adult Rats. *Brain Res. Mol. Brain Res.* **1993**, *18* (3), 195–200.

Pollack, M. H.; Otto, M. W.; Sachs, G. S.; Leon, A.; Shear, M. K.; Deltito, J. A.; Keller, M. B.; Rosenbaum, J. F. Anxiety Pathology Predictive of Outcome in Patients with Panic Disorder and Depression Treated with Imipramine, Alprazolam and Placebo. *J. Aff. Dis.* **1994**, *30*, 273–281.

Pollitt, J.; Young, J. Anxiety State or Masked Depression? A Study Based on the Action of Monoamine Oxidase Inhibitors. *Br. J. Psych.* **1971**, *119*, 143–149.

Pombo, S.; Figueira, M. L.; daCosta, N. F.; Ismail, F.; Yang, G.; Akiskal, K.; Akiskal, H. S. The Burden of Cyclothymia on Alcohol Dependence. *J. Aff. Dis.* **2013**, *153* (3), 1090–1096.

Pope Jr. HG, Hudson, J. I.; Jonas, J. M.; Bulimia in Men: A Series of Fifteen Cases. *J. Nerv. Mental Dis.* **1986**, *174*, 117–119.

Popoli, M.; Yan, Z.; McEwan, B. S.; Sanacora, G. The Stressed Synapse: The Impact of Stress and Glucocorticoids on Glutamate Transmission. *Nat. Rev. Neurosci.* **2012**, *13*, 22–37.

Porges, S. W. An Autonomic Mediator of Affect. In *The Development of Emotion Regulation and Dysregulation*; Garber, J., Dodge, K. A., Eds.; Cambridge University Press: Cambridge, 1996; pp 111–128.

Porges, S. W. The Polyvagal Theory: New Insights into Adaptive Reactions of the Autonomic Nervous System. *Cleve Clin. J. Med.* **2009**, *76* (Supp 2), S86–S90.

Porges, S. W. The Polyvagal Theory: Phylogenetic Substrates of a Social Nervous System. *Int. J. Psychophysiol.* **2001**, *42* (2), 123–146.

Porges, S. W.; Doussard-Roosevelt, J. A.; Portales, A. L.; Greenspan, S. I. Infant Regulation of the Vagal "Brake" Predicts Child Behavioral Problems: A Psychobiological Model of Social Behavior. *Dev. Psychobiol.* **1996**, *29* (8), 697–712.

Porter, R. J.; Bourke, C.; Gallagher, P. Neuropsychological Impairment in Major Depression: Its Nature, Origin and Clinical Significance. *Aust. NZ J. Psych.* **2007**, *41* (2), 115–128.

Posner, M. I.; Fan, J. *Attention as an Organ System: Topics in Integrative Neuroscience*; Cambridge University Press: Cambridge, 2008.

Propper, C.; Moore, G. A. The Influence of Parenting on Infant Emotionality: A Multi-Level Psychobiological Perspective. *Dev. Rev.* **2006,** *26,* 427–460.

Quirk, G. J. Prefrontal-Amygdala Interaction in the Regulation of Fear. In *Handbook of Emotion Regulation*; Gross, J. J., Ed.; Guilford Press: New York, 2007.

Quitkin, F. M.; Harrison, W.; Stewart, J. W.; McGrath, P. J.; Tricamo, E.; Ocepek-Welikson, K.; Rabkin, J. G.; Wager, S. G.; Nunes, E.; Klein, D. F. Response to Phenelzine and Imipramine in Placebo Nonresponders with Atypical Depression. *Arch. Gen. Psych.* **1991,** *48* (4), 319–323.

Quitkin, F. M.; McGrath, P. J.; Stewart, J. W.; Harrison, W.; Wager, S. G.; Nunes, E.; Rabkin, J. G.; Tricamo, E.; Markowitz, J.; Klein, D. F. Phenelzine and Imipramine in Mood Reactive Depressives: Further Delineation of the Syndrome of Atypical Depression. *Arch. Gen. Psych.* **1989,** *46,* 787–793.

Quitkin, F. M.; Rabkin, J. G.; Stewart, J. W.; McGrath, P. J.; Harrison, W. Study Duration in Antidepressant Research: Advantages of a 12-Week Trial. *J. Psych. Res.* **1986,** *20* (3), 211–216.

Quitkin, F. M.; Stewart, J. M.; McGrath, P. J.; Tricamo, E.; Rabkin, J. G.; Ocepek-Welikson, K.; Nunes, E.; Harrison, W.; Klein, D. F. Columbia Atypical Depression A Subgroup of Depressives with Better Response to MAOI Than to Tricyclic Antidepressants or Placebo. *Br. J. Psych.* **1993,** *163* (s21), 30–34.

Raabe, F. J.; Spengler, D. Epigenetic Risk Factors in PRSD and Depression. *Front Psych.* **2013,** *80.*

Racagni, G.; Popoli, M. Cellular and Molecular Mechanisms in the Long-Term Actions of Anti-Depressants. *Dial. Clin. Neurosci.* **2008,** *10,* 385–400.

Raison, C. L.; Borisov, A. S.; Woolwine, B. J.; Massung, B.; Vogt, G.; Miller, A. H. Interferon-Alpha Effects on Diurnal Hypothalamic-Pituitary-Adrenal Axis Activity: Relationship with Pro-Inflammatory Cytokines and Behavior. *Mol. Psych.* **2010,** *15,* 535–547.

Rao, U. DSM-5 Disruptive Mood Dysregulation Disorder. *Asian J. Psych.* **2014,** *11,* 119–123.

Ravindran, L. N.; Stein, M. B. Pharmacotherapy of PTSD: Premises, Principles, and Priorities. *Brain Res.* **2009,** *1293,* 4–39.

Regier, D. A.; Kuhl, E. A.; Kupfer, D. J. The DSM-5: Classification and Criteria Changes. *World Psych.* **2013,** *12* (2), 92–98.

Reilly-Harrington, N. A.; Knauz, R. O. Cognitive-Behavioral Therapy for Rapid Cycling Bipolar Disorder. *Cog Behav Prac* **2005,** *12* (1), 66–75.

Reimherr, F. W.; Marchant, B. K.; Strong, R. E.; Hedges, D. W.; Adler, L.; Spencer, T. J.; West, S. A.; Soni, P. Emotional Dysregulation in Adult ADHD and Response to Atomoxetine. *Biol. Psych.* **2005,** *58,* 125–131.

Reinhold, J. A.; Mandos, L. A.; Rickels, K.; Lohoff, F. W. Pharmacological Treatment of Generalized Anxiety Disorder. *Exp. Opin. Pharmacother.* **2011,** *12* (16), 2457–2467.

Repetti, R.; Taylor, S. E.; Seeman, T. Risky Families: Family Social Environments and the Mental and Physical Health of Offspring. *Psychol. Bull.* **2002,** *128,* 330–366.

Richarz, F. Ueber wesen und behandlung der melancholie mit aufregung (Melancholia agitans). *Allg. Ztschr. Psychiatr.* **1858,** *15,* 28–65.

Richter-Levin, G.; Akirav, I. Amygdala-Hippocampus Dynamic Interaction in Relation to Memory. *Mol. Neurobiol.* **2000**, *22*, 11–20.

Rickels, K.; Rynn, M. Pharmacotherapy of Generalized Anxiety Disorder. *J. Clin. Psych.* **2002**, *63* (Supp l14), 9–16.

Rifkin, A.; Quitkin, F.; Carillo, C.; Klein, D. F. Very High Dose Fluphenazine for Nonchronic Treatment-Refractory Patients. *Arch. Gen. Psych.* **1971**, *25*, 398–403.

Righetti-Veltema, M.; Bousquet, A.; Manzano, J. Impact of Post-Partum Depressive Symptoms on Mother and Her 18-Month Old Infant. *Eur. Ch. Adol. Psych.* **2003**, *12*, 75–83.

Rihmer, Z.; Akiskal, K. K.; Rihmer, A.; Akiskal, H. S. Current Research on Affective Temperaments. *Curr. Opin. Psych.* **2010**, *23*, 12–18.

Rihmer, Z.; Gonda, X. Antidepressant-Resistant Depression and Antidepressant-Associated Suicidal Behavior: The Role of Underlying Bipolarity. *Dep. Res. Treat.* **2011**, Art ID 906462, 5pp.

Ripoli, L. H. Pharmacologic Treatment for Borderline Personality Disorder. *Dialog. Clin. Neurosci.* **2013**, *15* (2), 13–224.

Ritchey, M.; Dolcos, F.; Eddington K. Strauman, T. J.; Cabeza, R. Neural Correlates of Emotional Processing in Depression: Changes with Cognitive Behavioral Therapy and Predictors of Response. *J. Psych. Res.* **2011**, *45* (5), 577–587.

Robinson, L. J.; Thompson, J. M.; Gallagher, P.; Goswami, U.; Young, A. H.; Ferrier, I. N.; Moore, P. B. A Meta-Analysis of Cognitive Deficits in Euthymic Patients with Bipolar Disorder. *J. Aff. Dis.* **2006**, *93* (1–3), 105–115.

Roceri, M.; Cirulli, F.; Pessina, C.; Peretto, P.; Racagni, G.; Riva, M. A. Postnatal Repeated Maternal Deprivation Produces Age-Dependent Changes of Brain-Derived Neurotrophic Factor Expression in Selected Rat Brain Regions. *Biol. Psych.* **2004**, *55* (7), 708–714.

Rodgers, R. J.; Cole, J. C. Anxiety Enhancement in the Murine Elevated Plus Maze by Immediate Prior Exposure to Social Stressors. *Phys. Behav.* **1993**, *53* (2), 383–388.

Rolls, E. T. *The Brain and Emotion*; Oxford University Press: Oxford, 1999.

Ronzoni, G.; del Arco, A.; Mora, F.; Segovia, G. Enhanced Noradrenergic Activity in the Amygdala Contributes to Hyperarousal in an Animal Model of PTSD. *Psychoneueroendocrinology* **2016**, *70*, 1–9.

Roozendaal, B.; McEwan, B. S.; Chattarji, B. S. Stress, Memory and the Amygdala. *Nat. Rev. Neurosci.* **2009**, *10*, 423–433.

Roozendaal, B.; McReynolds, J. R.; McGaugh, J. L. The Basolateral Amygdala Interacts with the Medial Prefrontal Cortex in Regulating Glucocorticoid Effects on Working Memory Impairment. *J. Neurosci.* **2004**, *24*, 1385–1392.

Roque, L.; Verissimo, M.; Fernandes, M.; Rebelo, A. Emotion Regulation and Attachment: Relationships with Children's Secure Base, During Different Situational and Social Contexts in Naturalistic Settings. *Inf. Behav. Dev.* **2013**, *36* (3), 298–306.

Rosenblat, J. D.; Carvalho, A. F.; Li, M.; Lee, Y.; Subramanieapillai, M.; McIntyre, R. S. Oral Ketamine for Depression: A Systematic Review. *J. Clin. Psych.* **2019**, *80* (3), 0–0.

Rosenthal, T. L.; Akiskal, H. S.; Scott-Strauss, A.; Rosenthal, R. H.; David, M. Familial and Developmental Factors in Characterological Depressions. *J. Aff. Dis.* **1981**, *3*, 183–192.

Ross, A.; Thomas, S. The Health Benefits of Yoga and Exercise: A Review of Comparison Studies. *J. Altern. Complem. Med.* **2010**, *6* (1), 3–12.

Ross, C. A.; Margolis, R. L. Research Domain Criteria: Strengths, Weaknesses, and Potential Alternatives for Future Psychiatric Research. *Mol. Neuropsych.* **2019,** *5* (4), 218–235.

Ross, R. Psychotic and Manic-like Symptoms During Stimulant Treatment of Attention Deficit Hyperactivity Disorder. *Am. J. Psych.* **2006,** *163* (7), 1149–1152.

Rosse, I. C. Clinical Evidences of Borderland Insanity. *J. Nerv. Mental Dis.* **1890,** *70,* 669–683.

Rothstein, J. D.; Jin, L.; Dykes-Hoberg, M.; Kunci, R. W. Chronic Inhibition of Glutamate Uptake Produces a Model of Slow Neurotoxicity. *Proc. Nat. Acad. Sci. USA* (PNAS) **1993,** *90,* 6591–6595.

Routledge, F. S.; Campbell, T. S.; McFetridge-Durdle, J. A.; Bacon, S. J. Improvements in Heart Rate Variability with Exercise Therapy. *Can. J. Cardiol.* **2010,** *26* (6), 303–312.

Rovai, L.; Maremmani, A. G. I.; Bacciardi, S.; Giazzarrini, D.; Pallucchini, A.; Spera, V.; Perugi, G.; Maremmani, I. Opposed Effects of Hyperthymic and Cyclothymic Temperament in Substance Use Disorder (Heroinor Alcohol-Dependent Patients). *J. Aff. Dis.* **2017,** *218,* 339–345.

Rowe, M.; Chuang, D. Lithium Neuroprotection: Molecular Mechanisms and Clinical Implications. *Exp. Rev. Mol. Med.* **2004,** *6* (21), 1–18.

Ruggero, C.; Zimmerman, M.; Chelminski, I.; Yound, D. Borderline Personality Disorder and the Misdiagnosis of Bipolar Disorder. *J. Psych. Res.* **2010,** *44* (6), 405–408.

Rule, R. R.; Shimamura, A. P.; Knight, R. T. Orbitofrontal Cortex and Dynamic Filtering of Emotional Stimuli. *Cogn. Affect. Behav. Neurosci.* **2002,** *2,* 264–270.

Rush, A. J.; Trivedi, M. H.; Wisniewski, S. R.; Nierenberg, A. A.; Stewart, J. W.; Warden, D.; Niederehe, G.; Thase, M. E.; Lavori, P. W.; Lebowitz, B. D.; McGrath, P. J.; Rosenbaum, J. F.; Sackeim, H. A.; Kupfer, D. J.; Luther, J.; Fava, M. Acute and Longer-Term Outcomes in Depressed Outpatients Requiring One or Several Treatment Steps: A STAR*D Report. *Am. J. Psych.* **2006,** *163* (11), 1905–1917.

Rybakowski, J. K. Lithium Treatment in the Era of Personalized Medicine. *Drug Dev. Res.* **2020,** Mar 24 (elect pub).

Salters-Pedneault, K.; Roemer, L.; Tull, M. T.; Rucker, L.; Mennin, D. S. Evidence of Broad Deficits in Emotion Regulation Associated with Chronic Worry and Generalized Anxiety Disorder. *Cog. Ther. Res.* **2006,** *30,* 469–480.

Sami, M. B.; Bhattacharyya, S. Are Cannabis-Using and Non-Using Patients Different Groups? Towards Understanding the Neurobiology of Cannabis Use in Psychotic Disorders. *J. Psychopharmacol.* **2018,** *32,* 825–849.

Sanacora, G.; Kendell, S. F.; Levin, Y.; Simen, A. A.; Fenton, L. R.; Coric, V.; Krystal, J. H. Preliminary Evidence of Riluzole Efficacy in Antidepressant-Treated Patients with Residual Depressive Symptoms. *Biol. Psych.* **2007,** *61,* 822–825.

Sanacora, G.; Treccani, G.; Popoli, M. Towards a Glutamate Hypothesis of Depression: An Emerging Frontier of Neuropsychopharmacology for Mood Disorders. *Neuropharmacology* **2012,** *62* (1), 63–77.

Sanchez, M. M.; Ladd, C. O.; Plotsky, P. M. Early Adverse Experiences as a Developmental Risk Factor for Later Psychopathology: Evidence from Rodent and Primate Models. *Dev. Psychopath.* **2001,** *13,* 419–450.

Sander, L. W. Polarity, Paradox and the Organizing Process in Development. In *Frontiers in Child Psychiatry*; Call, J., Galenson, E., Tyson, R., Eds.; Basic Books: New York, 2002.

Sanislow, C. A.; Quinn, K. J.; Sypher, I. NIMH Research Domain Criteria (RDoC). *The Encyclopedia of Clinical Psychology*, 1st ed. by Cautin, R. L., Lilienfeld, S. O. ©2015 John Wiley & Sons, Inc. Published 2015 by John Wiley & Sons, Inc, 2015.

Sapolsky, R. M. Glucocorticoid Toxicity in the Hippocampus: Temporal Aspects of Neuronal Vulnerability. *Br. Res.* **1985,** *359* (1–2), 300–305.

Sapolsky, R. M. Why Stress Is Bad for Your Brain. *Science* **1996,** *273,* 749–750.

Sapolsky, R. M.; Krey, L. C.; McEwan, B. S. The Neuroendocrinology of Stress and Aging: The Glucocorticoid Cascade Hypothesis. *Endocrine Rev.* **1986,** *7* (3), 284–301.

Sassi, R. B.; Nicoletti, M.; Brambilla, P.; Malinger, A. G.; Frank, E.; Kupfer, D. J.; Keshavan, M. S.; Soares, J. C. Increased Gray Matter Volume in Lithium-Treated Bipolar Disorder Patients. *Neurosci. Lett.* **2002,** *329* (2), 243–245.

Sato, T.; Bottlender, R.; Schroter, A.; Moller, H. J. Frequency of Manic Symptoms During a Depressive Episode and Unipolar 'Depressive Mixed State' as Bipolar Spectrum. *Acta Psych. Scand.* **2003,** *107* (4), 268–274.

Scassellati, C.; Bonvincini, C.; Faraone, S. V.; Gennarelli, M. Biomarkers and Attention-Deficit/Hyperactivity Disorder: A Systematic Review and Meta-Analysis. *J. Am. Acad. Child Adol. Psych.* **2012,** *51* (10), 1003–1019.

Schaffer, C. B.; Nordahl, T. E.; Schaffer, L. C.; Howe, J. Mood-Elevating Effects of Opioid Analgesics in Patients with Bipolar Disorder. *J. NeuroPsych. Clin. Neurosci.* **2007,** *19* (4), 449–452.

Schaller, J. L.; Behar, D. Treatment of a Case of Comorbid Bipolar Disorder and Attention-Deficit/Hyperactivity Disorder. *J. NeuroPsych. Clin. Neurosci.* **1998,** *10* (2), 235–236.

Scheuing, L.; Chiu, C. T.; Liao, H. M.; Chuang, D. M. Antidepressant Mechanism of Ketamine: Perspectives from Pre-Clinical Studies. *Front. Neurosci.* **2015,** *9* (240), 1–7.

Schildkraut, J. J. The Catecholamine Hypothesis of Affective Disorders: A Review of Supporting Evidence. *Am. J. Psych.* **1965,** *122* (5), 509–522.

Schindler, F.; Anghelescu, I. G. Lithium Versus Lamotrigine Augmentation in Treatment Resistant Unipolar Depression: A Randomized, Open-Label Study. *Int. Clin. Psychopharm.* **2007,** *22* (3), 179–182.

Schmitt, R.; Winter, D.; Niedtfield, I.; Herpertz, S. C.; Schmahl, C. Effects of Psychotherapy on Neuronal Correlates of Reappraisal in Female Patients with Borderline Personality Disorder. *Biol. Psych. Cog. Nerosci. Neuroim.* **2016,** *1* (6), 548–557.

Schnell, K.; Herpertz, S. C. Effects of Dialectic-Behavioral-Therapy on the Neural Correlates of Affective Hyperarousal in Borderline Personality Disorder. *J. Psych. Res.* **2007,** *41* (10), 837–847.

Schore, A. *Affect Regulation and the Origin of the Self: The Neurobiology of Emotional Development*; Erlbaum, NJ, 1994.

Schore, A. *Affect Regulation and the Origin of the Self: The Neurobiology of Emotional Development*; Routledge Classic Editions, 2016.

Schore, A. *Affect Regulation and the Repair of the Self*; Norton: New York, 2003.

Schore, J.; Schore, A. Regulation Theory and Affect Regulation in Psychotherapy: A Clinical Primer. *Smith Coll. Stud. Soc. Work* **2014,** *82* (2–3), 178–195.

Schou, M. Lithium in Psychiatric Practice and Prophylaxis. *J. Psych. Res.* **1968,** *6*, 67–95.

Schwarcz, R.; Pellicciari, R. Manipulation of Brain Kynurenines: Glial Targets, Neuronal Effects and Clinical Opportunities. *J. Pharmacol. Exp. Ther.* **2002,** *303*, 1–10.

Seckl, J. R. Prenatal Glucocorticoids and Long-Term Programming. *Eur. J. Endocrinol.* **2004,** *15* (s3), U49–U62.

Segerstrom, S. C.; Solberg Nes, L. Heart Rate Variability Reflects Self-Regulatory Strength, Effort and Fatigue. *Psych. Sci.* **2007,** *18* (3), 275–281.

Severus, E.; Bauer, M. Diagnosing Bipolar Disorder in DSM-5. *Int. J. Bipolar Dis.* **2013,** *1* (14), 1–3.

Sgoifo, A.; Koolhaas, J.; DeBoer, S.; Musso, E.; Stilli, D.; Buwalda, B.; Meerlo, P. Social Stress, Autonomic Activation, and Cardiac Activity in Rats. *Neurosci. Biobehav. Rev.* **1999,** *23* (7), 915–923.

Shaldubina, A.; Agam, G.; Belmaker, R. H. The Mechanism of Lithium Action: State of the Art Ten Years Later. *Prog Neuropsychopharm Biol. Psych.* **2001,** *25* (4), 855–866.

Shapiro, F. *Eye Movement Desensitization and Reprocessing (EMDR): Basic Principles, Protocols and Procedures,* 3rd ed.; Guilford Press: New York, 2018.

Sharma, V.; Khan, S.; Smith, A. Closer Look at Treatment Resistant Depression: Is It Due to a Bipolar Diathesis? *J. Aff. Dis.* **2005,** *84* (2–3), 251–257.

Shaw, P.; Stringaris, A.; Nigg, J.; Leibenluft, E. Emotion Dysregulation in Attention Deficit Hyperactivity Disorder. *Am. J. Psych.* **2014,** *171* (3), 276–293.

Shea, A.; Walsh, C.; MacMillan, H.; Steiner, M. Child Maltreatment and HPA Axis Dysregulation: Relationship to Major Depressive Disorder and Post Traumatic Stress Disorder in Females. *Psychoneuroendocrinology* **2005,** *30* (2), 162–178.

Shechter, A.; Kim, E. W.; St. Onge, M. P.; Westwood, A. J. Blocking Nocturnal Blue Light for Insomnia: A Randomized Controlled Trial. *J. Psych. Res.* **2018,** *96,* 196–202.

Shedler, J.; Beck, A.; Fonagy, P.; Gabbard, G. O.; Gunderson, J.; Kernberg, O.; Michaels, R.; Westen, D. Personality Disorders in DSM-5. *Am. J. Psych.* **2010,** *167* (9), 1026–1028.

Shen, G. H. C.; Alloy, L. B.; Abramson, L. Y.; Sylvia, L. G. Social Rhythm Regularity and the Onset of Major Affective Episodes in Bipolar Spectrum Individuals. *Bip. Dis.* **2008,** *10* (4), 520–529.

Shiffman, S.; Paty, J. A.; Gnys, M.; Kassel, J. A.; Hickox, M. First Lapses to Smoking: Within-Subjects Analysis of Real Time Reports. *J. Consult Clin. Psychol.* **1996,** *64* (2), 366.

Shirayama, M.; Shirayama, Y.; Iida, H.; Kato, M.; Kajimmura, N.; Watanabe, T.; Sekimoto, M.; Shirakawa, S.; Okawa, M.; Takahashi, K. The Psychological Aspects of Patients with Delayed Sleep Phase Syndrome (DSPS). *Sleep Med.* **2003,** *4* (5), 427–433.

Shisslak, C. M.; Perse, T.; Crago, M. Coexistence of Bulimia Nervosa and Mania: A Literature Review and Case Report. *Comp. Psych.* **1991,** *32,* 181–184.

Sicorello, M.; Schmahl, C. Emotion Dysregulation in Borderline Personality Disorder: A Fronto-Limbic Imbalance? *Curr. Opin. Psychol.* **2021,** *37,* 114–120.

Siegal, D. *The Developing Mind: How Relationships and the Brain Interact to Shape Who We Are;* The Guilford Press: New York, 2012.

Silver, J. M.; Yudofsky, S. C.; Slater, J. A.; Gold, R. K.; Katz Stryer, B. L.; Williams, D. T.; Wolland, H.; Endicott, J. Propranolol Treatment of Chronically Hospitalized Aggressive Patients. *J. NeuroPsych. Clin. Neurosci.* **1999,** *11* (3), 328–335.

Silvio, J. R.; Condemarin, R. Psychodynamic Psychiatrists and Psychopharmacology. *J. Am. Acad. Psychoanal. Dynam. Psych.* **2011,** *39* (1), 27–39.

Simon, B. Mind and Madness in Classical Antiquity. In *History of Psychiatry and Medical Psychology;* Wallace, E. R., Gach, J., Eds.; Springer: New York, 2008.

Sinha, R.; Li, C. S. R. Imaging Stress and Cue-Induced Drug and Alcohol Craving: Association with Craving and Clinical Implications. *Drug Alcohol Rev.* **2007**, *26* (1), 5–31.

Slade, A. The Implications of Attachment Theory and Research for Adult Psychotherapy: Research and Clinical Perspectives. In *Handbook of Attachment: Theory, Research, and Clinical Applications*; Cassidy, J., Shaver, P. R., Eds., The Guilford Press, 2008; pp 762–782.

Sloan, T.; Telch, M. J. The Effects of Safety-Seeking Behavior and Guided Threat Reappraisal on Fear Reduction During Exposure: An Experimental Investigation. *Behav. Res. Ther.* **2002**, *40* (3), 235–251.

Smith, D. J.; Ghaemi, S. N.; Craddock, N. The Broad Clinical Spectrum of Bipolar Disorder: Implications for Research and Practice. *J. Clin. Psychopharmacol.* **2008**, *22* (4), 397–400.

Sobanski, E.; Banaschewski, T.; Asherson, P.; Buitelaar, P.; Chen, W.; Franke, B.; Holtmann, M.; Krumm, B.; Sergeant, J.; Sonuga-Barke, E.; Stringaris, A.; Taylor, E.; Anney, R.; Ebstein, R. P.; Gill, M.; Miranda, A.; Mulas, F.; Oades, R. D.; Roeyers, H.; Rothberger, A.; Steinhausen H-C, Faraone, S. V. Emotional Lability in Children and Adolescents with Attention Deficit/Hyperactivity Disorder (ADHD): Clinical Correlates and Familial Prevalence. *J. Child Psychol. Psych.* **2010**, *51* (8), 915–923.

Solo, P. H.; George, A.; Nathan, R. S.; Schulz, P. M.; Perel, J. M. Paradoxical Effects of Amitriptyline on Borderline Patients. *Am. J. Psych.* **1986**, *143*, 1603–1605.

Sorensen, L.; Hugdahl, K.; Lundervold, A. J. Emotional Symptoms in Inattentive Primary School Children: A Population-Based Study. *J. Atten. Dis.* **2008**, *11*, 580–587.

Sotres-Bayon, F.; Bush, D. E.; LeDoux, J. E. Emotional Perseveration: An Update on Prefrontal-Amygdala Interaction in Fear Extinction. *Learn Mem.* **2004**, *11* (5), 525–535.

Souery, D.; Amsterdam, J.; deMontigny, C.; Lecrubier, Y.; Montgomery, S.; Lipp, O.; Racagni, G.; Zohar, J.; Mendlewicz, J. Treatment Resistant Depression: Methodological Overview and Operational Criteria. *Eur. Neuropsychopharm.* **1999**, *9*, 83–91.

Southam, E.; Kirby, D.; Higgins, G. A.; Hagan, R. M. Lamotrigine Inhibits Monoamine Uptake In Vitro and Modulates 5-Hydroxytryptamine Uptake in Rats. *Eur. J. Pharmacol.* **1998**, *358*, 19–24.

Souza, F.; Goodwin, G. Lithium Treatment and Prophylaxis in Unipolar Depression: A Meta-Analysis. *Br. J. Psych.* **1991**, *158*, 666–675.

Spangler, G.; Grossman, K. Individual and Physiological Correlates of Attachment Disorganization in Infancy. In *Attachment Disorganization*; Solomon, J., George, C., Eds.; Guilford Press: New York, 1997; pp 95–126.

Spangler, G.; Scieche, M. Emotional and Adrenocortical Responses of Infants to the Strange Situation: The Differential Function of Emotional Expression. *Int. J. Behav. Dev.* **1998**, *22*, 681–706.

Spear, L. P. The Adolescent Brain and Age-Related Manifestations. *Neurosci. Biobehav. Rev.* **2000**, *24*, 417–463.

Spence, S.; Courbasson, C. The Role of Emotional Dysregulation in Concurrent Eating Disorders and Substance Abuse Disorders. *Eating Behav.* **2012**, *13* (4), 382–385.

Spinazzola, J.; Ford, J.; Zucker, M, van der Kolk, B. A.; Silva, S.; Smith, S. F.; Blaustein, M. Survey Evaluates Complex Trauma Exposure, Outcome, and Intervention Among Children and Adolescents. *Psych. Ann.* **2005**, *35* (5), 433–439.

Spitz, R. *The First Year of Life: A Psychoanalytic Study of Normal and Deviant Development of Object Relations*; International Universities Press: New York, 1965.

Spitzer, R. L.; Endicott, J.; Robbins, E. *Research Diagnostic Criteria (RDC) for a Selected Group of Functional Disorders*, 3rd ed.; New York State Psychiatric Institute, Biometrics Research: New York, 1978.

Spitzer, R. L.; Endicott, J.; Wodruff, R. A.; Andreasen, N. Classification of Mood Disorders. In *Depression: Clinical, Biological and Psychological Perspectives*; Usdin, G., Ed.; Brunner/Mazel: New York, 1977.

Sroufe, L. A. Attachment Classification from the Perspective of Infant-Caregiver Relationships and Infant Temperament. *Ch. Dev.* **1985,** *56,* 1–14.

Sroufe, L. A. *Emotional Development: The Organization of Emotional Life in the Early Years*; Cambridge University Press: New York, 1996.

Sroufe, L. A.; Waters, E. Heart Rate as a Convergent Measure in Clinical and Developmental Research. *MerrillPalmer Quart.* **1979,** *23,* 3–27.

Staecevic, V.; Ulenhuth, E. H.; Kellner, R.; Pathak, D. Patterns of Comorbidity in Panic Disorder and Agoraphobia. *Psych. Res.* **1992,** *42* (2), 171–183.

Stansbury, K.; Gunnar, M. Adrenocortical Activity and Emotion Regulation. *Monogr. Soc. Res. Child Dev.* **1994,** *59* (2–3), 108–134.

Starcevic, V. Dysphoric About Dysphoria: Towards a Greater Conceptual Clarity of the Term. *Australasian Psych.* **2007,** *15* (1), 9–13.

Stein, G. S.; Hartshorn, S.; Jones, J.; Steinberg, D. Lithium in a Case of Severe Anorexia Nervosa. *Br. J. Psych.* **1982,** *140,* 526–528.

Stein, P. K.; Ehsani, A. A.; Domitrovich, P. P.; Kleiger, R. E.; Rottman, J. N. Effects of Exercise Training on Heart Rate Variability in Older Men. *Am. Heart J.* **1999,** *138* (5), 567–576.

Stern, D. N. *The Interpersonal World of the Human Infant*; Basic Books; New York, 1985.

Stevenson, J.; Haliburn, J.; Halivoc, S. Trauma, Personality Disorders, and Chronic Depression: The Role of the Conversational Model of Psychodynamic Therapy in Treatment Resistant Depression. *Psychoan. Psychother.* **2016,** *30* (1), 23–41.

Stickgold, R. EMDR: A Putative Neurobiological Mechanism of Action. *J. Clin. Psychol.* **2002,** *58* (1), 61–75.

Stickgold, R.; Scott, L.; Rittenhouse, C.; Hobson, A. Sleep-Induced Changes in Associative Memory. *J. Cog. Neurosci.* **1999,** *11* (2), 182–193.

Stone, M. H. Borderline Personality Disorder: History of the Concept. In *Borderline Personality Disorder*; Zanarini, M. C., Ed.; Taylor & Francis: New York, London, 2005; pp 1–18.

Stone, M. H. The Relationship Between Borderline Personality Disorder and Bipolar Disorder. *Am. J. Psych.* **2006,** *163* (7), 1126–1128.

Stone, M. H. Toward a Psychobiological Theory of Borderline Personality Disorder: Is Irritability the Red Thread That Runs Through Borderline Conditions? *Dissoc.: Prog. Dissoc. Disorders* **1988,** *1* (2), 2–15.

Strakowski, S. M.; Fleck, D. E.; Maj, M. Broadening the Diagnosis of Bipolar Disorder: Benefits vs. Risks. *World Psych.* **2011,** *10* (3), 181–186.

Stratigos, K. Frontline—Not Your Grandparents' Psychoanalysis. *J Am. Acad. Psychoanal. Dynam. Psych.* **2009,** *37* (2), 241–252.

Stringaris, A. Irritability in Children Adolescents: A Challenge for DSM-5. *Eur. Child. Adol. Psych.* **2011**, *20*, 61–66.

Stringaris, A.; Baroni, A.; Haimm, C.; Brotman, M.; Lowe, C. H.; Myers, F.; Rustgi, E.; Wheeler, W.; Kayser, R.; Towbin, K.; Leibenluft, E. Pediatric Bipolar Disorder Versus Severe Mood Dysregulation: Risk for Manic Episodes on Follow-up. *J. Am. Acad. Ch. Adol. Psych.* **2010**, *49* (4), 397–405.

Stringaris, A.; Cohen, P.; Pine, D. S.; Leibenluft, E. Adult Outcomes of Youth Irritability: A 20-Year Prospective Community-Based Study. *Am. J. Psych.* **2009**, *166*, 1048–1054.

Stringaris, A.; Zavos, H.; Leibenluft, E.; Maughan, B.; Eley, T. C. Adolescent Irritability: Phenotypic Associations and Genetic Links with Depressed Mood. *Am. J. Psych.* **2012**, *169* (1), 47–54.

Sudikoff, E. L.; Bertolin, M.; Lordo, D. N.; Kaufman DAS: Relationships Between Executive Function and Emotional Regulation in Healthy Children. *J. Neurol. Psychol.* **2015**, S (2), 1–8.

Suomi, S. J. Early Determinants of Behavior: Evidence from Primate Studies. *Br. Med. Bull.* **1997**, *53*, 170–184.

Surman, C. B. H.; Biederman, J.; Spencer, T.; Yorks, D.; Miller, C. A.; Petty, C. R.; Faraone, S. V. Deficient Emotional Self-Regulation and Adult Attention Deficit Hyperactivity Disorder: A Family Risk Analysis. *Am. J. Psych.* **2011**, *168* (6), 617–623.

Suveg, C.; Kendall, P. C.; Comer, J. S.; Robin, J. Emotion-Focused Cognitive-Behavior Therapy for Anxious Youth: A Multiple Baseline Evaluation. *J. Cont. Psychother.* **2006**, *36*, 77–85.

Suveg, C.; Zeman, J. Emotion Regulation in Children with Anxiety Disorders. *J. Clim. Ch. Adol. Psych.* **2004**, *33* (4), 750–759.

Svaldi, J.; Griepenstroh, J.; Tuschen-Caffer, B.; Ehring, T. Emotion Regulation Deficits in Eating Disorders: A Marker of Eating Pathology or General Psychopathology? *Psych. Res.* **2012**, *197* (1–2), 103–111.

Swendsen, J. D.; Tennen, H.; Carney, M. A.; Affleck, G.; Willard, A.; Hromi, A. Mood and Alcohol Consumption: An Experience Sampling Test of the Self-Medication Hypothesis. *J. Ab. Psychol.* **2000**, *109* (2), 198–204.

Szentagothai, J. The 'Module-Concept' in Cerebral Cortex Architecture. *Brain Res.* **1975**, *95*, 476–496.

Szyf, M.; McGowan, P.; Meany, M. The Social Environment and Epigenome. *Environ. Mol. Mutagen.* **2008**, *49* (1), 46–60.

Szyf, M.; Weaver, I.; Meaney, M. Maternal Care, the Epigenome, and Phenotypic Differences in Behavior. *Reprod. Toxicol.* **2007**, *24* (1), 9–19.

Talge, N. M.; Neal, C.; Glover, V. Antenal Maternal Stress and Long-Term Effects on Child Neurodevelopment: How and Why? *J. Ch. Psychol. Psych.* **2007**, *48*, 245–261.

Tanaka, K.; Watase, K.; Manabe, T.; Yamada, K.; Watanabe, M.; Takanashe, K.; Iwama, H.; Nishikawa, T.; Ichihara, N.; Kikuchi, T.; Okuyama, S.; Kawashima, N.; Hori, S.; Takimoto, M.; Wada, K. Epilepsy and Exacerbation of Brain Injury in Mice Lacking the Glutamate Transporter GLT-1. *Science* **1997**, *276* (5319), 1699–1702.

Tang, Y.; Tang, R.; Pasner, M. I. Mindfulness Meditation Improves Emotion Regulation and Reduces Drug Abuse. *Drug Alcohol. Dep.* **2016**, *163* (Supp 1), 513–518.

Tata, D. A.; Anderson, B. J. The Effects of Chronic Glucocorticoid Exposure on Dendritic Length, Synapse Numbers and Glial Volume in Animal Models: Implications for Hippocampal Volume Reductions in Depression. *Physiol. Behav.* **2010,** *99,* 186–193.

Tchanturia, K.; Doris, E.; Fleming, C. Effectiveness of Cognitive Remediation and Emotion Skills Training (CREST) for Anorexia Nervosa in Group Format: A Naturalistic Pilot Study. *Eur. Eat Disord. Rev.* **2013,** *22* (3), 200–205.

Teicher, M. H.; Glod, C.; Cole, J. O. Emergence of Intense Suicidal Preoccupation During Fluoxetine Treatment. *Am. J. Psych.* **1990,** *147* (2), 207–210.

Terao, T.; Nakano, H.; Inoue, Y.; Okamoto, T.; Nakamura, J.; Iwata, N. Lithium and Dementia: A Preliminary Study. *Prog. Neuropsychopharmacol. Biol. Psych.* **2006,** *30* (6), 1125–1128.

Terman, M.; Amira, L.; Terman, J. S.; Ross, D. Predictors of Response and Nonresponse to Light Treatment for Winter Depression. *Am. J. Psych.* **1996,** *153* (11), 1423–1429.

Thakura, R. G.; Ray, P.; Kanji, D.; Das, R.; Bisui, B.; Singh, O. P. Rapid Antidepressant Response with Ketamine: Is It the Solution to Resistant Depression? *Indian Psychol. Med.* **2012,** *34* (1), 56–60.

Thomas, L. A.; Brotman, M. A.; Muhrer, E. J.; Rosen, B. H.; Bones, B. L.; Reynolds, R. C.; Deveney, C. M.; Pine, D. S.; Leibenluft, E. Parametric Modulation of Neural Activity by Emotion in Youth with Bipolar Disorder, Youth with Severe Mood Dysregulation, and Healthy Volunteers. *Arch. Gen. Psych.* **2012,** *69* (12), 1257–1266.

Thomas, S. P.; Nandhra, H. S.; Jayaraman, A. Systematic Review of Lamotrigine Augmentation of Treatment Resistant Unipolar Depression (TRD). *J. Mental Health* **2010,** *19* (2), 168–175.

Thome, J.; Ehlis, A.; Fallgatter, A. J.; Krauel, K.; Lange, K. W.; Riederer, P.; Romanos, M.; Taurines, R.; Tucha, O.; Uzbekov, M.; Gerlach, M. Biomarkers for Attention-Deficit/ Hyperactivity Disorder. A Consensus Report of the WFSBP Task Force on Biological Markers and the World Federation of ADHD. *World J. Biol. Psych.* **2012,** *13* (5), 379–400.

Thompson, R. A. Emotion Regulation: A theme in search of a definition. In *The Development of Emotion Regulation and Dysregulation: Biological and Behavioral Aspects*; Fox, N.A., Ed. *Monogr. Soc. Res. Child Dev.* **1994,** *59* (2–3 Serial No. 240), 25–52.

Thompson, R. A.; Lewis, M. D.; Calkins, S. D. Reassessing Emotion Regulation. *Child Dev. Pers.* **2009,** *2* (3), 124–131.

Thompson, R. Emotion and Emotion Regulation: Two Sides of the Developing Coin. *Emot. Rev.* **2011,** *3* (1), 53–61.

Tomkins, S. S. Affect as Amplification: Some Modification in Theory. In *Emotion, Theory, Research and Experience*; Plutchik, R., Kellerman, H., Eds., Vol. I; Academic Press: New York, 1980.

Torgersen, S. Childhood and Family Characteristics in Panic and Generalized Panic Disorders. *Am. J. Psych.* **1986,** *143,* 630–632.

Tottenham, N.; Galvan, A. Stress and the Adolescent Brain: Amygdala-Prefrontal Cortex Circuitry and Ventral Striatum as Developmental Targets. *Neurosci. Biobehav. Rev.* **2016,** *70,* 217–227.

Tourian, L.; LeBoeuf, A.; Breton, J. J.; Cohen, D.; Gignac, M.; Labelle, R.; Guile, J. M.; Renaud, J. Treatment Options for the Cardinal Symptoms of Disruptive Mood Dysregulation Disorder. *J. Can. Acad. Ch. Adol. Psych.* **2015,** *24* (1), 41–54.

Trullas, R.; Skolnick, P. Functional antagonists at the NMDA receptor complex exhibit antidepressant actions. *Eur J Pharmacol* **1990**, *185*, 1–10.

Tsankova, N.; Renthal, W.; Kumar, A.; Nestler, E. Epigenetic Regulation in Psychiatric Disorders. *Nat. Rev. Neurosci.* **2007**, *8* (5), 355–376.

Tsao, J. C. I.; Lewin, M. R.; Craske, M. G. The Effects of Cognitive-Behavior Therapy for Panic Disorder on Comorbid Conditions. *J. Anx. Dis.* **1998**, *12* (4), 357–371.

Tsapakis, E. M.; Travis, M. J. Glutamate and Psychiatric Disorders. *Adv. Psych. Treat.* **2002**, *8* (3), 189–197.

Tull, M. T.; Stipelman, B. A.; Salters-Pedneault, K.; Gratz, K. L. An Examination of Recent Non-Clinical Panic Attacks, Panic Disorder, Anxiety Sensitivity, and Emotion Regulation Difficulties in the Prediction of Generalized Anxiety Disorder in an Analogue Sample. *J. Anx. Dis.* **2009**, *23* (2), 275–282.

Tunnard, C.; Rane, L. J.; Wooderson, S. C.; Markopoulou, K.; Poon, L.; Fekadu, A.; Juruena, M.; Cleare, A. J. The Impact of Childhood Adversity on Suicidality and Clinical Course in Treatment-Resistant Depression. *J. Aff. Dis.* **2014**, *152*–154, 122–130.

Tutter, A. Medication as Object. *J. Am. Psychoan. Ass.* **2006**, *54* (3), 781–804.

Uhleski, M. L.; Fuchs, P. N. Maternal Separation Stress Leads to Enhanced Emotional Responses to Noxious Stimuli in Adult Rats. *Behav. Br. Res.* **2010**, *212* (2), 208–212.

Umlauf, M. G.; Shattell, M. The Ecology of Bipolar Disorder: The Importance of Sleep. *Iss Mental Heal Nurs.* **2005**, *26* (7), 699–720.

Vahip, S.; Kesebir, S.; Alkan, M.; Yazici, O.; Akiskal, K. K.; Akiskal, H. S. Affective Temperaments in Clinically-Well Subjects in Turkey: Initial Psychometric Data on the TEMPS-A. *J. Aff. Dis.* **2005**, *85* (1–2), 13–125.

Van deer Kolk, B. A.; Fisler, R. E. Childhood Abuse and Neglect and Loss of Self-Regulation. *Bull. Mental Clin.* **1994**, *58* (2), 145–168.

Van den Bergh, B. R. H.; Van Calster, B.; Smits, T.; Van Huffel, S.; Lagae, L. Antenatal Maternal Anxiety Is Related to HPA-Axis Dysregulation and Self-Reported Depressive Symptoms in Adolescence: A Prospective Study on the Fetal Origins of Depressed Mood. *Neuropsychopharmacology* **2008**, *33*, 536–545.

van der Kolk, B. Developmental Trauma Disorder. *Psych. Ann.* **2005**, *35* (5), 401–408.

van der Kolk, B. *The Body Keeps the Score*; Penguin Books, 2014; p 153.

van der Kolk, B.; Spinazolla, J.; Blaustein, M.; Hopper, J.; Hopper, E.; Korn, D.; Simpson, W. B. A Randomized Clinical Trial of EMDR, Fluoxetine, and Pill Placebo in the Treatment of PTSD: Treatment Effects and Long-Term Maintenance. *J. Clin. Psych.* **2007**, *68* (1), 37–46.

van der Kolk, BA, Dreyfuss, D.; Michaels, M.; Shera, D.; Berkowitz, R.; Fisler, R.; Saxe, G. Fluoxetine in Posttraumatic Stress Disorder. *J. Clin. Psych.* **1994**, *55*, 517–522.

van Izjendoorn, H. W.; Schuengel, C.; Bakersmans-Kranenburg, M. J. Disorganized Attachment in Early Childhood: Meta-Analysis of Precursors, Concomitants, and Sequelae. *Dev. Psychopath.* **1999**, *11*, 225–249.

Van Lancker, D.; Cummings, J. L. Expletives: Neurobehavioral and Neurolinguistic Perspectives on Swearing. *Br. Res. Rev.* **1999**, *31* (1), 83–104.

van Loo, H. M.; Romeijn, J. W. Psychiatric Comorbidity: Fact or Artifact? *Theor. Med. Bioeth.* **2015**, *36* (1), 41–60.

Van Overwalle, F. Social Cognition and the Brain: A Meta-Analysis. *Human Brain Map.* **2009**, *30*, 829–858.

Vasantkumar, T.; Tanna, R.; Renningroth, P.; Woolson, R. F. Propranolol in the Treatment of Anxiety Neurosis. *Comp. Psych.* **1977**, *18* (4), 319–326.

Vasquez, G. H.; Tondo, L.; Mazzarini, L.; Gonda, X. Affective Temperaments in General Population: A Review and Combined Analysis from National Studies. *J. Aff. Dis.* **2012**, *139* (1), 18–22.

Vayas, A.; Chattarji, S. Modulation of Different States of Anxiety-Like Behavior by Chronic Stress. *Behav. Neurosci.* **2004**, *118*, 1450–1454.

Venkataraman, S.; Naylor, M.; King, C. Mania Associated with Bipolar Disorder in Adolescents. *J. Acad. Child Adol. Psych.* **1992**, *31* (2), 276–281.

Viaterna, M. H.; Takahashi, J. S.; Turek, F. W. Overview of Circadian Rhythms. *Nat. Inst. Alcohol Alcoholism* **2001**, *25*, 85–93.

Viau, V.; Sharma, S.; Plotsky, P. M.; Meaney, M. J. The Hypothalamic-Pituitary Adrenal Response to Stress in Handled and Non-Handled Rats: Differences in Stress-Induced Plasma Secretion Are Not Dependent Upon Increased Corticosterone Levels. *J. Neurosci.* **1993**, *13*, 1097–1105.

Virgin, Jr. C. E.; Ha, T. P. T.; Packan, D. R.; Tombaugh, G. C.; Yang, S. H.; Homer, H. C.; Sapolsky, R. M. Glucocorticoids Inhibit Glucose Transport and Glutamate Uptake in Hippocampal Astrocytes: Implications for Glucocorticoid Neurotoxicity. *J. Neurochem.* **1991**, *57* (4), 1422–1428.

Vyas, A.; Jadhav, S.; Chattarji, S. Prolonged Behavioral Stress Enhances Synaptic Connectivity in the Basolateral Amygdala. *Neurosciences* **2006**, *143*, 387–393.

Walden, T. A.; Smith, M. C. Emotion Regulation. *Motive Emot.* **1997**, *21*, 7–25.

Walker, M. P.; van der Helm, E. Overnight Therapy? The Role of Sleep in Emotional Brain Processing. *Psych. Bull.* **2009**, *135* (5), 731–748.

Walsh, M. A.; Brown, L. H.; Barrantes-Vidal, N.; Kwapil, T. R. The Expression of Affective Temperaments in Daily Life. *J. Aff. Dis.* **2013**, *145* (2), 179–186.

Wan, Y.; Xu, J.; Ma, D.; Zeng, Y.; Cibelli, M.; Maze, M. Postoperative Impairment of Cognitive Function in Rats: A Possible Role for Cytokine-mediated Inflammation in the Hippocampus. *Anesthesiol* **2007**, *106* (3), 436–443.

Waters, S. F.; Virmani, E. A.; Thompson, R. A.; Meyer, S.; Raikes, H. A.; Jochem, R. Emotion Regulation and Attachment: Unpacking Two Constructs and Their Association. *J. Psychopath. Behav. Asses.* **2010**, *32*, 37–47.

Watford, T. S.; Stafford, J. The Impact of Mindfulness on Emotion Dysregulation and Psychophysiological Reactivity Under Emotional Provocation. *Psych. Consc. Theor. Res. Prac.* **2015**, *2* (1), 90–109.

Waxmonsky, J.; Pelhan, W. E.; Gnagy, E.; Cummings, M. R.; O'Connor, B.; Majumdar, A.; Verley, J.; Hoffman, M. T.; Massetti, G. A.; Burrows-MacLean, L.; Fabiano, G. A.; Waschbusch, D. A.; Chacko, A.; Arnold, F. W.; Walker, K. S.; Garefino, A. C.; Robb, J. A. The Efficacy and Tolerability of Methylphenidate and Behavior-Modification in Children with Attention-Deficit/Hyperactivity Disorder and Severe Mood Dysregulation. *J. Ch. Adol. Psychopharmacol.* **2008**, *18* (6), 573–588.

Weaver, I. C.; LaPlante, P.; Weaver, S.; Parent, A.; Sharma, S.; Diorio, J.; Chapman, K. E.; Seckl, J. R.; Szyf, M.; Meaney, M. J. Early Environmental Regulation of Hippocampal Glucocorticoid Receptor Gene Expression: Characterization of Intracellular Mediators and Potential Genomic Target Sites. *Mol. Cell Endocrine* **2001**, *185* (1–2), 205–218.

Weinberg, M. K.; Olson, K. L.; Beeghly, M.; Tronick, E. Z. Making Up Is Hard to do, Especially for Mothers with High Levels of Depression and Their Infant Sons. *J. Ch. Psychol. Psych.* **2006**, *47*, 670–683.

Wesselmann, E. D.; Williams, K. D.; Hales, A. H. Vicarious Ostracism. *Front Human Neurosci.* **2013**, *7*, 153.

West, E. D.; Dally, P. J. Effects of Iproniazid in Depressive Syndromes. *Br. Med. J.* **1959**, *1* (5136), 1491–1494.

Weygandt, W. *Uber die Mischzustande des Manisch-Depressiven Irreseins*; J.F. Lehmann: Munich, 1899.

White, H. S.; Smith, M. D.; Wilcox, H. S. Mechanism of Action of Antiepileptic Drugs. *Int. Rev. Neurobiol.* **2007**, *81*, 85–110.

Wicki, W.; Angst, J. The Zurich Study. X. Hypomania in a 28to 30-Year-Old Cohort. *Eur. Arch. Psych. Clin. Neurosci.* **1991**, *40*:339–348.

Wiggins, J. L.; Brotman, M. A.; Adelman, N. E.; Kim, P.; Oakes, A. H.; Reynolds, R. C.; Chen, G.; Pine, D. S.; Leibenluft, E. Neural Correlates of Irritability in Disruptive Mood Dysregulation and Bipolar Disorders. *Am. J. Psych.* **2016**, *173*, 722–730.

Wildes J E, Ringham R M, Marcus, M. D. Emotion Avoidance in Patients with Anorexia Nervosa: Initial Test of a Functional Model. *Int. J. Eat Disord.* **2010**, *43* (5), 398–404.

Wilens, T. E.; Faraone, S. V.; Biederman, J.; Gunawardene, S. Does Stimulant Therapy of Attention-Deficit/Hyperactivity Disorder Beget Later Substance Abuse? A Meta-analytic Review of the Literature. *Pediatrics* **2003**, *111* (1), 179–185.

Williams, D. T.; Mehl, R.; Yudofsky, S.; Adams, D.; Roseman, B. The Effect of Propranolol on Uncontrolled Rage Outbursts in Children and Adolescents with Organic Brain Dysfunction. *Psychiatry* **1982**, *21* (2), 29–135.

Willis, T. *De Anima Brutorum*; Lyon, 1676.

Wilner, P. Depression: A Psychobiological Synthesis; Wiley: New York, 1985.

Wilson, A.; Cooperman, Y. Rejection Sensitive Dysphoria: Causes and Treatment. *Mental Health* **2021** Jun 3.

Wing Li, T. C. Psychodynamic Aspects of Psychotherapy. *J Am. Acad. Psychoanal. Dynam. Psych.* **2010**, *38* (4), 655–674.

Winokur, A.; March, V.; Mendels, J. Primary Affective Disorders in Relatives of Patients with Anorexia Nervosa. *Am. J. Psych.* **1980**, *137*, 695–698.

Winokur, G.; Clayton, P. J.; Reich, T. *Manic-Depressive Illness*; St. Louis CV Mosby, St. Loius, 1969.

Wirz-Justice, A.; Terman, M. Commentary on "Blue-Blocking Glasses as Additive Treatment for Mania: A Randomized Placebo-Controlled Trial". *Bip. Dis.* **2016**, *18* (4), 383–384.

Witter, D. P.; Ramnaraine, L.; Shapiro, M. A. A Possible Case of Bipolar Disorder Unmasked by Dextromethorphan in a 16-yo Adolescent. *Adol Psych* **2015**, *5* (4), 273–276.

Wittling, W.; Roschmann, R. Emotion-Related Hemisphere Asymmetry: Subjective Emotional Responses to Laterally Presented Films. *Cortex* **1993**, *29* (3), 431–448.

Woerner, M. G.; Correll, C. U.; Alvir, J. M. J.; Greenwald, B.; Delman, H.; Kane, J. M. Incidence of Tardive Dyskinesia with Risperdal or Olanzapine in the Elderly: Results from a 2-Year Prospective Study in Antispychotic-Naïve Patients. *Neuropsychopharmacology* **2011**, *36*, 1738–1746.

Wolf, O. T.; Dziobek, I.; McHugh, P.; Sweat, V.; deLeon, M. J.; Javier, E.; Convit, A. Subjective Memory Complaints in Aging Are Associated with Elevated Cortisol Levels. *Neurobiol. Aging* **2005**, *26* (10), 1357–1363.

Wolf, S. A.; Steiner, B.; Akpinarli, A.; Kammertoens, T.; Nassenstein, C.; Braun, A.; Blankenstein, T.; Kempermann G: CD4-Positive T Lymphocytes Provide a Neuroimmunological Link in the Control of Adult Hippocampal Neurogenesis. *J. Immunol.* **2009**, *182* (7), 3979–3984.

World Health Organization. *Mental Disorders: Glossary and Guide to their Classification in Accordance with the Ninth Revision of the International Classification of Diseases*; WHO: Geneva, 1978.

World Health Organization. *The ICD-10 Classification of Mental and Behavioral Disorders*; World Health Organization: Geneva, 1992.

Wozniak, J.; Biederman, J.; Kiely, K.; Ablon, J. S.; Faraone, S. V.; Mundy, E.; Mennin, D. Mania-Like Symptoms Suggestive of Childhood-Onset Bipolar Disorder in Clinically Referred Children. *J. Am. Acad. Ch. Adol. Psych.* **1995**, *34*, 867–876.

Wright, K. A.; Everson-Hock, E. S.; Taylor, A. H. The Effects of Physical Activity on Physical and Mental Health among Individuals with Bipolar Disorder: A Systematic Review. *Mental Health Phys. Act* **2009**, *2* (2), 89–94.

Yatnam, L. N.; Goldstein, J. M.; Vieta, E.; Bowden, C. L.; Grunze, H.; Post, R. M.; Suppes, T.; Calabrese, J. R. Atypical Antipsychotics in Bipolar Disorder: Potential Mechanisms of Action. *J. Clin. Psych.* **2005**, *66* (Supp 5), 40–48.

Yehuda, R.; Lehrner, A. Intergenerational Transmission of Trauma Effects: Putative Role of Epigenetic Mechanisms. *World Psych.* **2018**, *17* (3), 248–257.

Yirmiya, R.; Endotoxin Produces a Depressive-Like Episode in Rats. *Brain Res.* **1996**, *711*, 163–174.

Yorke, J. The Significance of Human–Animal Relationships as Modulators of Trauma Effects in Children: A Developmental Neurobiological Perspective. *Early Ch. Dev. Care* **2010**, *180* (5), 559–570.

Yudofsky, S.; Williams, D.; Gorman, J. Propranolol in the Treatment of Rage and Violent Behavior in Patients with Chronic Brain Syndromes. *Am. J. Psych.* **1981**, *138* (2), 18–220.

Yuskel, C.; Ongur, D. Magnetic Resonance Spectroscopy Studies of Glutamate-Related Abnormalities in Mood Disorders. *Biol. Psych.* **2010**, *68*, 785–794.

Zakzanis, KK, Leach, L, Kaplan, E: On the Nature and Pattern of Neurocognitive Function in Major Depressive Disorder. *Neuropsych. Neuropsychol. Behav. Neurol.* **1998**, *11*, 111–119.

Zanarini, M. C.; Frankenburg, F. R.; Hennen, J.; Silk, K. H. The Longitudinal Course of Borderline Psychopathology: 6-Year Prospective Follow-Up of the Phenomenology of Borderline Personality Disorder. *Am. J. Psych.* **2003**, *16* (2), 274–283.

Zanarini, M. C.; Gunerson, J. G.; Frankenberg, F. R.; Cognitive Features of Borderline Personality Disorder. *Am. J. Psych.* **1990**, *147* (57), 57–63.

Zarate Jr, C. A.; Brutsche, N. E.; Ibrahim, L.; Franco-Chaves, J.; Diazgranados, A.; Cravchik, A.; Selter, J.; Marquardt, C. A.; Liberty, V.; Luckenbaugh, D. A. Replication of Ketamine's Antidepressant Efficacy in Bipolar Depression: A Randomized Controlled Add-On Trial. *Biol. Psych.* **2012**, *71*:11, 939–946.

Zarate Jr, C. A.; Du, J.; Quiroz, J.; Gray, N. A.; Denicoff, K. D.; Singh, J.; Charney, D.; Manji, H. K. Regulation of Cellular Plasticity Cascades in the Pathophysiology and

Treatment of Mood Disorders: Role of the Glutamatergic System. *Ann. NY Acad. Sci.* **2003,** *1003* (1), 273–291.

Zarate Jr, C. A.; Quiroz, J. A.; Singh, J. B.; Denicoff, K. D.; De Jesus, G.; Luckenbaugh, D. A.; Charney, D. S.; Manji, H. K. An Open-Label Trial of the Glutamate-Modulating Agent Riluzole in Combination with Lithium for the Treatment of Bipolar Depression. *Biol. Psych.* **2005,** *57,* 430–432.

Zarate Jr. CA, Singh, J. B.; Carlson, P. J.; Brutsche, N. E.; Ameli, R.; Luckenbaugh, M. A.; Charney, D. S.; Manji, H. K. A Randomized Trial of an N-Methyl-D-aspartate Antagonist in Treatment-Resistant Major Depression. *Arch. Gen. Psych.* **2006,** *63* (8), 856–864.

Zarate, C. Riluzole in Psychiatry. *Exp. Opin. Drug Metab. Toxicol.* **2008,** *4* (9), 1223–1234.

Zarrow, M. X.; Campbell, P. S.; Denenberg, V. H. Handling in Infancy: Increased Levels of the Hypothalamic-Corticotropin Releasing Factor (CRF) Following Exposure to a Novel Situation. *Proc. Soc. Exp. Biol. Med.* **1972,** *356,* 141–143.

Zeanah, C. H.; Fox, N. A. Temperament and Attachment Disorders. *J. Clin. Ch. Adol. Psychol.* **2004,** *33* (1), 32–41.

Zelazo, P. D.; Cunningham, W. A. Executive Function: Mechanisms Underlying Emotion Regulation. In *Handbook of Emotion Regulation*; Gross, J. J., Ed., Guilford Press: New York, 2007.

Zhang, B.; Zhang, Y.; Wu, W.; Xu, T.; Yin, Y.; Zhang, J.; Huang, D.; Li, W. Chronic Glucocorticoid Exposure Activates BK-NLRP1 Signal Involving in Hippocampal Neuron Damage. *J. Neuroinflamm.* **2017,** *14* (139), 1–13.

Zhong, J.; Lee, W. H. Lithium: A Novel Treatment for Alzheimer's Disease? *Exp. Opin. Drug Saf.* **2007,** *6* (4), 375–383.

Zotev, V.; Kreuger, F.; Phillips, R.; Alvarez, R. P.; Kyle-Simmons, W.; Bellgowan, P.; Drevets, W. C.; Bodurka, J. Self-Regulation of Amygdala Activation Using Real-Time fMRI Neurofeedback. *PLoS One* **2011,** *6* (9), e24522.

Zotev, V.; Phillips, R.; Young, K. D.; Drevets, W. C.; Bodurka, J. Prefrontal Control of the Amygdala During Real-Time fMRI Neurofeedback Training of Emotion Regulation. *PLoS One* **2013,** *8* (11), e79184

Zou, D. Y.; Wu, Y. L.; Yao, W. X.; Cao, Y.; Wu, C. F.; Tanaka, M. Effect of MK-801 and Ketamine on Hydroxyl Radical Generation in the Posterior Cingulate and Retrosplenial Cortex of Free-Moving Mice, as Determined by In Vivo Microdialysis. *Pharmacol. Biochem. Behav.* **2007,** *86,* 1–7.

Index